Advances in Novel Precast Concrete Structures

Advances in Novel Precast Concrete Structures

Guest Editors

Dong-Zhi Guan

Zhangfeng Zhu

Jian Sun

Lianglong Song

Sen Yang

MDPI Basel • Beijing • Wuhan • Barcelona • Belgrade • Novi Sad • Cluj • Manchester

Guest Editors

Dong-Zhi Guan
School of Civil Engineering
Southeast University
Nanjing
China

Zhangfeng Zhu
College of Civil Engineering
Nanjing Tech University
Nanjing
China

Jian Sun
School of Civil Engineering
Southeast University
Nanjing
China

Lianglong Song
College of Civil and
Transportation Engineering
Hohai University
Nanjing
China

Sen Yang
Department of Civil
Engineering
Shanghai University
Shanghai
China

Editorial Office
MDPI AG
Grosspeteranlage 5
4052 Basel, Switzerland

This is a reprint of the Special Issue, published open access by the journal *Buildings* (ISSN 2075-5309), freely accessible at: https://www.mdpi.com/journal/buildings/special_issues/0CBX915WZB.

For citation purposes, cite each article independently as indicated on the article page online and as indicated below:

Lastname, A.A.; Lastname, B.B. Article Title. *Journal Name* **Year**, *Volume Number*, Page Range.

ISBN 978-3-7258-5899-6 (Hbk)
ISBN 978-3-7258-5900-9 (PDF)
https://doi.org/10.3390/books978-3-7258-5900-9

Cover image courtesy of Dongzhi Guan

Contents

buildings

MDPI

Article

Modified Energy-Based Design Method of the Precast Partially Steel-Reinforced Concrete Beam–CFST Column Eccentrically Braced Frame

Fugui Hou, Weiguang Chong, Yu Lin *, Xijun He and Guanglei Zhang

College of Civil Engineering, Nanjing Forestry University, Nanjing 210037, China; zhangguanglei@njfu.edu.cn (G.Z.)
* Correspondence: linyuzero05@njfu.edu.cn

Abstract

The eccentrically braced frame (EBF) is a typical structural system used in high-rise buildings. Current related design methods focus on the concrete and steel structures rather than on the complex composite structure. In addition, they tend to overlook the contribution of the energy-dissipation unit and its corresponding additional influence on the structure. In this study, a precast composite EBF structure is selected as a case study, including the partially steel-reinforced concrete (PSRC) beam and the concrete-filled steel tubular (CFST) column. A modified energy-based design method is proposed to leverage the excellent seismic performance of the precast composite EBF structure. The multi-stage energy-dissipation mechanism and the additional influence of the eccentric braces are systematically considered through the energy distribution coefficient and the layout of dampers. A case study of a 12-floor, three-bay precast composite EBF structure is conducted using a series of nonlinear time-history analyses. Critical seismic responses, including the maximum inter-story drift ratio, residual inter-story drift ratio, and peak acceleration, are systematically analyzed to evaluate the effectiveness of the proposed design theory. The distribution coefficient is recommended to range from 0.70 to 0.80 to balance the energy-dissipation contribution between the frame and the eccentric braces. In terms of the damper layout, the energy-dissipation contribution of the eccentric brace should differ among the lower, middle, and upper floors.

Keywords: precast composite structure; eccentrically braced frame; energy balance; seismic response; additional influence; multi-stage energy dissipation

1. Introduction

Owing to the advantages of steel and concrete, the composite structure is gradually being accepted worldwide. As steel typically exhibits significant hysteretic behavior, it serves as an important energy-dissipation unit in the structure [1–3]. Considering that bond slip results in the pinching effect in traditional reinforced concrete (RC) components [4,5], researchers have proposed an embedded steel configuration at the ends of beams to achieve seismic performance comparable to that of steel structures [6,7]. The partially steel-reinforced concrete (PSRC) beam has been demonstrated as feasible in several forms of section steel in frame structures [8,9]. For frame structures, the concrete filled steel tubular (CFST) column is considered a suitable member to pair with the PSRC beam due to the ease of steel connection installation [10–12]. Experimental results of substructure tests have consistently supported the seismic performance of these structures. However, owing to the high bearing capacity and compact cross-section of the CFST column [13–15], its lateral stiffness is relatively limited, and hence, frame structures using CFST columns are typically applied as multi-story buildings.

To broaden the application of the PSRC beam–CFST column frame, an additional lateral force-resisting system is necessary. The braced frame structure is considered an evolution of frame structures in high-rise building design, including the concentrically braced frame (CBF) [16,17] and the eccentrically braced frame (EBF) [18,19]. The EBF structure can achieve the advantages of both the pure frame and the CBF in the event of the frequent earthquake (FE) and the design basis earthquake (DBE). In addition, it can absorb the excessive large earthquake energy through the damper and protect the frame from damage [20]. As the energy-dissipating unit can be easily replaced, the EBF structure possesses excellent post-earthquake repairability. The Y-shaped layout is one of the typical configurations in the EBF structure [21,22], which is recommended to work with a shear panel damper (SPD). Lin et al. [23] pointed out that the PSRC beam–CFST column EBF structure exhibits a two-stage failure mode, namely the failure of the SPD and the beam plastic hinge.

As for the traditional design method of the EBF system, the cross-sectional dimensions of the structural members are usually determined based on the assumption of elastic theory [24]. To overcome this shortcoming, an energy-based theory is developed to accurately quantify the input and absorbed earthquake energy [25,26]. Based on the transformation of structural energy dissipation, the base shear force and the lateral force distribution can be calculated. Chao et al. [27] proposed a distribution pattern for the lateral force to uniformly distribute the interstory drift ratio (IDR), which has been demonstrated to be valid in both the frame and CBF structures. Considering the ductile seismic mechanism during the maximum considered earthquake (MCE), most design methods only take into account the positive effect of plastic deformation of the energy-dissipating unit. The additional load from the eccentric brace to the frame is considered through designing the brace and frame member in sequence together with an amplification factor for the frame member. Consequently, the frame tends to be too stiff to achieve a necessary yielding IDR for the eccentric brace. On one hand, it is difficult for the brace to reach the design yielding force, and the corresponding energy-dissipation capacity cannot function as expected [28–31]. On the other hand, no matter which form of the connection configuration is used, the overestimated additional load will cause waste in construction materials due to excessive safety factors [32,33]. Therefore, it is important to design the energy-dissipation unit with respect to the expected additional load and post-damage degradation of the frame [23,34,35]. Xiong et al. [36] recommended an energy distribution coefficient for the eccentric brace in an EBF steel structure. It is worth noting that the RC structure tends to cause a serious pinching effect in the hysteretic behavior [37,38]. The concrete-based EBF structure may be characterized by a dramatic difference from the steel structure.

In this study, a modified energy-based design method is proposed for the application of the precast composite EBF structure in high-rise buildings. The multi-stage energy-dissipation mechanism and the additional influence of the eccentric brace are considered systematically through the distribution coefficient for the energy dissipation and the layout of the damper. A 12-floor, three-bay precast PSRC beam–CFST column EBF structure is designed. Through a series of nonlinear time-history analyses, the effectiveness of the modified method is evaluated through critical seismic responses, including the maximum interstory drift ratio (MIDR), the residual interstory drift ratio (RIDR), and the peak acceleration.

2. Design Method

To effectively predict structural energy absorption, the seismic input energy is considered fixed, depending on the target IDR during the MCE. To ensure structural safety, the total energy dissipation of the whole structure should exceed the expected input energy, and the energy should be distributed rationally among structural components, especially under the maximum considered earthquake, so that each element can contribute effectively [25,26,39]. As shown in Figure 1, the design process for the precast composite EBF structure consists of a total of nine steps.

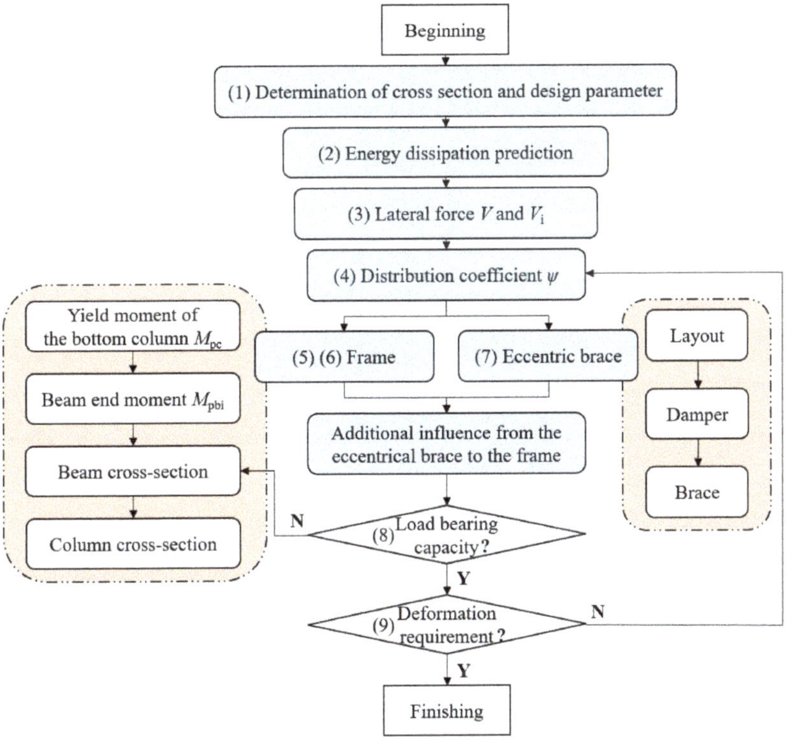

Figure 1. Design process.

(1) Determination of the cross-sectional dimensions and design parameters

The cross-sectional dimensions of the PSRC beam and the CFST column can be estimated based on the floor height and span. A series of seismic design parameters should be determined, including the fundamental period T, the yield IDR θ_y, the plastic IDR θ_p, the ultimate IDR θ_u, the seismic influence coefficient α_1, and the damping ratio ζ.

(2) Determination of the energy dissipation

Considering that the structure is in an elastic–plastic stage during the MCE, the input earthquake energy is equal to the energy dissipated by all structural members. To quantify the energy absorption, the dissipated energy is calculated as the work of the earthquake-equivalent static loading on the target displacement. The formula is as follows:

$$E_e + E_p = \gamma_s E_I \tag{1}$$

where E_e, E_p, and E_I denote the elastic energy absorption, the plastic energy absorption, and the input earthquake energy, respectively. The relationship among the three indices is plotted in Figure 2. The term γ_s is the modification factor for the input energy, accounting for the structural damping; K_s and α_s are determined as the initial lateral stiffness and post-yield lateral stiffness hardening ratio, respectively; and V_y, Δ_y, and Δ_u represent the design base shear force, the yield displacement, and the ultimate displacement during the MCE, respectively. Correspondingly, V_e and Δ_e represent the base shear force and elastic displacement for the equivalent elastic structure. Let the ductility reduction factor be $R_\mu = V_e/V_y$, and the displacement ductility coefficient $\mu_s = \Delta_u/\Delta_y$. Newmark and Hill [40] proposed Formula (2) to define the relationship among R_μ, μ_s, and T. Therefore, the energy dissipation can be calculated through Formulas (3)–(5).

$$\gamma_s = \frac{2\mu_s - 1 + \alpha_s(\mu_s - 1)^2}{R_\mu^2} \tag{2}$$

$$E_I = \frac{1}{2}M\left(\frac{T}{2\pi}\alpha_1 g\right)^2 \tag{3}$$

$$E_e = \frac{1}{2}M\left(\frac{T}{2\pi}\cdot\frac{V}{W}\cdot g\right)^2 \tag{4}$$

$$E_p = \frac{WT^2 g}{8\pi^2}\left[\gamma_s\alpha_1 - \left(\frac{V}{W}\right)^2\right] \tag{5}$$

where M and W denote the total mass and weight of the structure and V represents the structural design base shear.

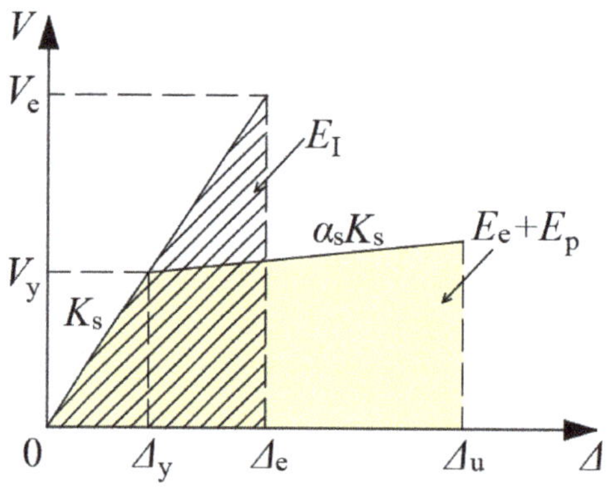

Figure 2. Energy balance.

(3) Determination of the lateral force

To utilize the excellent seismic performance of the EBF structure, the distribution pattern for the lateral force proposed by Chao et al. [27] is adopted to maintain the IDR uniform during the structural plastic stage. The lateral force distribution coefficient β_i for the ith floor can be represented as follows:

$$\beta_i = \frac{V_i}{V_n} = \left(\frac{\sum_{j=i}^{n} W_j h_j}{W_n h_n}\right)^{0.75T^{-0.2}} \tag{6}$$

where W_j and W_n denote the representative gravity load of the jth floor and the top floor, respectively; V_j and V_n represent the shear force of the jth floor and the top floor, respectively; and h_j and h_n denote the distance from the jth floor or the top floor to the bottom, respectively. The analytical model is shown in Figure 3. Then, the shear force of the ith floor can be obtained as follows:

$$V_i = (\beta_i - \beta_{i+1})V_n = (\beta_i - \beta_{i+1})\left(\frac{W_n h_n}{\sum_{j-i}^{n} W_j h_j}\right)^{0.75T^{-0.2}} \tag{7}$$

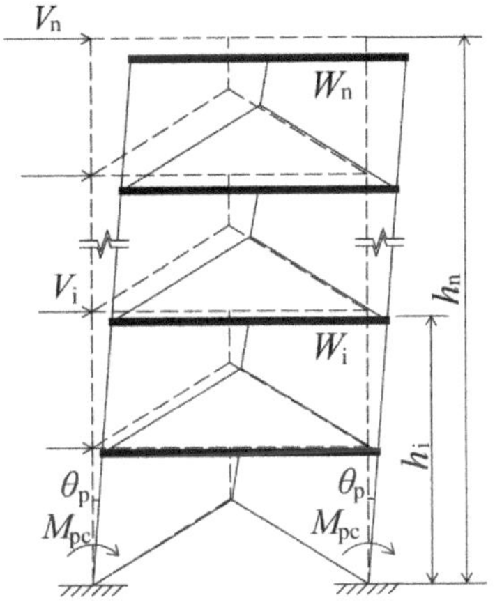

Figure 3. Analytical model.

According to θ_p, the plastic energy dissipation corresponding to the IDR can be obtained as follows:

$$E_p = \sum_{i=1}^{n} V_i h_i \theta_p \tag{8}$$

Substituting the above equation into Equation (5), the structural design base shear can be calculated as follows:

$$V = \frac{-\lambda + \sqrt{\lambda^2 + 4\gamma_s \alpha_1^2}}{2} W \tag{9}$$

$$\lambda = \left(\frac{8\pi^2\theta_p}{T^2 g}\right)\left(\sum_{i=1}^{n}(\beta_i - \beta_{i+1})h_i\right)\left(\frac{W_n h_n}{\sum_{j=i}^{n} W_j h_j}\right)^{0.75T^{-0.2}} \tag{10}$$

(4) Determination of the distribution coefficient ψ for the energy dissipation

The index can be considered the ratio of the energy dissipation of the damper to the total energy absorbed by the structural system. For the composite structure, on one hand, the excessive contribution of the damper will underestimate the energy-dissipation capacity of the frame itself. On the other hand, a damper with a large energy absorption can transfer a relatively large additional force to the frame. Additionally, the concrete damage and bond slip potentially result in a degradation of the seismic performance. Therefore, the value for ψ is recommended within 0.70–0.80, which will be discussed in Section 3.4.

(5) Design of the frame beam

Although the total energy should be dissipated through the plastic hinge deformation in the beam member to prevent the global structural failure, it is difficult to avoid all the column members from participating in the energy dissipation. From a construction cost perspective, the limited contribution of the bottom column is considered feasible under the

MCE. As shown in Formula (11), the energy dissipation of the frame is considered equal to the work done by both the beam plastic hinge and the bottom column.

$$(1-\psi)\sum_{i=1}^{n} V_i h_i \theta_p = \sum_{i=1}^{n} 2\eta M_{pbi}\theta_p + 2M_{pc}\theta_p \tag{11}$$

where M_{pbi} and M_{pc} denote the yield moment of the beam end at the ith floor and the bottom column, respectively. Note that the scenario where all beam ends yield under the MCE condition does not align with the actual seismic damage patterns of frame structures. Reference [36] introduces an energy-dissipation reduction coefficient η to account for the contribution of the partial beam plastic hinges. The recommended value is set between 0.5 and 1.0. According to the structural yield mechanism, M_{pc} can be calculated as follows:

$$2M_{pc}\theta = \frac{1}{2}\varphi V h_1 \theta \tag{12}$$

where θ is the rotation of the bottom column. φ represents the over-strength of the bottom column. As for the PSRC beam–CFST column, φ can be taken as 1.20 according to the experimental result of six corresponding joint specimens [8]. Additionally, according to Reference [11], the PSRC beam can be initially designed as the steel-reinforced concrete beam. Then, the cross-sectional dimensions of the embedded section steel can be determined according to the constructional requirements, which depend on the following two considerations:

(1) There is weakness around the connection area between the embedded section and the CFST column. According to AISC 358-16 [41] with FEMA-350 [42], the connection should be set with a distance of a + 0.5b to the column side, where a = (0.5~0.75)b_f and b = (0.65~0.85)h. The terms b_f and h represent the width and height of the embedded section, respectively;

(2) The section steel should be inserted into the reinforced beam with adequate embedded length to avoid local failure. Guo et al. [43] suggest that the embedded length should be no less than 200 mm. Therefore, the distance is set as (a + b + 200) from the end of the embedded section to the column edge.

(6) Design of the frame column

The moment and axial force of the side and middle columns are shown in Figure 4 as an isolated unit. Therefore, the shear forces can be expressed as follows:

$$V_{cn,S} = \frac{\sum_{i=1}^{n} \xi_i M_{pbi} + \frac{h_c}{2}\sum_{i=1}^{n} V_{pbi} + M_{pc}}{\sum_{i=1}^{n}(\beta_i - \beta_{i+1})h_i} \tag{13}$$

$$V_{cn,M} = \frac{\sum_{i=1}^{n} \xi_i \left(M_{pbi,L} + M_{pbi,R}\right) + \frac{h_c}{2}\sum_{i=1}^{n}\left(V_{pbi,L} + V_{pbi,R}\right) + M_{pc}}{\sum_{i=1}^{n}(\beta_i - \beta_{i+1})h_i} \tag{14}$$

where h_c represents the height of the column cross-section and ξ_i denotes the amplification factor for the moment of the beam end; the amplification factor can be set between 1.0 and 1.1, according to Reference [44].

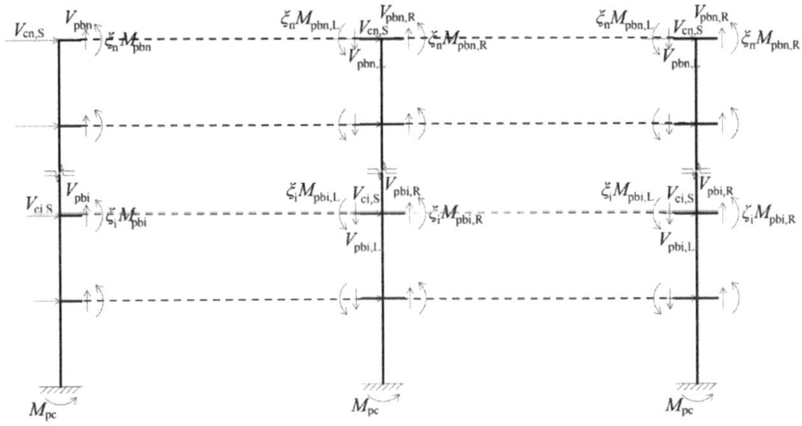

Figure 4. Isolated unit.

Then, the load-bearing capacity of the column can be calculated as follows:

$$M_{ci,S} = \sum_{j=i}^{n} \xi_j M_{pbj} + \frac{h_c}{2}\sum_{j=i}^{n} V_{pbj} + \sum_{j=i}^{n}(h_j - h_{j-1})V_{cj,S} \quad (15)$$

$$M_{ci,M} = \sum_{j=i}^{n} \xi_j \left(M_{pbj,L} + M_{pbj,R}\right) + \frac{h_c}{2}\sum_{j=i}^{n} \left(V_{pbj,L} + V_{pbj,R}\right) + \sum_{j=i}^{n}(h_j - h_{j-1})V_{cj,M} \quad (16)$$

$$N_{ci,S} = \sum_{j=i}^{n} V_{pbj} + \sum_{j=i}^{n} N_{Wj,S} \quad (17)$$

$$N_{ci,M} = \sum_{j=i}^{n} \left(V_{pbj,L} - V_{pbj,R}\right) + \sum_{j=i}^{n} N_{Wj,M} \quad (18)$$

where $M_{ci,S}$ and $M_{ci,M}$ denote the moment of the end of the side column and middle column, respectively; $N_{ci,S}$ and $N_{ci,M}$ represent the axial force of the end of the side column and middle column, respectively; and $N_{wj,S}$ and $N_{wj,M}$ denote the vertical load transferred from the jth floor to the side column and middle column, respectively.

(7) Design of the eccentric brace system

Different forms of SPDs can be used as the energy-dissipation unit in the eccentric brace system. Reference [45] recommends a modified strip model to predict the hysteretic behavior and critical mechanical properties of SPD. Then, the cross-sectional dimensions of the SPD, namely the height, width, and thickness of the shear panel, can be determined according to the total energy-dissipation demands calculated from the distribution coefficient, ψ. The formula is listed as follows:

$$\psi \sum_{i=1}^{n} V_i h_i \theta_p = \sum_{i=1}^{n} W_{i,damper} \quad (19)$$

where $W_{i,damper}$ represents the energy absorbed by the damper in the ith floor with respect to the target displacement. As the eccentric brace remains elastic during the whole process, its dimensions can be calculated according to the ultimate shear force of the damper.

(8) Determination of the additional influence of the eccentric brace system

The additional influence mainly manifests as the moment and axial force on the beam and the column. With respect to the layout of the eccentric brace, namely symmetrical and asymmetrical, the degree of the additional influence tends to be different between the side and middle columns. Therefore, the load-bearing capacity of the frame member should be compared with the sum of the internal force calculated in steps (5) and (6) and the additional influence based on the ultimate force of the damper.

(9) Check the deformation requirement

The time-history analysis should be conducted on the precast composite EBF structure. The critical seismic response should meet the coded deformation requirement, including the MIDR, the RIDR, and the peak acceleration.

3. Case Study

3.1. Benchmark Model

To verify the aforementioned design method, a 12-floor, three-bay precast PSRC beam–CFST column EBF office structure is designed for a building on a site with an intensity 8 earthquake (see Figure 5). Heights of the first floor and other floors are set as 4000 mm and 3300 mm, respectively. The seismic design group and site class are Group III and Class II according to CB50011-2010 [24]. Dead loads are taken as 5.00 kN/m^2 for both the standard floor and the roof. Live loads of the standard floor and the roof are set as 2.00 kN/m^2 and 0.45 kN/m^2, respectively. The factors of the representative gravity load are taken as 1.0 and 0.5 for the dead load and the live load [46]. The shear panel damper is made of Q235B, while other steel members are all made of Q355B. Strength grades of the concrete and the reinforcement are taken as C40 and HRB400, respectively. According to ASCE 7-16 [46], the fundamental period of the structure is set as 1.20 s. Other design parameters are summarized in Table 1.

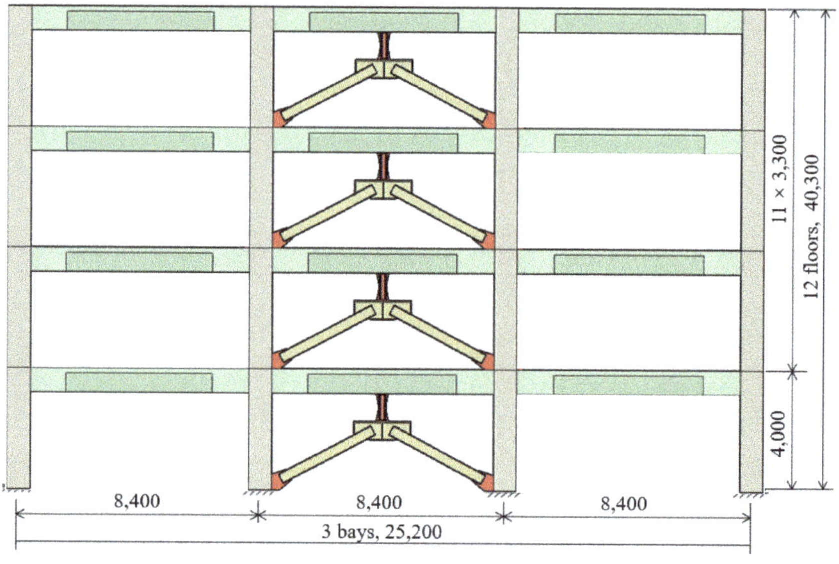

Figure 5. Dimensions of the benchmark model (unit: mm).

Table 1. Design parameter.

Design Parameter	Value	Design Parameter	Value
T	1.20 s	α_1	0.325
θ_y	0.005	R_μ	2.5
θ_p	0.015	μ_s	2.5
θ_u	0.020	γ_s	0.688
α_s	0.135	ζ	0.05

3.2. Numerical Model

Reference [23] experimentally investigates the seismic performance of the PSRC beam–concrete encased CFST column frame with a duplex assembled I-shaped SPD. Therefore, the composite EBF structure is selected as the actual seismic system in this section. Nowadays, many advanced techniques have been used to comprehensively simulate the seismic response of a frame structure using a 3D solid finite element method [47,48]. However, considering the analysis efficiency and hardware dependency, the numerical model (see Figure 6) is established as a fiber beam element model through OpenSees [49] to characterize the hysteretic behavior of the EBF system. Considering that the force-based element could achieve ideal accuracy with few elements [50], the beam and the column models are represented by the forceBeamColumn elements with five integration points set along the longitudinal direction. The cross-section of each element consists of unconfined concrete, confined concrete, and steel through the built-in Concrete01 and Steel02 materials. As the plastic damage and the confinement effect significantly influence the simulation accuracy [51–53], compressive strengths of the confined tube concrete and core beam concrete are regarded as 1.57 times and 1.19 times higher than the compressive strength of the unconfined concrete, respectively. Material properties are listed in Table 2. The pretension method is characterized by one truss element and two elasticBeamColumn elements in each beam. The second-order effect of gravity is considered by introducing the P-delta coordinate transformation. The scissor model, which consists of one zeroLength element and four elasticBeamColumn elements, is adopted to characterize the shear behavior of the beam-column joint zone. The rotational constitutive relationship of the core zeroLength element is determined according to the superposed model [54]. Rayleigh damping is typically assumed for the first two modes of the building.

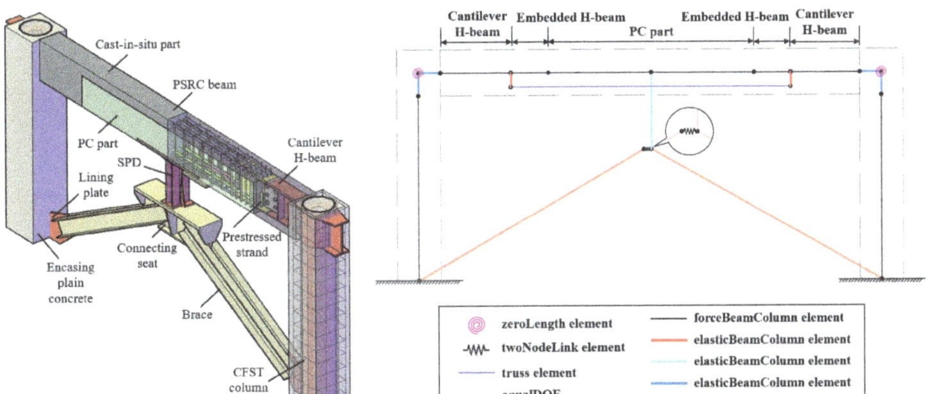

(**a**) (**b**)

Figure 6. Elements in the numerical model: (**a**) structural concept; (**b**) simulation.

Table 2. Material properties.

Material	Type	Nominal Diameter/Thickness (mm)	Yield Strength (Mpa)	Ultimate Strength (Mpa)	Elastic Modulus (Mpa)
Steel	Stirrup	6	614.1	709.8	208.1
	Longitudinal bar	10	542.5	635.2	205.3
		14	441.4	615.0	201.9
	Shear plate in SPD	8	299.2	443.5	202.0
	Tube	7	430.0	562.3	204.8
	Prestressed strand	10.8	1633.1	1751.3	203.1
Unconfined concrete	Encasing concrete	-	-	30.3	29.9
	Precast concrete	-	-	38.0	32.1
	In situ concrete	-	-	39.6	32.5

For the eccentric brace, as the brace maintains elasticity throughout the process, the elasticBeamColumn element is used to represent the load-transfer mechanism. With respect to the duplex-configured I-shaped SPD, the modified strip model is used to characterize the hysteretic behavior according to Reference [45] and is represented by twoNodeLink elements consisting of the parallel material combining Steel01 and Hysteretic materials. With respect to the additional moment from the damper to the beam, an elasticBeamColumn element is adopted with the same height as the damper.

The numerical seismic performance is plotted in Figure 7 from aspects of both the hysteresis curve and the energy-dissipation curve. As there is little difference between the numerical model and the tested specimen, the model establishment can be demonstrated to be valid.

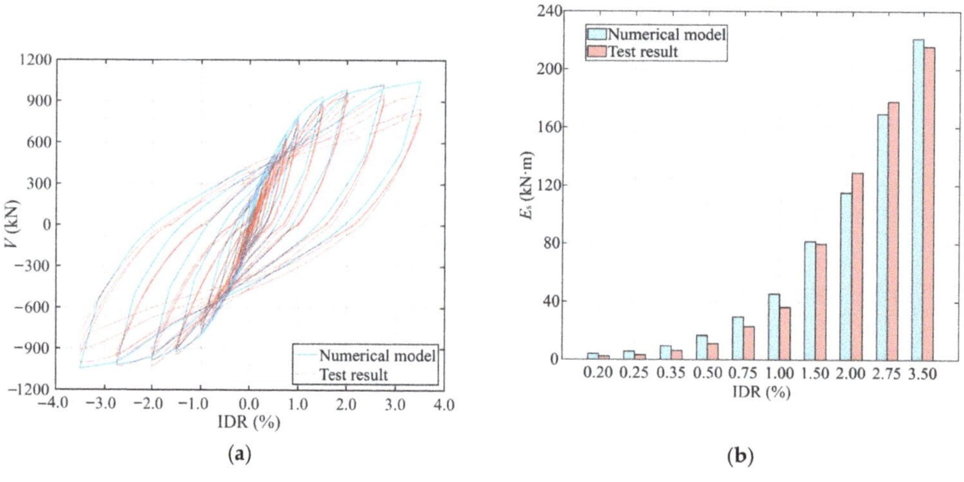

Figure 7. Comparison of the numerical model and test results: (**a**) hysteresis curve; (**b**) energy-dissipation capacity.

3.3. Selection of the Ground Motion

As the environment of the case is similar to Site Class D in the United States, the ground motion suggested by FEMA-P695 [55] is used in the nonlinear time-history analysis. Twenty-two groups of ground motions are obtained from the PEER database [56] and listed in Table 3. To reflect different intensities of the FE, DBE, and MCE, the corresponding peak ground accelerations (PGAs) are modulated to 70 gal, 200 gal, and 400 gal [24], respectively.

Table 3. Ground motion.

No.	Name	Monitoring Station	Magnitude	Seismic Component	PGA (g)
GM1	Northridge	Beverly Hills-Mulhol	6.7	MUL009	0.416
GM 2	Northridge	Canyon Country-WLC	6.7	LOS000	0.410
GM 3	Duzce, Turkey	Bolu	7.1	BOL090	0.822
GM 4	Hector Mine	Hector	7.1	HEC090	0.338
GM 5	Imperial Valley	Delta	6.5	H-DLT262	0.238
GM 6	Imperial Valley	El Centro Array#11	6.5	H-E11140	0.364
GM 7	Kobe, Japan	Nishi-akashi	6.9	NIS090	0.503
GM 8	Kobe, Japan	Shin-Osaka	6.9	SHI000	0.243
GM 9	Kocaeli, Turkey	Duzce	7.5	DZC180	0.312
GM 10	Kocaeli, Turkey	Arcelik	7.5	ARC000	0.219
GM 11	Landers	Yermo Fire Station	7.3	YER270	0.244
GM 12	Landers	Coolwater	7.3	CLW-TR	0.417
GM 13	Loma Prieta	Capitola	6.9	CAP090	0.443
GM 14	Loma Prieta	Gilroy Array#3	6.9	G03000	0.555
GM 15	Manjil, Iran	Abbar	7.4	ABBAR-L	0.515
GM 16	Superstition Hills	El Centro Imp. Co.	6.5	B-ICC090	0.238
GM 17	Superstition Hills	Poe Road	6.5	B-POE360	0.300
GM 18	Cape Mendocino	Rio Dell Overpass	7.0	RIO270	0.385
GM 19	Chi-chi, Taiwan	CHY101	7.6	CHY101-E	0.340
GM 20	Chi-chi, Taiwan	TCU045	7.6	TCU045-N	0.507
GM 21	San Fernando	LA-Hollywood Stor	6.6	PEL090	0.210
GM 22	Friuli, Italy	Tolmezzo	6.5	A-TMZ270	0.315

3.4. Determination of the Distribution Coefficient for the Energy Dissipation

To investigate the effective value of the distribution coefficient, ψ, the range recommended in Reference [36] for the steel structure is expanded to 0.60–0.90. Maintaining the total energy dissipation identical across all models, a total of seven tested models (ψ = 0.60, 0.65, 0.70, 0.75, 0.80, 0.85, 0.90) are designed according to the method proposed in Section 2. The details of the cross-sectional dimensions are listed in Appendix A. Each model is analyzed under 22 ground motion records listed in Table 3 with the modulated PGAs for the FE, DBE, and MCE.

3.4.1. Maximum Interstory Drift Ratio

The distribution of the MIDR for each model under various seismic actions is plotted in Figure 8. To accurately analyze the overall structural response under 22 ground motion records, both the mean value (μ) and the mean value plus standard deviation ($\mu + \sigma$) are displayed as the solid line and the dashed line, respectively. In general, the mean values of the MIDR are less than the IDR limit required according to Reference [24], namely 0.4% for the FE, 1.0% for the DBE, and 2.0% for the MCE. With the increase in the floor level, the MIDR initially increases and then decreases. The peak MIDR is found to appear from the second floor to the sixth floor.

Since the value of ψ denotes the design contribution of the displacement-based damper, a limited IDR inevitably has a negative influence on the energy dissipation of the EBF system, especially for the upper floor (e.g., the floor higher than the eighth floor). When ψ ranges from 0.70 to 0.80, the average MIDR under the MCE varies from 0.75% to 1.25%. The drift distribution remains uniform, with no localized weak stories. For ψ = 0.60, the main structure absorbs a higher proportion of energy, leading to slightly larger deformations (e.g., an average drift of 0.91% under the MCE). The maximum IDR tends to occur on the sixth and seventh floors. With ψ exceeds 0.85, the distribution of the IDR becomes gradually uneven with the earthquake intensity increasing. Therefore, the energy distribution coefficient ψ is recommended to be within 0.70–0.80 to achieve a uniform MIDR.

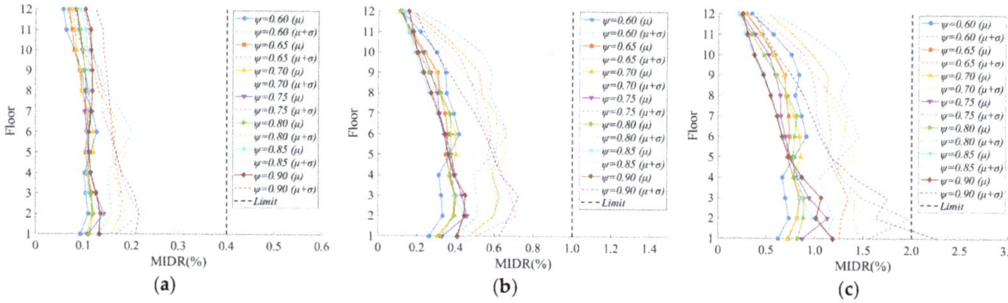

Figure 8. Maximum interstory drift ratio: (**a**) FE; (**b**) DBE; (**c**) MCE.

3.4.2. Residual Interstory Drift Ratio

The RIDR is an important seismic performance indicator to evaluate the post-earthquake structural reparability. Reference [57] suggests the RIDR of a braced frame structure should be less than 0.50%. The distribution of RIDR of each model under various seismic actions is plotted in Figure 9. It can be found that RIDRs of most curves are less than the required limit, except for the dashed curve of ψ at 0.85 and 0.90 in Figure 9c. The phenomenon results in the excessive contribution of the damper with a large ψ. Since its energy dissipation results from the plastic deformation of the shear panel, the corresponding excessive restoring force prevents the frame from returning to the initial state. Therefore, the energy distribution coefficient ψ is recommended to be less than 0.85 to achieve acceptable post-earthquake repairability.

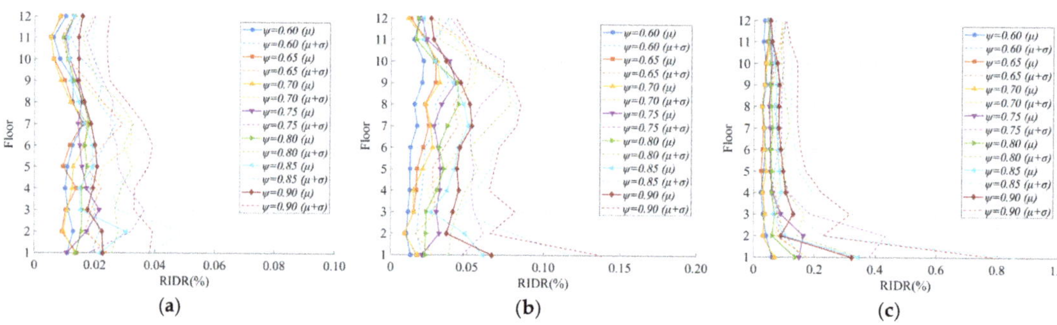

Figure 9. Residual interstory drift ratio: (**a**) FE; (**b**) DBE; (**c**) MCE.

3.4.3. Peak Acceleration

The peak acceleration of each model is displayed in Figure 10. Regardless of the intensity of the input earthquake, there are limited differences in curves with respect to floors below the seventh floor. Note that the peak acceleration in a common frame usually rises with the increase in the floor level. For the FE curve, peak accelerations of the upper floors are significantly decreased, except for the top floor owing to the whipping effect. Therefore, it can be concluded that the eccentric brace system is conducive to controlling the acceleration of the floors close to the top floor. The influence of the distribution coefficient can be observed in curves of the DBE and the MCE with the floor above the eighth floor. The phenomenon indicates that the EBF system gradually functions through plastic deformation. For the DBE curve, there is a dramatic difference between ψ = 0.60 and other values. For the

MCE curve, the tendency of curves can be sorted into four groups, namely 0.60, 0.65 to 0.70, 0.75 to 0.80, and 0.85 to 0.90. An increase in ψ can dramatically enhance the contribution of the damper, and hence, control the seismic response of the upper floor. Considering that curves of $\psi \geq 0.65$ characterize a similar peak acceleration in the top floor, the range can be regarded as suitable with respect to the seismic performance and the cost of damper.

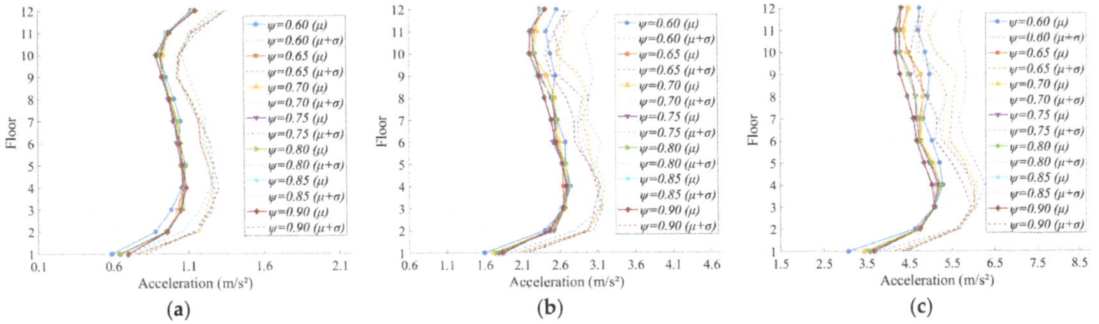

Figure 10. Peak acceleration: (**a**) FE; (**b**) DBE; (**c**) MCE.

3.5. Layout of Dampers

According to Figure 8, it can be found that the distribution of the MIDR varies with the change in the distribution coefficient ψ and the floor level. A small ψ results in a uniform MIDR in all floors, which is consistent with the assumed expectation proposed by Chao et al. [27]. As for a large ψ, the seismic response tends to appear in a local uniform style. For example, under different intensities, the MIDR is mainly concentrated in the middle and lower parts, e.g., the second, third, and fifth floors, while the peak acceleration is concentrated in the fourth and top floors. The MIDR and RIDR achieve relatively small values from the eighth floor to the top floor. The phenomenon mainly results from a totally consistent layout of dampers along each floor. To stress the uniformity, the layout of the damper should be adapted according to the floor levels. Therefore, the building is divided into three categories for the convenience of engineering design, namely the lower floors (first to third), the middle floors (fourth to seventh), and the upper floors (eighth and twelfth). Considering the value of ψ recommended in Section 3.4, only tested models with a ψ ranging from 0.70 to 0.80 are updated in this section. The damper configuration is redesigned as shown in Table A4. The energy-dissipation capacity of dampers in lower, middle, and upper floors accounts for 30%, 35%, and 35% of the total energy dissipated in the eccentric brace system, respectively. The beam and the column are identical to those listed in Tables A1 and A2. The different analysis results are listed in Table 4 with respect to the average value and the coefficient of variation (CoV). It can be found that there is little difference in the average value between the original and updated models. However, note that the CoV significantly decreases, especially for the cases under the DBE and MCE.

Figures 11–13 compare the seismic performance indicators before and after the model update, including the MIDR, RIDR, and peak acceleration. Under the FE, it should be mentioned that the updated layout of the damper leads to a significant increase in the upper floors, especially when ψ is equal to 0.75 or 0.80. However, as the corresponding values are relatively small and far from the limit, the amplification effect can be regarded as acceptable.

Table 4. Comparison of seismic responses.

ψ	Seismic Intensity	Value Type	MIDR		RIDR		Peak Acceleration	
			Original	Updated	Original	Updated	Original	Updated
0.70	FE	Average	0.10	0.10	0.011	0.012	0.98	0.98
		CoV	15.37%	12.65%	33.89%	26.14%	12.85%	12.77%
	DBE	Average	0.31	0.31	0.02	0.021	2.43	2.43
		CoV	30.14%	26.67%	35.91%	30.63%	11.19%	11.26%
	MCE	Average	0.69	0.69	0.039	0.039	4.65	4.65
		CoV	25.96%	23.07%	29.55%	28.27%	9.99%	10.15%
0.75	FE	Average	0.11	0.11	0.015	0.016	0.98	1.01
		CoV	14.18%	20.16%	21.81%	24.36%	12.73%	14.95%
	DBE	Average	0.31	0.31	0.03	0.028	2.42	2.45
		CoV	32.93%	26.11%	22.02%	20.51%	10.38%	10.93%
	MCE	Average	0.68	0.66	0.074	0.064	4.58	4.61
		CoV	35.65%	27.46%	53.94%	52.23%	9.51%	9.26%
0.80	FE	Average	0.11	0.11	0.015	0.017	0.98	1.02
		CoV	9.45%	20.50%	17.83%	28.75%	13.18%	14.82%
	DBE	Average	0.3	0.31	0.029	0.03	2.42	2.45
		CoV	31.78%	26.48%	31.9%	24.79%	11.05%	11.44%
	MCE	Average	0.67	0.67	0.063	0.06	4.58	4.6
		CoV	32.73%	26.50%	37.6%	35.69%	9.89%	9.75%

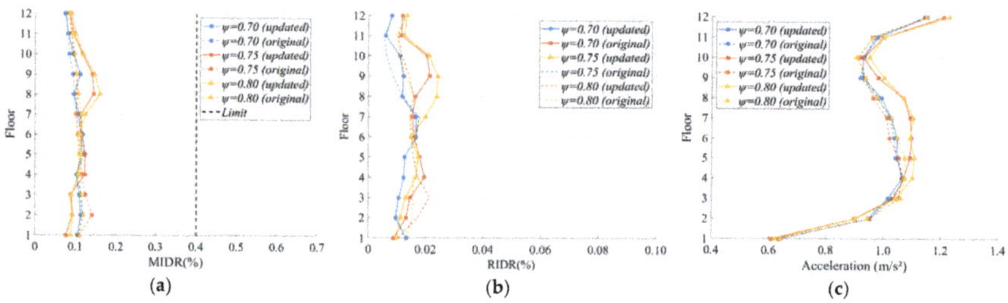

Figure 11. Seismic responses under FEs: (**a**) MIDR; (**b**) RIDR; (**c**) peak acceleration.

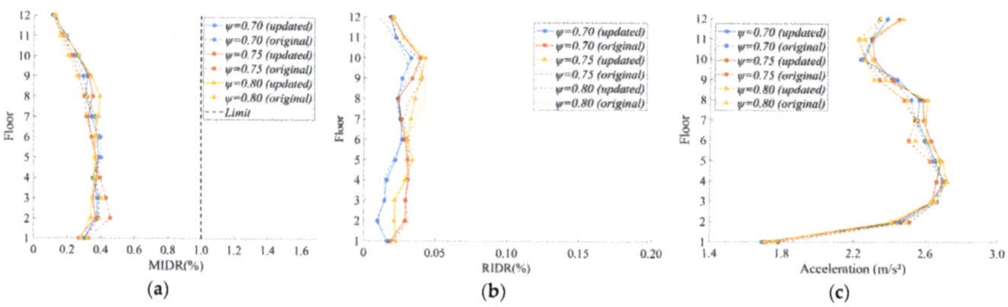

Figure 12. Seismic responses under DBEs: (**a**) MIDR; (**b**) RIDR; (**c**) peak acceleration.

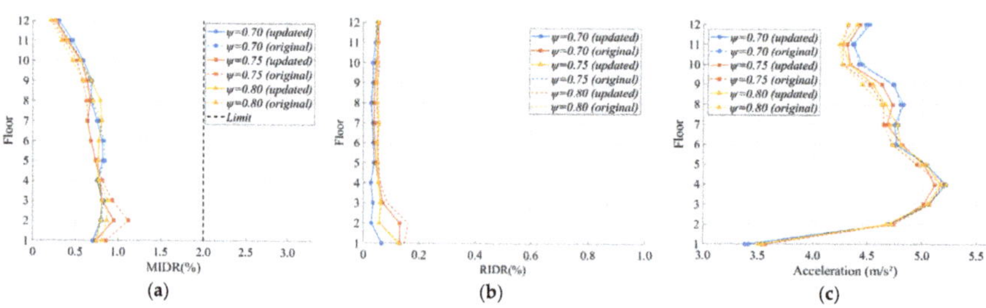

Figure 13. Seismic responses under MCEs: (**a**) MIDR; (**b**) RIDR; (**c**) peak acceleration.

As for the DBE, MIDR curves (see Figure 12a) of the updated models exhibit a trend of generally vertical development, except for the top three floors. In the same range, a larger RIDR can be observed in curves of updated models than those of original models (see Figure 12b). The phenomenon results from an increased energy absorption contribution of the plastic deformation of the damper, which potentially prevents the unchanged frame from returning to its initial state. According to Figure 12c, as curves are close to each other, it can be concluded that the layout has little influence on the peak acceleration.

For the MIDR curve under the MCE (see Figure 13a), an obvious decrease in the lower floors can be observed together with a slight increase in the middle and upper floors. Taking curves with $\psi = 0.75$ as an example, although the peak MIDR appears on the second floor in both original and updated models, the value decreases from 1.13% to 0.95%. There is no doubt that the drop of 15.93% can dramatically relieve the negative influence of the P-delta effect, especially for the frame member in the lower floors. The result demonstrates the effectiveness of increasing the energy-dissipation capacity around the bottom of the structure. Given the minimal difference observed between the original and updated models (see Figure 13b,c), the adjustment in the damper layout can be identified as the dominant contributor to mitigating the MIDR.

Figure 14 plots the comparison between the updated and original models of the case with $\psi = 0.75$, where dashed lines denote the average value of the corresponding sample. Generally, the MIDR does not change sharply between each pair of adjacent floors under the DBE and MCE, which meets the assumption proposed in Step (4). It can be found that the MIDRs of the middle floors are close to the average value for the updated model in the FE, DBE, and MCE, while a similar phenomenon is only observed in the original sample under the MCE. This phenomenon results from the excessive stiffness of the SPD designed for the middle and high floors of the original model. Therefore, the regional energy distribution strategy benefits the participation of all eccentric braces.

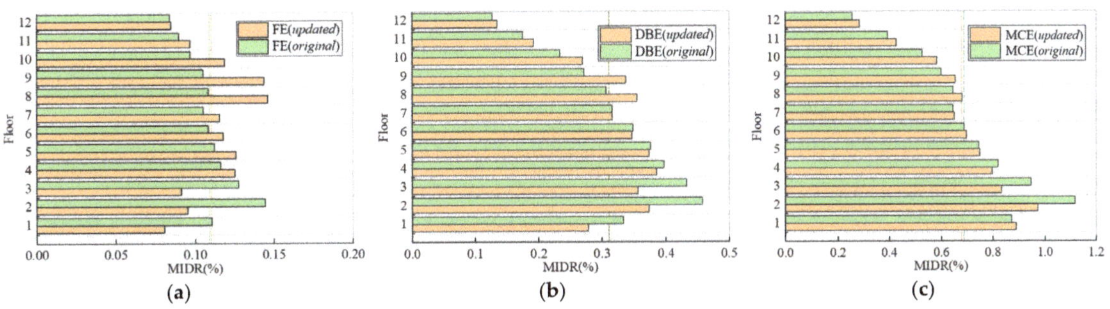

Figure 14. MIDR with $\psi = 0.75$: (**a**) FE; (**b**) DBE; (**c**) MCE.

4. Conclusions

This study presents a modified energy-based design method for the precast PSRC beam–CFST column EBF structure. The multi-stage energy-dissipation mechanism is systematically considered from the perspective of structural plastic energy absorption. The additional influence of the damper in the eccentric brace is accounted for, addressing the limitations in previous analysis. Based on the case study of a 12-floor, three-bay precast composite EBF structure, the following conclusions can be drawn:

(1) The traditional energy-based design method has been modified through a parallel design process that integrates both the eccentric brace and the frame members. Owing to both the energy absorption and additional load of the eccentric brace, the MIDR, the

RIDR, and the peak acceleration consistently meet the code requirements, demonstrating the feasibility of the proposed method.

(2) To represent the multi-stage energy-dissipation mechanism, the distribution coefficient ψ is introduced to balance the energy absorption between the frame and the eccentric brace. In terms of limiting the seismic response and construction cost, a range of 0.70 to 0.80 is recommended for ψ.

(3) The uniformity of the cross-sectional dimensions of the damper negatively impacts the overall seismic response. The excessive stiffness of the SPD on the upper floors leads to a reduction in the contribution of the energy absorption.

(4) A simplified layout, which divides the structure into the lower, middle, and upper floor range, is proposed to optimize the energy-dissipation contribution of the damper. This layout helps reduce the MIDR on the lower floors under the MCE while maintaining a uniform MIDR distribution across the middle floors.

Although the modified energy-based design method is primarily proposed for the precast PSRC beam–CFST column EBF structure in this study, it is expected to be applicable to other composite EBF structure with a steel beam or a steel-reinforced concrete beam. The damper layout method is a simplified approach designed for convenience in engineering practice. To further optimize the seismic performance of the precast EBF structure from a research perspective, the energy-dissipation contribution of each eccentric brace can be individually taken into account with respect to the floor level through a fragility analysis.

Author Contributions: Conceptualization, Y.L.; methodology, Y.L.; formal analysis, F.H and W.C; investigation, F.H., X.H., and G.Z.; writing—original draft preparation, F.H. and W.C.; writing—review and editing, F.H., X.H., and G.Z.; supervision, Y.L.; project administration, Y.L.; funding acquisition Y.L. All authors have read and agreed to the published version of the manuscript.

Funding: This research was funded by the National Natural Science Foundation of China (grant no. 52308257), the Basic Research Program of Jiangsu (grant no. BK20230399), the China Postdoctoral Science Foundation (grant no. 2023M741730), and the Natural Science Foundation of the Higher Education Institutions of Jiangsu Province (grant no. 23KJB560019).

Data Availability Statement: The original contributions presented in the study are included in the article. Further inquiries can be directed to the corresponding authors.

Acknowledgments: The authors gratefully acknowledge the financial support provided by the National Natural Science Foundation of China, the Natural Science Foundation of Jiangsu Province, the China Postdoctoral Science Foundation, and the Natural Science Foundation of the Higher Education Institutions of Jiangsu Province.

Conflicts of Interest: The authors declare no conflicts of interest.

Appendix A

For both the side column and the middle column, the outer diameter and the thickness of the steel tube are provided sequentially in Table A1. The thickness of the fireproof concrete encasing the tube and the embedded section steel is taken as 65 mm, according to [23].

An H-beam section is used as the embedded section steel for the beam. Table A2 lists the H-beam height, H-beam width, web thickness, and flange thickness sequentially.

Table A1. Cross-sectional dimensions for the column.

Floor	Distribution Coefficient ψ													
	0.60		0.65		0.70		0.75		0.80		0.85		0.90	
	Side	Middle	Side	Middle	Side	Middle	Side	Middle	Side	Middle	Side	Middle	Side	Middle
12	430 × 8	529 × 8	430 × 8	430 × 8	430 × 8	430 × 8	430 × 8	430 × 8	430 × 8	430 × 8	430 × 8	430 × 8	430 × 8	430 × 8
11	430 × 8	529 × 8	430 × 8	430 × 8	430 × 8	430 × 8	430 × 8	430 × 8	430 × 8	430 × 8	430 × 8	430 × 8	430 × 8	430 × 8
10	430 × 8	529 × 8	430 × 8	430 × 8	430 × 8	430 × 8	430 × 8	430 × 8	430 × 8	430 × 8	430 × 8	430 × 8	430 × 8	430 × 8
9	430 × 8	529 × 8	430 × 8	430 × 8	430 × 8	430 × 8	430 × 8	430 × 8	430 × 8	430 × 8	430 × 8	430 × 8	430 × 8	430 × 8
8	430 × 8	630 × 8	430 × 8	529 × 8	430 × 8	529 × 8	430 × 8	430 × 8	430 × 8	430 × 8	430 × 8	430 × 8	430 × 8	430 × 8
7	430 × 8	630 × 8	430 × 8	529 × 8	430 × 8	529 × 8	430 × 8	529 × 8	430 × 8	430 × 8	430 × 8	430 × 8	430 × 8	430 × 8
6	430 × 8	630 × 8	430 × 8	529 × 8	430 × 8	529 × 8	430 × 8	529 × 8	430 × 8	529 × 8	430 × 8	430 × 8	430 × 8	430 × 8
5	529 × 8	630 × 8	430 × 8	630 × 8	430 × 8	529 × 8	430 × 8	529 × 8	430 × 8	529 × 8	430 × 8	430 × 8	430 × 8	430 × 8
4	529 × 8	720 × 10	430 × 8	630 × 8	430 × 8	630 × 8	430 × 8	529 × 8	430 × 8	529 × 8	430 × 8	430 × 8	430 × 8	430 × 8
3	529 × 8	720 × 10	430 × 8	630 × 8	430 × 8	630 × 8	430 × 8	529 × 8	430 × 8	529 × 8	430 × 8	529 × 8	430 × 8	430 × 8
2	529 × 8	720 × 10	529 × 8	630 × 8	529 × 8	630 × 8	430 × 8	529 × 8	430 × 8	630 × 8	430 × 8	529 × 8	430 × 8	529 × 8
1	529 × 8	720 × 10	529 × 8	630 × 8	529 × 8	630 × 8	430 × 8	630 × 8	430 × 8	630 × 8	430 × 8	529 × 8	430 × 8	529 × 8

Table A2. Cross-sectional dimensions for the beam.

Floor	Distribution Coefficient ψ						
	0.60	0.65	0.70	0.75	0.80	0.85	0.90
12	450 × 300 × 11 × 18	400 × 300 × 10 × 16	400 × 300 × 10 × 16	400 × 300 × 10 × 16	400 × 300 × 10 × 16	400 × 300 × 10 × 16	350 × 300 × 9 × 14
11	450 × 300 × 11 × 18	400 × 300 × 10 × 16	400 × 300 × 10 × 16	400 × 300 × 10 × 16	400 × 300 × 10 × 16	400 × 300 × 10 × 16	350 × 300 × 9 × 14
10	450 × 300 × 11 × 18	450 × 300 × 11 × 18	450 × 300 × 11 × 18	400 × 300 × 10 × 16	400 × 300 × 10 × 16	400 × 300 × 10 × 16	400 × 300 × 10 × 16
9	450 × 300 × 11 × 18	450 × 300 × 11 × 18	450 × 300 × 11 × 18	400 × 300 × 10 × 16	400 × 300 × 10 × 16	400 × 300 × 10 × 16	400 × 300 × 10 × 16
8	500 × 300 × 11 × 18	450 × 300 × 11 × 18	450 × 300 × 11 × 18	450 × 300 × 11 × 18	400 × 300 × 10 × 16	400 × 300 × 10 × 16	400 × 300 × 10 × 16
7	500 × 300 × 11 × 18	450 × 300 × 11 × 18	450 × 300 × 11 × 18	450 × 300 × 11 × 18	400 × 300 × 10 × 16	400 × 300 × 10 × 16	400 × 300 × 10 × 16
6	500 × 300 × 11 × 18	500 × 300 × 11 × 18	450 × 300 × 11 × 18	450 × 300 × 11 × 18	400 × 300 × 10 × 16	400 × 300 × 10 × 16	400 × 300 × 10 × 16
5	500 × 300 × 11 × 18	500 × 300 × 11 × 18	450 × 300 × 11 × 18	450 × 300 × 11 × 18	400 × 300 × 10 × 16	400 × 300 × 10 × 16	400 × 300 × 10 × 16
4	600 × 300 × 12 × 20	500 × 300 × 11 × 18	500 × 300 × 11 × 18	450 × 300 × 11 × 18	450 × 300 × 11 × 18	400 × 300 × 10 × 16	400 × 300 × 10 × 16
3	600 × 300 × 12 × 20	500 × 300 × 11 × 18	500 × 300 × 11 × 18	450 × 300 × 11 × 18	450 × 300 × 11 × 18	400 × 300 × 10 × 16	400 × 300 × 10 × 16
2	600 × 300 × 12 × 20	500 × 300 × 11 × 18	500 × 300 × 11 × 18	450 × 300 × 11 × 18	450 × 300 × 11 × 18	400 × 300 × 10 × 16	400 × 300 × 10 × 16
1	600 × 300 × 12 × 20	500 × 300 × 11 × 18	500 × 300 × 11 × 18	450 × 300 × 11 × 18	450 × 300 × 11 × 18	400 × 300 × 10 × 16	400 × 300 × 10 × 16

The duplex assembled I-shaped SPD is designed according to Reference [45]. Table A3 sequentially lists the height, width, and thickness of each shear panel. Note that each duplex-configured I-shaped SPD consists of two identical shear panels.

Table A3. Cross-sectional dimensions for the damper.

Floor	Distribution Coefficient ψ						
	0.60	0.65	0.70	0.75	0.80	0.85	0.90
12	374-235-8	374-250-8	374-265-8	374-280-8	374-290-8	374-305-8	374-315-8
11	374-235-8	374-250-8	374-265-8	374-280-8	374-290-8	374-305-8	374-315-8
10	374-235-8	374-250-8	374-265-8	374-280-8	374-290-8	374-305-8	374-315-8
9	374-235-8	374-250-8	374-265-8	374-280-8	374-290-8	374-305-8	374-315-8
8	374-235-8	374-250-8	374-265-8	374-280-8	374-290-8	374-305-8	374-315-8
7	374-235-8	374-250-8	374-265-8	374-280-8	374-290-8	374-305-8	374-315-8
6	374-235-8	374-250-8	374-265-8	374-280-8	374-290-8	374-305-8	374-315-8
5	374-235-8	374-250-8	374-265-8	374-280-8	374-290-8	374-305-8	374-315-8
4	374-235-8	374-250-8	374-265-8	374-280-8	374-290-8	374-305-8	374-315-8
3	374-235-8	374-250-8	374-265-8	374-280-8	374-290-8	374-305-8	374-315-8
2	374-235-8	374-250-8	374-265-8	374-280-8	374-290-8	374-305-8	374-315-8
1	374-235-8	374-250-8	374-265-8	374-280-8	374-290-8	374-305-8	374-315-8

Appendix B

Table A4. Cross-sectional dimensions for the damper, considering the layout.

Floor	Distribution Coefficient ψ		
	0.70	0.75	0.80
12	328-245-7	328-260-7	374-260-8
11	328-245-7	328-260-7	374-260-8
10	328-245-7	328-260-7	374-260-8
9	328-245-7	328-260-7	374-260-8
8	328-245-7	328-260-7	374-260-8
7	374-280-8	374-280-8	374-300-8
6	374-280-8	374-280-8	374-300-8
5	374-280-8	374-280-8	374-300-8
4	374-280-8	374-280-8	374-300-8
3	374-280-8	374-340-8	422-340-9
2	374-280-8	374-340-8	422-340-9
1	374-280-8	374-340-8	422-340-9

References

1. Liu, Y.; Ma, Z.; Lv, X. Seismic performance study of a novel steel truss energy dissipation coupling beam. *J. Constr. Steel Res.* **2024**, *216*, 108556. [CrossRef]
2. Jia, M.; Chen, Y.; Jin, P. Development and validation analysis of a steel-lead hybrid dual-yield BRB for multi-stage seismic energy dissipation. *Thin Wall. Struct.* **2024**, *199*, 111822. [CrossRef]
3. Chen, J.; Abbas, N.; Sun, J.; Furuta, T.; Wei, Y.; Xiong, H. An innovative energy-dissipation angle bracket for CLT structures: Experimental tests and numerical analysis. *Eng. Struct.* **2024**, *314*, 118381. [CrossRef]
4. Zhang, W.; Yang, X.; Lin, J.; Lin, B.; Huang, Y. Experimental and numerical study on the torsional behavior of rectangular hollow reinforced concrete columns strengthened by CFRP. *Structures* **2024**, *70*, 107690. [CrossRef]
5. Zhang, Z.; Guo, F.; Gao, J.; Deng, E.; Kong, J.; Zhang, L. Seismic performance of an innovative prefabricated bridge pier using rapid hardening ultra-high performance concrete. *Structures* **2025**, *74*, 108558. [CrossRef]
6. Li, B.; Kulkarni, S.A.; Leong, C.L. Seismic performance of precast hybrid-steel concrete connections. *J. Earthq. Eng.* **2009**, *13*, 667–689. [CrossRef]

7. Yang, K.H.; Seo, E.A.; Hong, S.H. Cyclic flexural tests of hybrid steel–precast concrete beams with simple connection elements. *Eng. Struct.* **2016**, *118*, 344–356. [CrossRef]
8. Feng, S.; Guan, D.; Guo, Z.; Liu, Z.; Li, G.; Gong, C. Seismic performance of assembly joints between HSPC beams and concrete-encased CFST columns. *J. Constr. Steel Res.* **2021**, *180*, 106572. [CrossRef]
9. Zhang, J.; Zhang, T.; Yuan, W. Development and experimental seismic study of bearing-two-level yielding energy dissipation beam-column-slab joints for precast concrete frames. *J. Build. Eng.* **2024**, *86*, 108779. [CrossRef]
10. Guo, L.; Wang, J.; Wang, W.; Hu, Z. Experimental study and analytical evaluation on seismic performance of CECFST moment resisting frame with bolted connections. *Eng. Struct.* **2022**, *259*, 114074. [CrossRef]
11. Feng, S.; Guan, D.; Ni, L.; Lin, Y.; Liu, Z.; Guo, Z.; Li, G. Experimental study on seismic behavior of joints connecting precast H-steel reinforced concrete beams and concrete-filled steel tube columns. *J. Build. Eng.* **2022**, *45*, 103444. [CrossRef]
12. Lin, Y.; Guan, D.; Guo, Z.; Zhu, M.; Yang, H. Seismic performance on PSPC beam-concrete encased CFST column frame with a built-in reduced beam section. *Case Stud. Constr. Mater.* **2024**, *21*, e03966. [CrossRef]
13. Chen, J.; Chan, T.M.; Chung, K.F. Design of square and rectangular CFST cross-sectional capacities in compression. *J. Constr. Steel Res.* **2021**, *176*, 106419. [CrossRef]
14. Zhang, Y.; Ren, C.; Yi, J.; Wei, Y.; Li, G.; Zhao, H. Axial compression performance of concrete-filled steel tubular columns with damaged BFRP jackets. *Constr. Build. Mater.* **2024**, *442*, 137640. [CrossRef]
15. Rong, C.; Peng, Y.; Shi, Q.; Wang, P. Eccentric compression performance of concrete filled steel tube slotted columns: Experiment and simulation analysis. *Structures* **2025**, *74*, 108580. [CrossRef]
16. Zhang, R.; Qiu, C.; Wang, W. Peak and residual deformation-based seismic design for multi-story hybrid concentrically braced frames. *J. Build. Eng.* **2023**, *67*, 106075. [CrossRef]
17. Li, H.; Zhang, W. Seismic energy and axial column demands of steel concentrically braced frames. *Structures* **2024**, *64*, 106586. [CrossRef]
18. Chen, Z.P.; Zhu, S.; Yu, H.; Wang, B. Development of novel SMA-based D-type self-centering eccentrically braced frames. *Eng. Struct.* **2022**, *260*, 114228. [CrossRef]
19. Ulger, T.; Karabulut, M.; Mert, N. Lateral load performance of a reinforced concrete frame with pultruded GFRP box braces. *J. Struct. Eng. Appl. Mech.* **2022**, *5*, 40–49. [CrossRef]
20. Mortazavi, P.; Kwon, O.S.; Christopoulos, C. Pseudo-dynamic hybrid simulations of steel eccentrically braced frames equipped with cast steel replaceable modular yielding links. *Earthq. Eng. Struct. Dyn.* **2023**, *52*, 3622–3648. [CrossRef]
21. Li, T.; Su, M.; Sui, Y.; Ma, L. Real-time hybrid simulation on high strength steel frame with Y-shaped eccentric braces. *Eng. Struct.* **2021**, *226*, 111369. [CrossRef]
22. Liu, B.; Lu, Y.; Li, W.; Li, J.; Zhao, J.; Wang, S.; Ni, G.; Meng, Q. Study on Seismic Behavior of Different Forms of Eccentrically Braced Steel Frames. *Buildings* **2024**, *14*, 2064. [CrossRef]
23. Lin, Y.; Wang, Y.; Wu, W.; Yang, S.; Guo, Z.; Guan, D.; Li, G. Experimental study on PSPC beam–Concrete encased CFST column frame installed with novel steel panel dampers. *Eng. Struct.* **2023**, *288*, 116211. [CrossRef]
24. *GB50011-2010*; Code for Seismic Design of Buildings. China Architecture & Building Press: Beijing, China, 2016. (In Chinese)
25. Ye, L.; Cheng, G.; Qu, Z.; Lu, X. Study on energy-based seismic design method and application on steel braced frame structures. *J. Build. Struct.* **2012**, *33*, 36–45. (In Chinese)
26. Ye, L.; Miao, Z.; Cheng, G.; Ma, Q.; Lu, X. Study on the energy-based seismic design method of building structures. *Eng. Mech.* **2014**, *31*, 1–12+20. (In Chinese) [CrossRef]
27. Chao, S.; Goel, S.; Lee, S. A seismic design lateral force distribution based on inelastic state of structures. *Earthq. Spectra* **2012**, *23*, 547–569. [CrossRef]
28. Ali, A.; Zhang, C.; Bibi, T.; Sun, L. Experimental investigation of sliding-based isolation system with re-centering functions for seismic protection of masonry structures. *Structures* **2024**, *60*, 105871. [CrossRef]
29. Cao, X.Y.; Shen, D.; Ji, K.; Qu, Z.; Wang, C.N. Recovery resilience framework of replaceable AB-BRB for seismic strengthening during the aftershock stage. *Thin Wall. Struct.* **2024**, *205*, 112389. [CrossRef]
30. Huang, L.; Hu, J.; Zeng, B.; Xie, Q. Cyclic behavior of variable friction dampers with two-stage energy dissipation and stiffness. *J. Constr. Steel Res.* **2025**, *226*, 109290. [CrossRef]
31. Zhang, Y.; Xia, B.; Guo, X.; Sun, B.; Hu, D.; Wei, Y. Genetic algorithm-enhanced Housner intensity measure for seismic vulnerability analysis of reinforced concrete column-steel beam (RCS) frame structure. *Soil Dyn. Earthq. Eng.* **2025**, *193*, 109320. [CrossRef]
32. Li, T.; Su, M.; Guo, J. A plastic design method based on multi-objective performance for high-strength steel composite K-shaped eccentrically braced frame. *J. Constr. Steel Res.* **2022**, *198*, 107562. [CrossRef]
33. Ouyang, X.; Zhang, Y.; Ou, X.; Shi, Y.; Liu, S.; Fan, J. Seismic fragility analysis of buckling-restrained brace-strengthened reinforced concrete frames using a performance-based plastic design method. *Structures* **2022**, *43*, 338–350. [CrossRef]
34. Zhai, Z.; Liu, Y.; Guo, W.; Mercan, O.; Chen, H.; Zhou, F. A seismic resilient design method for structures equipped with two-level yielding dampers, accounting for extremely rare earthquakes. *Eng. Struct.* **2023**, *294*, 116797. [CrossRef]

35. Yan, X.; Shu, G.; Rahgozar, N.; Alam, M.S. Seismic design and performance evaluation of hybrid braced frames having buckling-restrained braces and self-centering viscous energy-dissipative braces. *J. Constr. Steel Res.* **2024**, *213*, 108359. [CrossRef]
36. Xiong, J.; Chen, W.; Hu, S.; Wang, X.; Zeng, S. Study of an optimum design method for Y-shaped EBFs based on energy balance theory and energy dissipation coefficient. *J. Nanchang Univ. (Nat. Sci.)* **2019**, *43*, 284–295. (In Chinese)
37. Cao, X.Y.; Feng, D.C.; Wu, G.; Wang, Z. Experimental and theoretical investigations of the existing reinforced concrete frames retrofitted with the novel external SC-PBSPC BRBF sub-structures. *Eng. Struct.* **2022**, *256*, 113982. [CrossRef]
38. Yu, X.; Li, Z.; Yang, A.; Li, Y.; Lu, D.; Dai, K. Time-Dependent Seismic Fragility of Coastal RC Frames Considering Effect of Distance from Coastline. *Buildings* **2025**, *15*, 737. [CrossRef]
39. Habibi, A.; Chan, R.; Albermani, F. Energy-based design method for seismic retrofitting with passive energy dissipation systems. *Eng. Struct.* **2013**, *46*, 77–86. [CrossRef]
40. Newmark, N.; Hall, W. *Earthquake Spectra and Design*; Earthquake Engineering Research Institute: El Cerrito, CA, USA, 1982.
41. *ANSI/AISC 358-16*; Prequalified Connections for Special and Intermediate Steel Moment Frames for Seismic Applications, Including Supplement No. 1. American Institute of Steel Construction: Chicago, IL, USA, 2016.
42. FEMA-350. *Recommended Seismic Design Criteria for New Steel Moment-Frame Buildings*; Federal Emergency Management Agency: Washington, DC, USA, 2000.
43. Guo, X.N.; Gao, S.Y.; Bui, T.; Luo, X. Experimental study on static performance of embedded channel joints of precast concrete beams. *J. Tongji Univ. (Nat. Sci.)* **2017**, *45*, 1258–1264. (In Chinese)
44. Leelataviwat, S.; Goel, S.C.; Stojadinović, B. Toward performance-based seismic design of structures. *Earthq. Spectra* **1999**, *15*, 435–461. [CrossRef]
45. Lin, Y.; Yang, S.; Guan, D.; Guo, Z. Modified strip model for indirect buckling restrained shear panel dampers. *J. Constr. Steel Res.* **2020**, *175*, 106371. [CrossRef]
46. ASCE. *Minimum Design Loads and Associated Criteria for Buildings and Other Structures, ASCE/SEI 7-16 [S.]*; American Society of Civil Engineers: Reston, VA, USA, 2016.
47. Zhang, C.; Shu, J.; Zhang, H.; Ning, Y.; Yu, Y. Estimation of load-carrying capacity of cracked RC beams using 3D digital twin model integrated with point clouds and images. *Eng. Struct.* **2024**, *310*, 118126. [CrossRef]
48. Long, X.; Iyela, P.M.; Su, Y.; Atlaw, M.M. Numerical predictions of progressive collapse in reinforced concrete beam-column sub-assemblages: A focus on 3D multiscale modeling. *Eng. Struct.* **2024**, *315*, 118485. [CrossRef]
49. *OpenSees, Open System for Earthquake Engineering Simulation*; Computer Program. University of California: Berkeley, CA, USA, 2025. Available online: http://opensees.berkeley.edu (accessed on 5 January 2025).
50. Neuenhofer, A.; Filippou, F.C. Geometrically nonlinear flexibility-based frame finite element. *J. Struct. Eng.-ASCE* **1998**, *124*, 704–711. [CrossRef]
51. Niu, Y.; Wang, W.; Su, Y.; Jia, F.R. Plastic damage prediction of concrete under compression based on deep learning. *Acta Mech.* **2024**, *235*, 255–266. [CrossRef]
52. Han, L.H. *Concrete Filled Steel Tubular Structures*: Theory and Practice, 3rd ed.; China Science Pubilshing: Beijing, China, 2016. (In Chinese)
53. Feng, D.C.; Wang, Z.; Wu, G. Progressive collapse performance analysis of precast reinforced concrete structures. *Struct. Des. Tall Spec.* **2019**, *28*, e1588. [CrossRef]
54. Fukumoto, T.; Morita, K. Elastoplastic behavior of panel zone in steel beam-to-concrete filled steel tube column moment connections. *J. Struct. Eng.-ASCE* **2005**, *131*, 1841–1853. [CrossRef]
55. FEMA-P695. *Quantification of Building Seismic Performance*; Federal Emergency Management Agency: Washington, DC, USA, 2009.
56. PEER. *PEER Strong Ground Motion Databse, University of California, Berkeley*. 2025. Available online: https://peer.berkeley.edu/peer-strong-ground-motion-databases (accessed on 10 January 2025).
57. Erochko, J.; Christopoulos, C.; Tremblay, R.; Choi, H. Residual drift response of SMRFs and BRB frames in steel buildings designed according to ASCE 7-05. *J. Struct. Eng.-ASCE* **2011**, *137*, 589–599. [CrossRef]

buildings

MDPI

Article

Vibration Performances of a Full-Scale Assembled Integral Two-Way Multi-Ribbed Composite Floor

Liang Gong [1], Yan Feng [2,*], Wenbin Zhang [1], Ming Xu [2,*] and Xiangqiang Zeng [3]

1 School of Architecture Engineering, Nanjing Vocational Institute of Transport Technology, Nanjing 211188, China
2 Key Laboratory of RC&PC Structures of Ministry of Education, Southeast University, Nanjing 210096, China
3 Shandong Shijia Yuanda Technology Development Co., Ltd., Qingdao 224000, China
* Correspondence: 103009030@seu.edu.cn (Y.F.); xuming@seu.edu.cn (M.X.)

Abstract

The static performances of an assembled integral two-way multi-ribbed composite floor system have been studied experimentally and numerically, while the dynamic characteristics and comfort analysis under a human load have not been investigated. In this article, a 9.2 m × 9.2 m floor system, composed of 16 precast panels and integrated into a whole structure through six wet joints, was designed and tested under pedestrian loads. Dynamic performances related to its natural frequencies, vibration mode shapes, and maximum acceleration were analyzed. Theoretical formulas were proposed to predict its natural frequency and maximum acceleration under a single-person load. It was found that the dynamic behavior of this innovative floor system meets the requirements of GB50010-2010 and ISO 2631. Elastic plate theory could be applied to predict the natural frequency and acceleration, with the bending stiffness obtained from the experiment. Some design and dynamic test suggestions for this floor system and similar structures are proposed based on a parametric analysis.

Keywords: assembled integral structure; two-way multi-ribbed composite floor; vibration mode shapes; natural frequencies; acceleration–time history curve; theoretical formula

1. Introduction

The floor system, accounting for nearly 40% of material usage and construction time, is a critical component of modern buildings. In response to the growing demand for accelerated construction timelines and sustainable building practices, prefabricated floor systems have become increasingly essential [1,2].

Compared to cast-in-situ floors, the main challenges associated with assembled concrete floors are the reduced stiffness and increased displacement due to the joints. As a result, innovative assembled or assembled integral floor systems, such as the Vierendeel-sandwich-plate floor system [3], the PK prestressed composite slab floor system [4], the assembled monolithic hollow-ribbed floor [5], the prefabricated PC floor system [6], and other new systems [7–11] were proposed and investigated through experimental and theoretical methods. It was found that lightweight hollow floors with reliable wet joints are one of the best solutions for modern large-span floor systems [12–15].

To fully display the advantages of hollow floors and combine with the techniques of assembled structures, an assembled integral two-way multi-ribbed composite floor system was proposed by Zeng and Xu [16,17]. By dividing the floor system into precast

https://doi.org/10.3390/buildings15091551

2.05 m × 2.05 m panels with multiple ribs connecting with wet joints and couplers, the hollow ratio could reach 60%, and the construction could be significantly enhanced by 40%. Static experiments and numerical simulations were conducted to verify its performance regarding stiffness and crack resistance [16,17], and satisfactory results were reached.

However, with the increase in the span, the vibration under human loads might be prominent because of its lower vertical stiffness. A dynamic performance should be conducted to check whether its natural frequency and acceleration under human loads meet the requirements of design codes, and further explore the numerical method to predict its dynamic behaviors.

Research has been conducted worldwide on dynamic behaviors for assembled integral floor systems, mainly focusing on experimental and numerical methods. Tilden [18] and Fuller [19] were among the first researchers to experimentally quantify the dynamic load effects of individuals and groups, respectively. Other researchers, such as Greimann and Klaiber [20], Tuan and Saul [21], and many others [22–24], extended the method to obtain other dynamic responses from other loads, including dancing, sports, and crowd harmonic loads. However, the experimental method is mainly based on field tests, not analytical or numerical methods. The finite element method (FEM) is the most commonly used method to predict human-induced vibrations. Nie et al. [25] investigated the dynamic characteristics of a long-span floor under crowd-induced rhythmic excitation by using a stochastic vibration approach combined with a FEM for modal analysis. Liu [26] studied three types of materials with different properties (orthogonally anisotropic materials, strip-shaped unidirectional materials, and isotropic materials), analyzing the vibration modes of these materials on the same hollow-core slab. They found the optimal material parameters for each mode and identified the orthogonally anisotropic material as the best choice. Wen et al. [27] carried out experimental research on the vibration comfort of a new type of precast hollow-cross rib floor system. They proposed a finite-element analysis method, and the results showed that the floor system has good vibration comfort performance. But the main problems for assembled integral floor systems are that a large number of elements and a lot of interfacial simulation treatment may slow down the calculation efficiency and increase the calculation time.

As is concluded above, it is found that the experimental methods for dynamic tests are quite mature, but there are still some problems related to the dynamic calculation under human loads with a FEM. In this paper, to enhance the calculation efficiency, a finite difference method (FDM) is used to predict its vibration modes, frequencies, and peak acceleration under a human load. To further verify the effectiveness, a 9.2 m × 9.2 m floor system, composed of 16 precast panels and integrated into a whole structure through six wet joints, was designed and tested under a pedestrian load. Dynamic performances related to its natural frequency, vibration modes, and acceleration tests under a pedestrian load were analyzed.

2. Experimental Investigation

2.1. Introduction of the Innovative Floor System

The innovative floor system consists of four main components: a precast ribbed bottom slab (PRBS), lightweight infills and a cast-in-situ upper slab (CUS), as shown in Figure 1. The three different parts above and adjacent bottom slab are integrated into a reliable system through three methods (see Figure 2): (1) The connection between the PRBS and the CUS is realized by the shear strength of the rough interface, cast-in-situ joints, and stirrups. (2) The lightweight infills are fixed on the precast bottom slab through the positioning rebars and recessed cavities. (3) The rebars of the adjacent PRBSs are connected by couplers for squeezing the splicing of rebars and cast-in-situ joints where the shear keys are averagely arranged alongside the outsides of the PRBS.

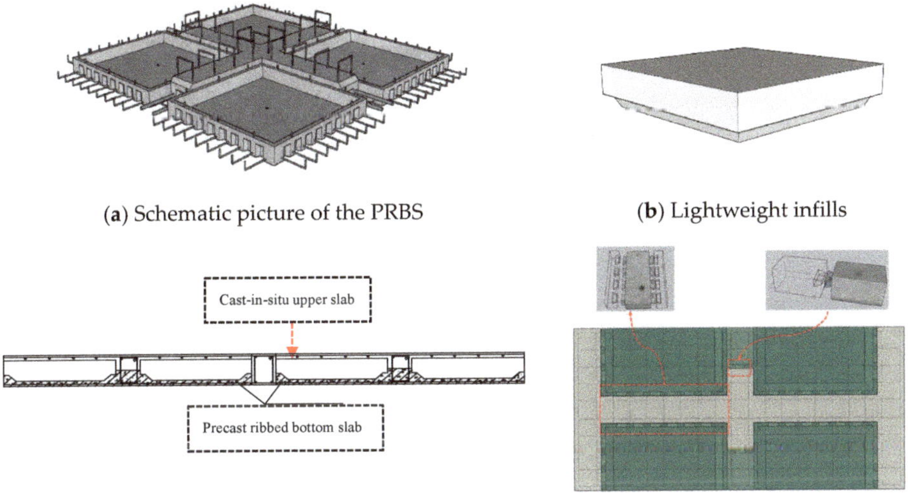

(**a**) Schematic picture of the PRBS (**b**) Lightweight infills

(**c**) Cast-in-situ concrete topping (**d**) Wet joint with spliced rebar

Figure 1. Precast panels of the assembled integral two-way multi-ribbed composite floor system.

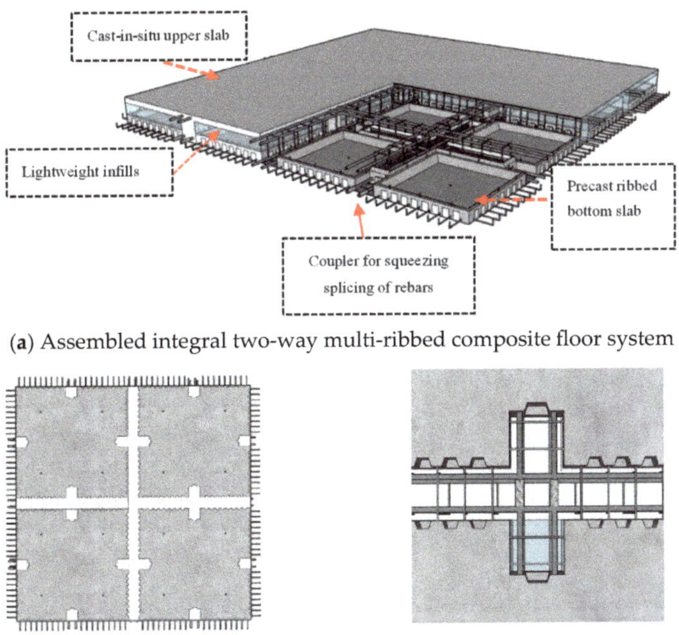

(**a**) Assembled integral two-way multi-ribbed composite floor system

(**b**) Bottom view (**c**) Details of the joint

Figure 2. Schematic picture of the assembled integral two-way multi-ribbed composite floor system.

2.2. Specimen Preparation

The specimen, as shown in Figures 3 and 4, consists of 16 PRBs and 64 lightweight infills. The total size of the specimen is 9.6 m × 9.6 m with a joint width of 200 mm and a floor height of 300 mm. It is supported by four beams and four columns, designed by the GB50010-2010 "Code for design of concrete structures" [28]. The beam is 400 × 800 mm (width × height) reinforced with 8Φ16, while the column is 400 × 800 mm reinforced with 16Φ16. Detailed information on the prefabricated PRBs is illustrated in Figure 3g or references [16,17].

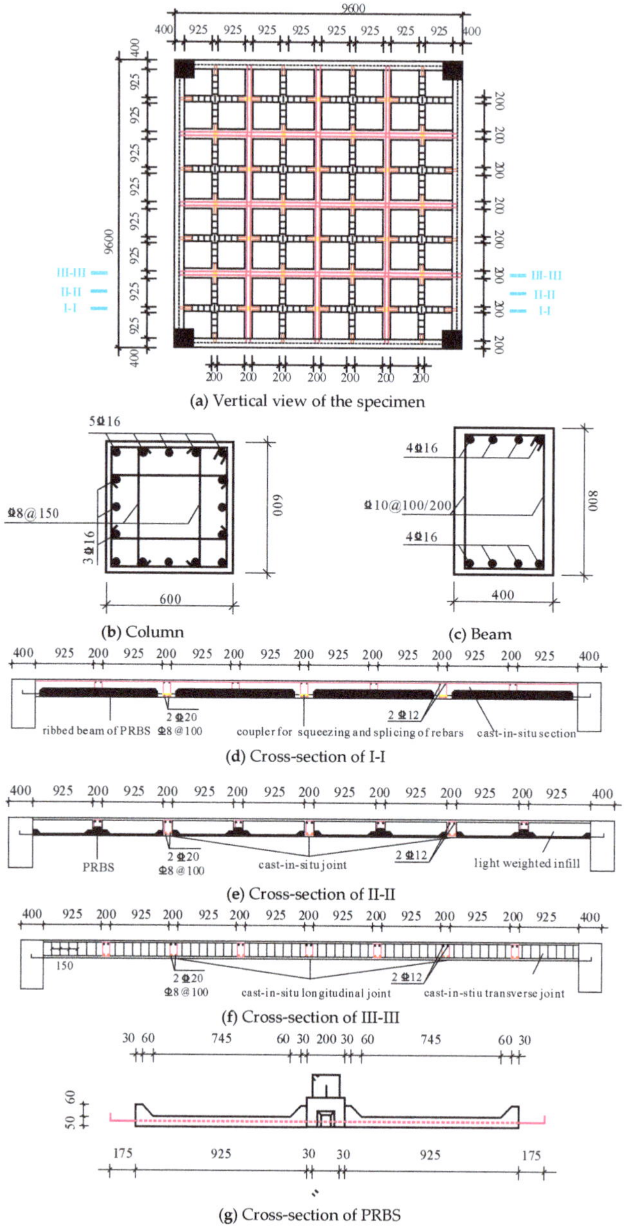

Figure 3. Detailed sizes of the assembled integral two-way multi-ribbed composite floor system.

(a) Assembled picture of 16 precast panels (b) Vibration tests

Figure 4. Test specimen.

All 16 PRBSs were manufactured in factories, while the beams, columns, CUS, and joints were poured and cured at the laboratory of civil engineering of the Southeast University in China. The longitudinal rebars of adjacent PRBS panels were connected using mechanical splicing couplers (highlighted in yellow in Figure 3). The cast-in-situ concrete topping and wet joints were integrated with RC beams and columns to form a unified floor system.

2.3. Material Properties

The rebars adopted in the experiment were all HRB400, according to the Chinese standard GB50010-2010 "Code for design of concrete structures" [28]. The diameters were 6 mm, 12 mm and 20 mm, respectively. The concrete used in the cast-in-situ columns, beams, PRBC, CUS, and cast-in-situ joints was all C35. Rebars with a diameter of 20 mm were sleeved into the coupler and connected by cold extrusions.

Rebars, couplers, and concrete were tested according to Chinese standard GB 228.1-2021 "Metallic materials—Tensile tests—Part 1: methods of test under room temperature" [29], JG/T 163-2013 "Coupler for rebar mechanical splicing" [30], and GB50010-2010 [28], respectively. Mechanical properties of the rebar, coupler, and concrete were all tested in the civil engineering laboratory of Southeast University, Nanjing, China, with hydraulic machines (MTS, the US). Detailed mechanical properties are collected in Tables 1 and 2.

Table 1. Mechanical properties of the rebars and coupler.

Materials	Diameters of Rebar [mm]	Yield Stress [MPa]	Ultimate Stress [MPa]	Young's Modulus [MPa]
Rebars	20	420.31	620.06	217,696
	12	451.24	616.58	209,643
	6	436.73	580.46	213,487
Coupler	20	414.57	608.02	176,867

Table 2. Mechanical properties of the concrete.

Positions of Concrete	Cubic Compressive Stress [MPa]	Young's Modulus [MPa]
Cast-in-situ section	36.75	30,043
Precast panels	39.47	29,567

$$Z = \begin{Bmatrix} X \\ \dot{X} \end{Bmatrix}, H = \begin{Bmatrix} 0 & I \\ -M^{-1}K & -M^{-1}C \end{Bmatrix}, F = \begin{Bmatrix} 0 \\ M^{-1} \end{Bmatrix} \tag{3}$$

$$x(t+1) = Ax(t) + Bu(t) \tag{4}$$

$$y(t) = Dx(t) + Eu(t) \tag{5}$$

where, M, C, K, and X are the mass, damping, stiffness, and displacement matrices of a multi-DOF system. $u(t)$ is a random vibration from the earth, which could be considered as white noise. $x(t)$, $y(t)$ are the n-dimensional and m-dimensional state vectors at discrete sampling times t, respectively. n is the number of system model orders and m is the number of measurement meters. A and B are the state and matrix of motion equation in a discrete form. D is the observation matrix and E is the feedthrough (or direct transmission) matrix while N is the number of samples. To obtain A, the following three matrices are constructed:

$$U_p = \begin{bmatrix} u(0) & u(1) & \cdots & u(N-p-1) & u(N-p) \\ u(1) & u(2) & & u(N-p) & u(N-p+1) \\ & \vdots & \ddots & & \vdots \\ u(p-2) & u(p-1) & \cdots & u(N-3) & u(N-2) \\ u(p-1) & u(p) & & u(N-2) & u(N-1) \end{bmatrix} \tag{6}$$

$$Y_p = \begin{bmatrix} y(0) & y(1) & \cdots & y(N-p-1) & y(N-p) \\ y(1) & y(0) & & y(N-p) & y(N-p+1) \\ & \vdots & \ddots & & \vdots \\ y(p-2) & y(p-1) & \cdots & y(N-3) & y(N-2) \\ y(p-1) & y(p) & & y(N-2) & y(N-1) \end{bmatrix} \tag{7}$$

$$Y_f = \begin{bmatrix} y(p) & y(p+1) & \cdots & y(N-2) & y(N-1) \\ y(1) & y(0) & & y(p-1) & y(p-2) \\ & \vdots & \ddots & & \vdots \\ y(p-2) & y(p-1) & \cdots & y(0) & y(1) \\ y(p-1) & y(p-2) & & y(1) & y(0) \end{bmatrix} \tag{8}$$

Based on reference [31,32], we know $U_p \approx O_f X_f + input\ terms$. To eliminate the influence of input terms, the following step in Equation (10) is conducted, and $\overline{Y_f}$ is obtained. Then, we perform a singular value decomposition on $\overline{Y_f}$. Solving Equations (11) and (12), A could be obtained. λ_i stands for the eigenvalue of A, and the corresponding eigenvectors of A are the vibration shaking modes. And from Equation (5), the natural frequency ω_i and damping ratio ξ_i could be calculated as presented in Equation (13).

$$Z_p = [U_p; Y_p] \tag{9}$$

$$\overline{Y_f} = Y_f * \left(I - {Z_p}^T\left(Z_p {Z_p}^T\right)^{-1} Z_p\right) = U \sum V^T. \tag{10}$$

$$O_p = U(:, 1, n) \sum (1:n, 1:n)^{1/2} \tag{11}$$

$$O_p = \begin{bmatrix} D & DA & \cdots & DA^{p-1} \end{bmatrix} \tag{12}$$

$$\frac{ln(\lambda_i)}{\Delta t} = \alpha_i + \beta_i \boldsymbol{i} \tag{13}$$

$$\omega_i = \sqrt{{\alpha_i}^2 + {\beta_i}^2} \qquad \xi_i = \frac{\alpha_i}{\omega_i} \tag{14}$$

3. Experimental Results

3.1. Vibration Modes, Frequencies, and Damping Ratios

The first three vibration modes of the full-scale specimen are shown in Figure 6. Additionally, the damping ratios and frequencies of each vibration mode are collected in Table 3. It is observed that the first frequency of this assembled integral floor system is 11.718 Hz with a damping ratio of 3.304%. The first vibration mode is symmetric, and the deflection at the middle point is greater than in the other parts. The second frequency is 27.314 Hz, with a relatively small damping ratio of 0.855%. The second vibration mode shape is antisymmetric. For the third mode, the frequency is 43.510 Hz, and the damping ratio is 0.793%.

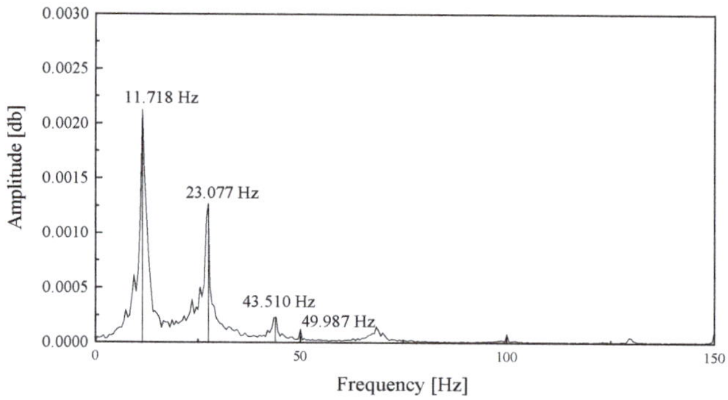

Figure 6. Amplitude–frequency curve.

Table 3. Summary of the vibration mode shapes, frequencies, and damping ratios.

NO of Vibration Modes	Frequencies [Hz]	Damping Ratio [%]	Vibration Mode Shapes
1	11.718	3.304	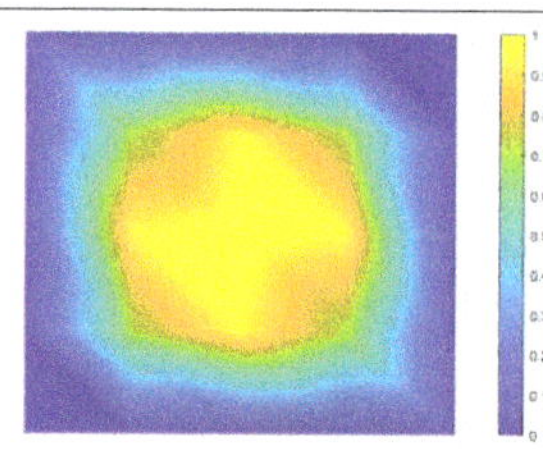
2	27.314	0.855	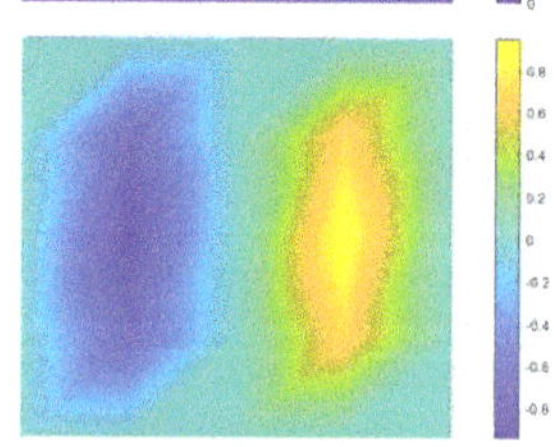
3	43.510	0.793	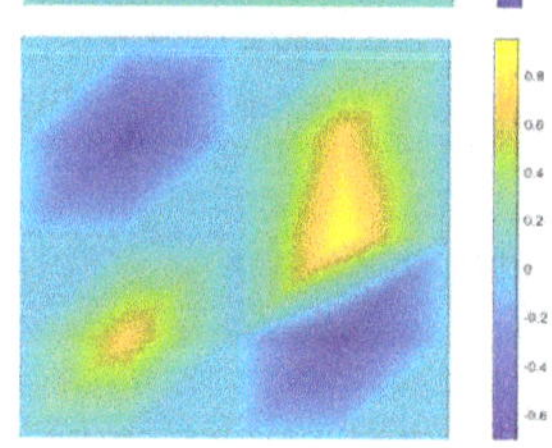

3.2. Acceleration Response of the Single-Person Tests

The acceleration-time history curves of the specimen under single-person walking are presented in Figure 7. Even though we want to control the walking frequency at 2 Hz, the average frequency in the experiment is about 1.83 Hz. This would not influence the comfort analysis as a larger frequency will generate greater acceleration. Also, for the numerical simulation, we can directly use the actual interval of each step.

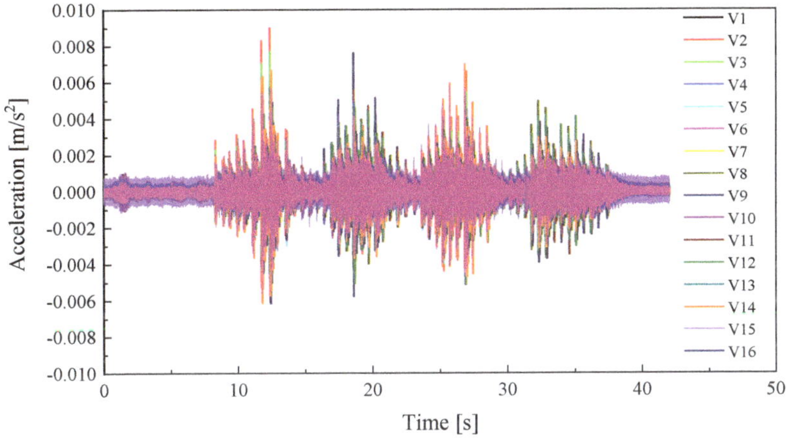

Figure 7. Acceleration time-history curve of one-person walking at 2 Hz.

Figure 8 presents the acceleration time-history curve of V11 under single-person walking. The wave is similar to a free vibration due to the 3.304% damping ratio, and there is no obvious wave superposition. In the two back-and-forth processes, the acceleration waves are of similar shape, and the three acceleration waves are also similar. The maximum acceleration of 15 m was all less than 0.01 m/s^2, meeting the vibration serviceability requirements of GB 50010-2010 and ISO 2631 [33]. Under the same load, the accelerations of V2, V14, and V3 were greater than the others. The first frequency is 11.718 Hz, five times larger than the load frequency. Thus, the conclusion that no resonance exists under one-person loads could be roughly reached.

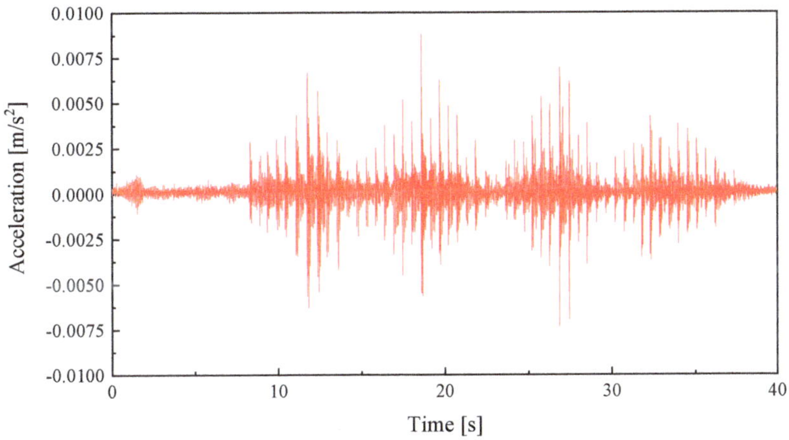

Figure 8. Acceleration time-history curve of V11 under single-person walking.

3.3. Acceleration Response of the Multi-Person Tests

The acceleration-history curves of the specimen under multi-person loads are presented in Figure 9. Compared with the time-history curves of a one-person load, the maximum acceleration of 15 m was much greater. It reached 0.15 m/s^2 for multi-person walking and 0.45 m/s^2 for multi-person walking, larger than the specification requirements of vibration serviceability of GB 50010-2010 and ISO 2631. Figure 10 presents the acceleration time-history curve of V11 under six-person walking. The waves are no longer like a free vibration, which means the six accelerations may add up in the walking process.

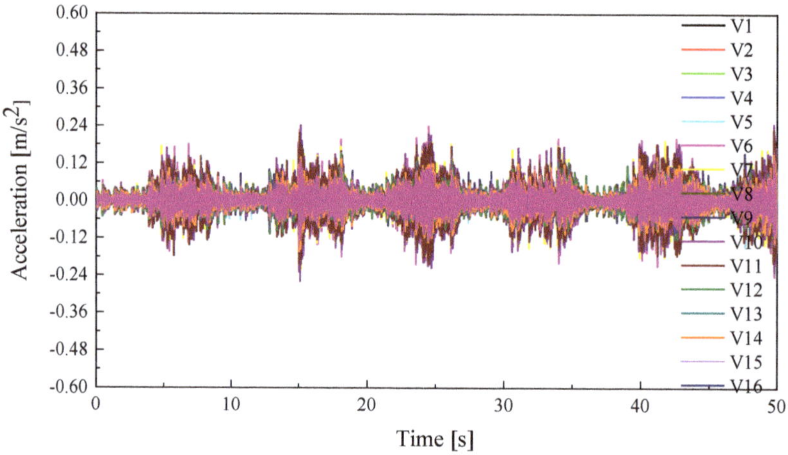

Figure 9. Acceleration time-history curve of six-person walking at 2 Hz.

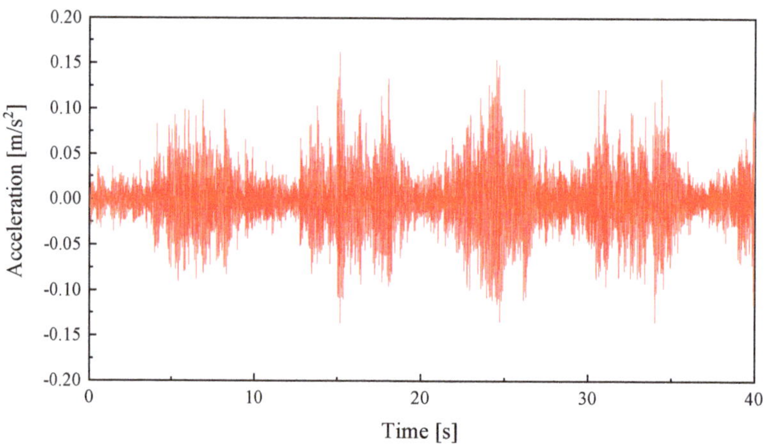

Figure 10. Acceleration time-history curve of V11 under six-person walking.

4. Theoretical Derivation

4.1. Simplification of Beam–Plate Connection and Boundary Conditions

To derive the analytical solution for this floor system, some simplifications or assumptions must be made. The main assumption contains two aspects, i.e., the properties of the floor, beam-plate connection, and boundary conditions:

The floor system is considered as an orthogonally anisotropic elastic shell, where the differences between the hollow part and ribbed beam part are neglected. This assumption is inherited from some previous studies on assembled floor systems.

The influence of beams is neglected. As mentioned, the beam height is 800 mm, about 1/11.5 of the span, and the distance of the stirrups is 100 mm along the beam, which is considered to be stiff enough to ignore its bending deflections under human loads. Furthermore, even though some simplifications were made, satisfactory results were still reached, demonstrating the efficiency of this simplification.

The boundary condition is idealized as fixed ends for the four sides. From Figure 3, it could be observed that there are only tensile rebars at the bottom of the beam–plate connections. According to GB 50010-2010, this design could be considered as fixed ends.

In this way, the innovative floor system could be considered as a simply supported shell on four sides. The boundary conditions are x = 0, y = 0, M = 0 at the four sides. In mathematical form, boundary conditions are presented as follows:

$$\text{x} = 0 \text{ or x} = 9.2\text{ m} \qquad w = 0,\ \frac{\partial^2 w}{\partial x^2} = 0 \tag{15}$$

$$\text{x} = 0 \text{ or y} = 9.2\text{ m} \qquad w = 0,\ \frac{\partial^2 w}{\partial x^2} = 0 \tag{16}$$

4.2. Natural Frequency and Vibration Modes

In reference [16,17], the static performance of this assembled integral two-way multi-ribbed composite floor system could be divided into elastic, elastic–plastic, and damaged stages. Compared to the live loads of the elastic limit, the pedestrian loads were minimal. Therefore, the elastic mechanics method could be used in this section to calculate the vibration modes and frequency. The equations controlling its vibration are shown as follows:

$$D\left[\frac{\partial^4 w}{\partial x^4} + 2\frac{\partial^4 w}{\partial x^2 \partial y^2} + \frac{\partial^4 w}{\partial y^4}\right] + m\frac{\partial^2 w}{\partial t^2} = 0 \tag{17}$$

where E is the modulus of the concrete, w is the out-of-plane deflection of the floor system, y is the in-plane vertical coordination of the floor system, x is the in-plane horizontal coordination of the floor system, D is the bending stiffness of the floor system, and m is the average mass of the floor system. Equation (14) could be transformed to a fourth-order partial difference equation. The actual stiffness of this full-scale 9.2 m × 9.2 m integral two-way multi-ribbed composite floor system can be determined from static experiments.

Using separation of variables in the form $w(x,t) = \sum_{i=1}^{N} \phi_i(x,y)\theta_i(t)$, where θ_i is only related to t, ϕ_i is the shape function only related x and y. Equation (17) could be transformed to Equation (18). In Equation (18), only when $D\left[\frac{\partial^4}{\partial x^4} + 2\frac{\partial^4}{\partial x^2 \partial y^2} + \frac{\partial^4}{\partial y^4}\right]\phi_i$ and $m\frac{\partial^2}{\partial t^2}\theta_i$ equal a same constant, this equation could be solved. Introducing this constant as ${\omega^2}_i$, we then get Equation (19). Solving Equation (19), the shape function and natural frequencies could be obtained as Equations (20) and (21).

$$D\left[\frac{\partial^4}{\partial x^4} + 2\frac{\partial^4}{\partial x^2 \partial y^2} + \frac{\partial^4}{\partial y^4}\right]\phi_i(x,y) + m\frac{\partial^2}{\partial t^2}\theta_i = 0 \tag{18}$$

$$D\left[\frac{\partial^4}{\partial x^4} + 2\frac{\partial^4}{\partial x^2 \partial y^2} + \frac{\partial^4}{\partial y^4}\right]\phi_i(x,y) = -m\frac{\partial^2}{\partial t^2}\theta_i = {\omega_i}^2 \tag{19}$$

$$W_{ij}(x,y) = A_{ij} sin\frac{i\pi x}{a} sin\frac{i\pi y}{b} \tag{20}$$

$$f_{ij} = \frac{\pi}{2a^2}\left(i^2 + j^2\right)\sqrt{\frac{D}{m}} \tag{21}$$

where, $W_{ij}(x, y)$ is the vibration mode and f_{ij} is the frequency, i and j indicate the number of half-waves in the x and y direction.

It can be found from the figures above that the vibration mode shapes of the analytical and experimental results are almost identical (see Figures 11–13). The absolute errors of the first three modes are 4%, 11%, and 122%, respectively, as collected in Table 4. It can be concluded that Equations (17) and (18) could be used to predict the frequencies and vibration mode shapes of this assembled integral two-way multi-ribbed composite floor.

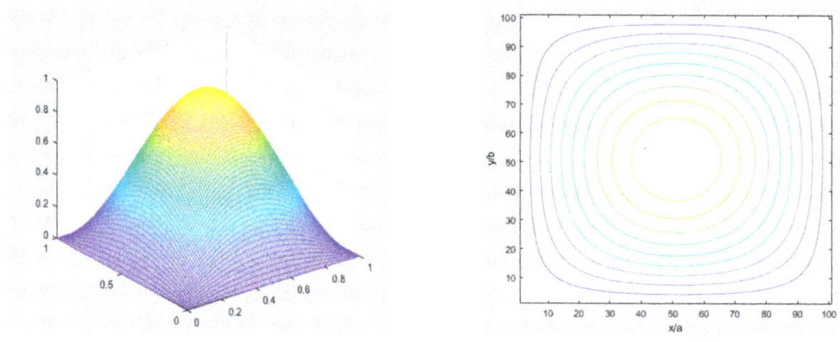

(**a**) 3-D (**b**) 2-D

Figure 11. 1st vibration mode with f_1 = 12.20 Hz.

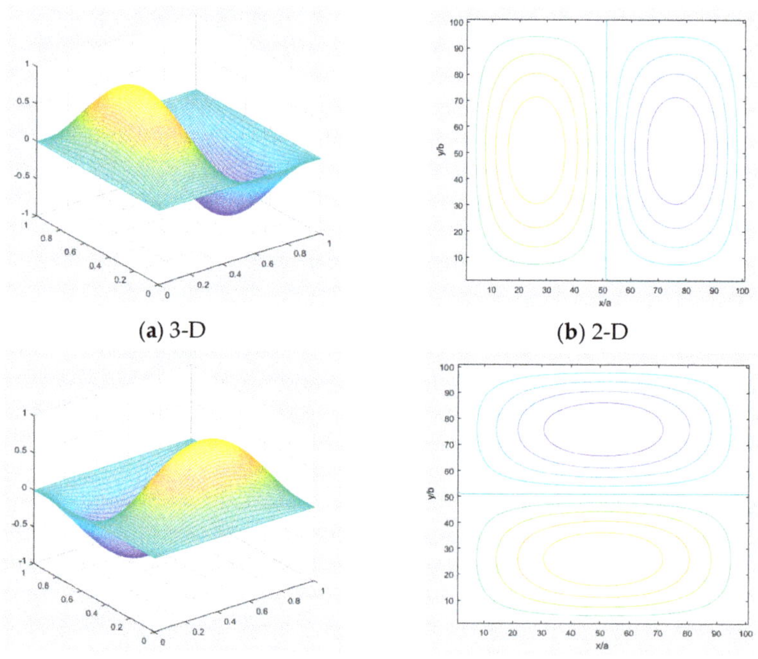

(**a**) 3-D (**b**) 2-D (**c**) 3-D (**d**) 2-D

Figure 12. 2nd vibration mode with f_2 = 30.49 Hz.

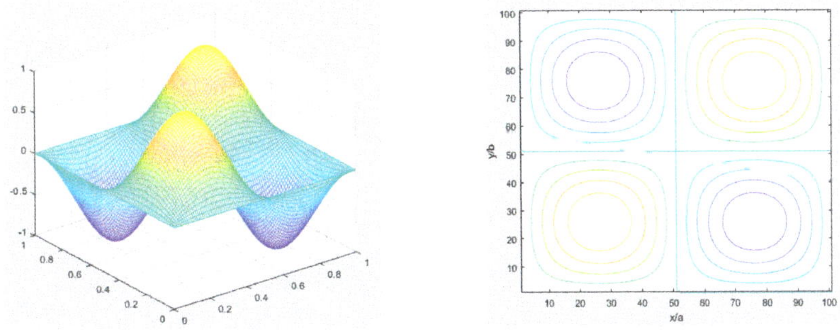

(a) 3-D (b) 2-D

Figure 13. 3rd vibration mode with f_3 = 48.78 Hz.

Table 4. Comparison of experiment, simulation, and GB50010 results.

Frequencies [Hz]	First Order	Errors [%]	Second Order	Errors [%]	Third Order	Errors [%]
Test	11.718	-	27.314	-	43.510	-
Analytical method	12.20	+4.11%	30.49	+11.63	48.78	+12.11%

4.3. Calculation of Acceleration Under Human Loads

Since the equations in Section 4.1 prove that the vibration mode shapes and frequencies could be well predicted by elastic theory, the vibrations under human loads could also be calculated. Taking into consideration the human loads, the vibration equations could be revised as Equation (8). The human load is modeled by using a single-footfall force loading on the footprints [34,35]. Assuming the foot force is exerted on the spot (x_i, y_i), then the single-footfall force $F(t)$ could be expressed as:

$$D\left[\frac{\partial^4 w}{\partial x^4}+2\frac{\partial^4 w}{\partial x^2 \partial y^2}+\frac{\partial^4 w}{\partial y^4}\right]+m\frac{\partial^2 w}{\partial t^2}+C\frac{\partial w}{\partial t}=F(t)\delta(x-x_i)\delta(y-y_i) \tag{22}$$

$$F(t)=G\sum_{k=1}^{3}B_k\sin\left(\frac{k\pi}{t_e}t\right) \tag{23}$$

$$B_1=-\frac{0.0698}{t_e}+1.211;\ B_2=\frac{0.10528}{t_e}-0.1284;\ B_3=\frac{0.3002}{t_e}-0.1534 \tag{24}$$

where $F(t)$ is the force from a single footfall, δ is the Dirac function, G is the weight of a human body, B_k is the Fourier coefficient, f_s is the walking step rate, and T_e is the duration of a single footstep, while C is the damping matrix.

When the step frequency equals 2 Hz, and T_e is 0.5 s, the single-footfall force could be obtained, as illustrated in Figure 14. The curve of the single-footfall force is characterized by two peaks, which represent the moments that the first-foot contacts the floor with "heel strike" and the other leaves with "toe off". There is an overlap between the two contacts of "heel strike" and "toe off" during the walking process. Thus, the duration of the single-footfall T_e will be greater than the period of human walking $t_s = 1/f_s$. The loading scheme considering the overlap periods is shown in Figure 15 [36,37].

In the multi-person load calculation, the same single-footfall model in Figure 14 is assumed for each volunteer. In the experiment, six people are walking at the same speed. Thus, the multi-person load could be expressed as the six single-footfall forces. However, it is necessary to note that the simulation using the multi-person load above is not so satisfactory. It should belong to the simulation method of the crowd load, which will be studied in the following article.

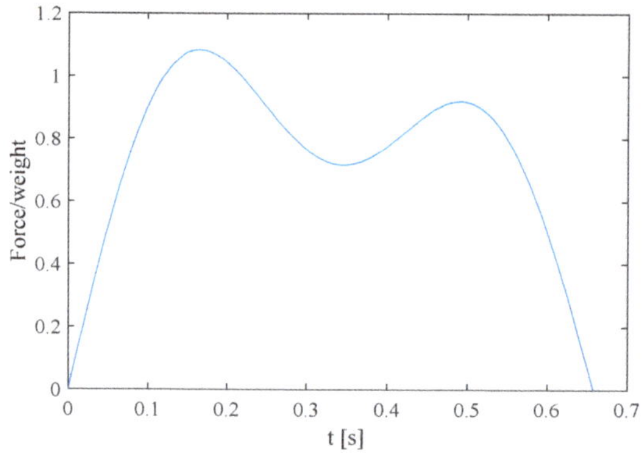

Figure 14. Single foot force–time curve.

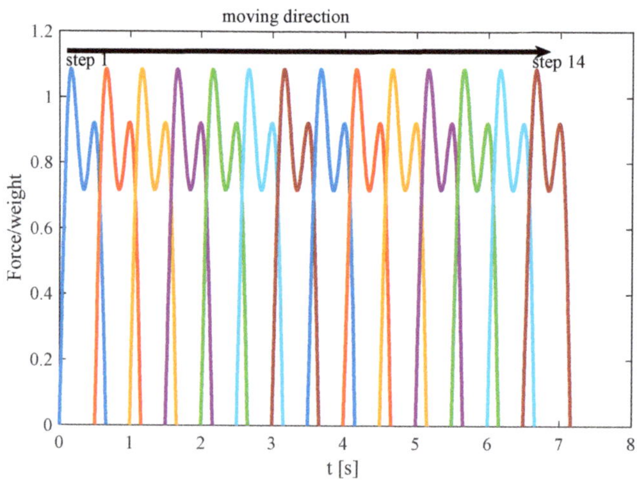

Figure 15. Moving force exerted on the specimen.

By using the single-footfall curve, the human load could be simplified to 14 point loads with an interval of 0.54 s. Experimental results show that the damping ratio is 3%, and the first frequency of the floor system is 11.72 Hz, five times larger than the human-load frequency. Thus, it can be assumed that the energy of each step could be partly dissipated by the free vibration of the floor system at the interval of two steps. In other words, each step could be seen as a single impact force on the floor. This assumption could also be proven through the time–acceleration history curves from the vibration tests.

Equation (22) is solved numerically by the finite difference method (FDM); the discretized forms are presented as Equation (25)–(29) and solved by the Newton–Raphson method.

$$\frac{\partial^4 w}{\partial x^4} = \frac{w_{i+2,j} - 4w_{i+1,j} + 6w_{i,j} - 4w_{i-1,j} + w_{i-2,j}}{\Delta x^4} \tag{25}$$

$$\frac{\partial^4 w}{\partial y^4} = \frac{w_{i,j+2} - 4w_{i,j+1} + 6w_{i,j} - 4w_{i,j-1} + w_{i,j-2}}{\Delta x^4} \tag{26}$$

$$\frac{\partial^4 w}{\partial x^2 \partial y^2} = \frac{w_{i+1,j+1} - 2w_{i+1,j} + w_{i+1,j} - 2w_{i,j+1} + 4w_{i,j} - 2w_{i,j-1} + w_{i-1,j+1} - 2w_{i-1,j} + w_{i-1,j-1}}{\Delta x^2 \Delta y^2} \tag{27}$$

$$\frac{\partial^2 w}{\partial t^2} = \frac{w_{t+1} - 2w_t + w_{t-1}}{\Delta t^2} \tag{28}$$

$$\frac{\partial w}{\partial t} = \frac{w_{t+1} - w_t}{\Delta t} \tag{29}$$

The acceleration–time history curves of single-person loads are presented in Figure 16 and compared with the experimental results. Since the layout of all acceleration meters is symmetric, only the acceleration time–history curves of V11, V9 and V13 are illustrated in Figure 16. It was found that they are of similar shape under single-person walking and running, and the peak acceleration calculated by the method above is about 0.0081 m/s^2. The single-person peak acceleration obtained from the vibration test is 0.075 m/s^2, where the absolute error is less than 8%.

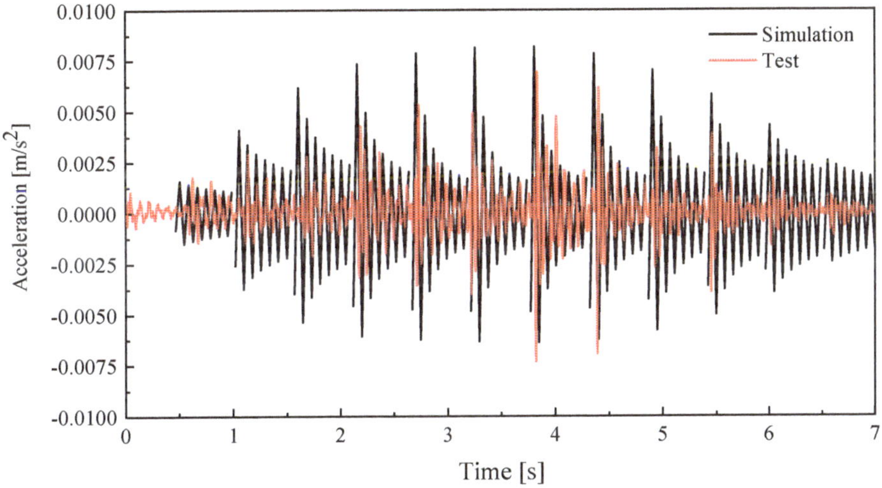

(**a**) V11

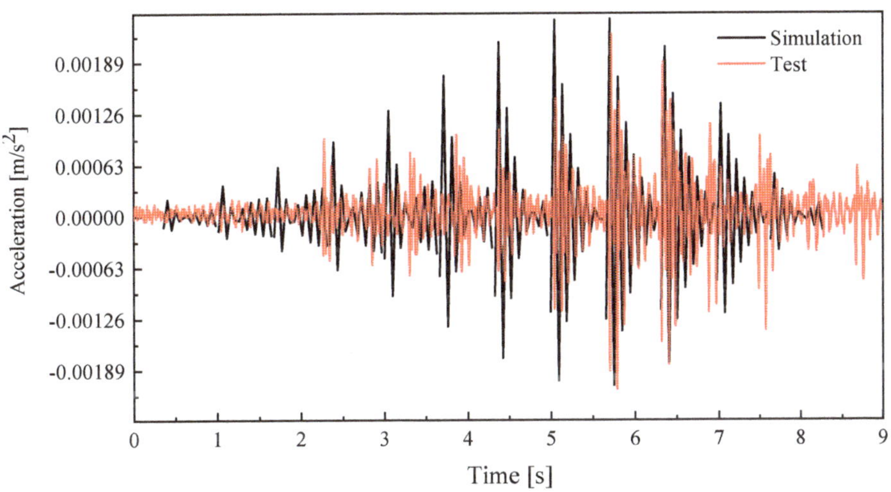

(**b**) V9

Figure 16. *Cont.*

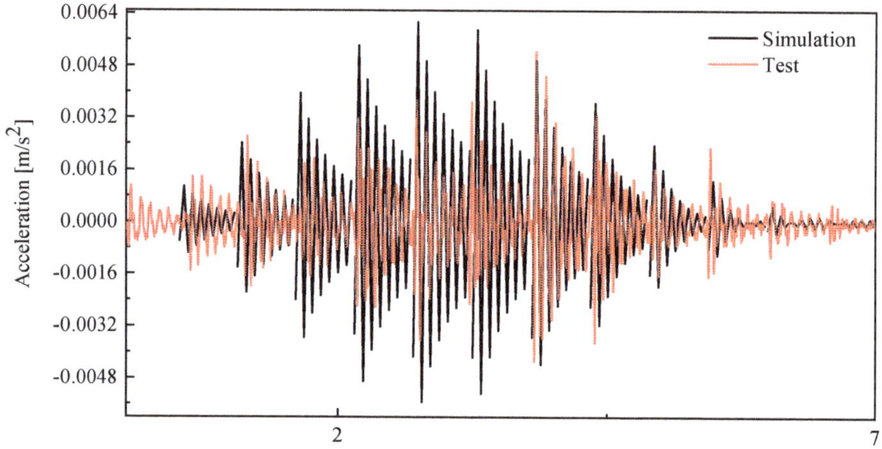

(c) V13

Figure 16. Comparison of acceleration time history curve between experimental and numerical results under a single-person load.

According to JGJ 3-2010 "Technical specification for concrete structures of tall buildings" [38], the maximum acceleration of floor systems can be calculated by Equation (30), where g is the gravity acceleration and F_p is the contact force calculated by Equation (22). The maximum acceleration predicted by JGJ 3-2010 is 0.0295 m/s^2 with an absolute error of 60%. This specification is used for cast-in-situ concrete floor systems, which are stiffer than this innovative floor system. Furthermore, the damping ratio for cast-in-situ concrete floor systems is assumed to be 0.05, while the test indicates the first-order damping ratio for this floor is 0.033, which is also a main reason for the larger accelerations.

$$a_{max} = \frac{F_p}{\xi_i m} g \quad (30)$$

4.4. Parametric Analysis

When calculating the natural frequency, it is evident that the bending stiffness D is very important. In our experimental study, it is straightforward to obtain D through a static experiment. When applied to other floors with different sizes, the finite element method should be employed to determine the bending stiffness of the innovative floor system.

Based on the framework above, influence factors, such as damping ratio, single-person weight, step frequency and bending stiffness, are studied, and details of these parameters are illustrated in Table 5. The relationships between these parameters and the maximum accelerations are illustrated in Figure 17.

Table 5. Parametric analysis of single-person loads.

Stepping Frequencies [Hz]	Weight of a Single Person [N]	Damping Ratio [%]	Bending Stiffness [D]
1.25	600	3	0.9
1.5	700	4	1.0
1.75	800	5	1.1
2	900	6	1.2

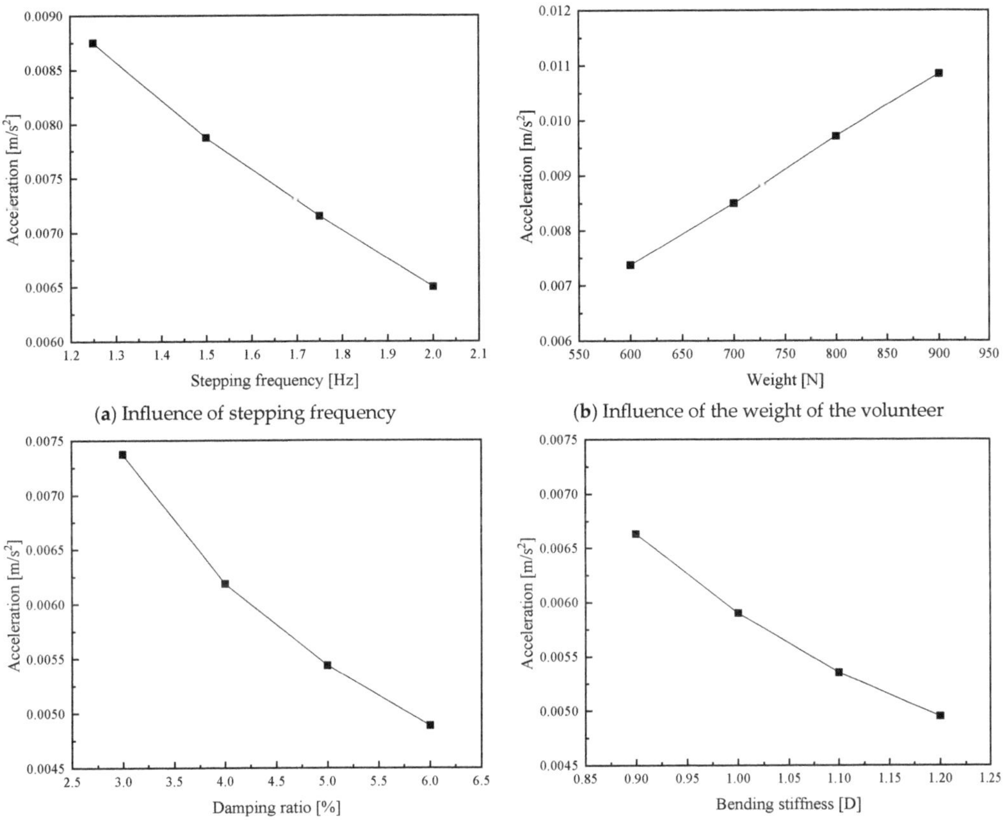

(**a**) Influence of stepping frequency
(**b**) Influence of the weight of the volunteer
(**c**) Influence of damping ratio
(**d**) Influence of bending stiffness

Figure 17. Parametric analysis of the assembled integral two-way multi-ribbed composite floor system under a single-person load.

It is observed that the weight and stepping frequency on the maximum acceleration is linear, while the influence of the damping ratio and bending stiffness is nonlinear. Increasing the bending stiffness and the damping ratio can both decrease the vibration of this innovative floor system. But increasing the bending stiffness always means a thicker plate or higher beam, which would increase the costs of construction. Thus, it is recommended that some construction strategies be adopted to enhance the damping ratio and control the vibration of this innovative floor system.

5. Discussion and Further Application

5.1. Error Discussion

Even though satisfactory results were reached, there are still some errors in the numerical simulation. The errors mainly come from two aspects, which are the incorrect prediction of single-person loads and the calculation error of the bending stiffness from the theoretical derivation.

For the first error, the single foot force–time curve is presented in Figure 14, where the point loads from each step are considered as fixed values. In real cases, the foot force depends on the height of each leg lifted while walking, which makes it unable to be captured by a fixed function.

For the latter one, the error mainly comes from two aspects: (1) over-simplification of boundary conditions; the beam around the four sides should be taken into consideration for more accurate prediction; (2) the whole floor system is simulated with a uniform shell, which cannot fully capture the different vibrations when stepping on the ribbed beam or the hollow lightweight infills.

5.2. Further Application

The method proposed in this article could be applied not only to predict the natural frequency and maximum acceleration for similar assembled integral floor systems, but also to walls and decks with ribbed beams. But the key problem is to check if these components could be simulated by a uniform plate with fixed ends, especially the interface behavior between the precast panels and cast-in-situ parts.

Design parameters, such as the thickness of the PRB, height and distance of the ribbed beam, are the key factors that directly change the bending stiffness of this innovative assembled integral two-way multi-ribbed composite floor system. With the increase in thickness of the PRB and the height of the ribbed beam, the frequency would increase correspondingly due to a larger bending stiffness, and the maximum acceleration would decrease. However, in this paper, we simulate the floor system as a uniform shell with fixed ends, where the bending stiffness could be obtained from static tests. When applied to other spans or different designs, the relationship between the bending stiffness and design parameters is not clear. A 3d finite element method should be the potential method, but there might be too many solid elements slowing down the calculation efficiency and precision. Thus, a simplified model is needed, and this study will be the key part of our next article.

The damping ratio is the key to controlling the vibration of the floor system with the same bending stiffness. For this type of assembled integral two-way multi-ribbed composite floor system, the ribbed beam distance and thickness of the whole floor system would significantly enhance the stiffness, through which the maximum acceleration could also be controlled. Considering the costs, it is recommended that some construction strategies be adopted to enhance the damping ratio and control the vibration of this innovative floor system.

For the same vibration tests, it is recommended to put acceleration meters at both the top and bottom of the test specimen. Furthermore, it is recommended that more volunteers than the design code requires are needed, since the relationship between the maximum acceleration and multi-person load is nonlinear. Single-person, multi-person walking and running should also be a necessary choice for further study of the dynamic behaviors of floor systems.

6. Conclusions

This paper presented the dynamic performances of a 9.2 m × 9.2 m assembled integral two-way multi-ribbed composite floor system, both experimentally and theoretically. The natural frequency, vibration mode shapes, and acceleration–time history curves were analyzed, and simplified formulas were derived to predict their dynamic performances. The main conclusions are presented as follows:

1. The dynamic behavior of this innovative floor system meets the requirements of GB50010-2010 and ISO 2631. The first-order vibration frequency of the 9.2 m × 9.2 m floor is 11 Hz, larger than 3 Hz. The maximum acceleration for a single person is 0.07 m/s^2, smaller than the limited value of 0.35 m/s^2. For multiple persons, the maximum acceleration is 0.45 m/s^2.
2. Elastic plate theory could be applied to predict the natural frequency and acceleration, with the bending stiffness obtained from the experiment. The errors of elastic vibration theory in calculating the first three vibration mode shapes and natural frequencies are

approximately 10%. And the absolute error of the acceleration time–history curve is less than 10%, demonstrating its efficiency in predicting the vibration behaviors of this innovative assembled integrated two-way multi-ribbed composite floor system.
3. The bending stiffness and damping ratio are the key factors influencing the vibration acceleration of this innovative floor system. Considering the costs, it is recommended that some construction strategies be adopted to enhance the damping ratio and control the vibration of this innovative floor system.
4. For similar vibration tests, it is recommended to put acceleration meters at both the top and bottom of the test specimen. Furthermore, it is recommended that more volunteers than the design code requires are needed, since the relationship between the maximum acceleration and multi-person load is nonlinear.

Author Contributions: Conceptualization, M.X. and L.G.; methodology, Y.F.; software, W.Z.; validation, M.X., Y.F. and L.G.; formal analysis, L.G.; investigation, Y.F.; resources, W.Z.; data curation, L.G.; writing—original draft preparation, W.Z.; writing—review and editing, L.G.; visualization, X.Z.; supervision, M.X.; project administration, X.Z.; funding acquisition, M.X. and X.Z. All authors have read and agreed to the published version of the manuscript.

Funding: The study was supported by the Basic Science (Natural Science) Research Major Project of Higher Education Institutions in Jiangsu Province (23KA560006), whose support is gratefully acknowledged.

Data Availability Statement: The XLSX data used to support the findings of this study may be accessed by emailing the corresponding author, who can be contacted at 103009030@seu.edu.cn.

Conflicts of Interest: Author Xiangqiang Zeng was employed by the company Shandong Shijia Yuanda Technology Development Co., Ltd. The remaining authors declare that the research was conducted in the absence of any commercial or financial relationships that could be construed as a potential conflict of interest. The authors declare that there are no conflicts of interest regarding the publication of this paper.

Symbols and Notation List

a	in-plane horizontal length of the floor	$\boldsymbol{A}$	state matrix of a multi-DOF system
a_{max}	maxmimum acceleration under human load	$\boldsymbol{B}$	state matrices matrix of a multi-DOF system
b	in-plane vertical length of the floor	$\boldsymbol{C}$	damping matrix of a multi-DOF system
Bk	Fourier coefficient	$\boldsymbol{D}$	observation matrix
D	bending stiffness of the floor system	$\boldsymbol{E}$	feedthrough (or direct transmission) matrix
f_{ij}	natural frequency, i and j indicate the number of half-waves in the x and y direction.	$\boldsymbol{K}$	stiffness matrix of a multi-DOF system
fs	walking step rate	$\boldsymbol{M}$	mass matrix of a multi-DOF system
$F(t)$	force from a single footfall	$\boldsymbol{X}$	displacement of a multi-DOF system
m	average mass of the floor system	$\boldsymbol{u(t)}$	white noise from the earth
G	pedestrian weight	xk	acceleration and velocity of a multi-DOF system due to the white noise
g	gravity acceleration	yk	observed acceleration, velocity and displacement of a multi-DOF system
t	time	$\delta(x)$	Dirac function.
Te	duration of a single footstep	ωi	ith angular frequency of a multi-DOF system
w	out-of-plane deflection of the floor system	ξi	ith damping ratio of a multi-DOF system
W_{ij}	vibration mode shape		
x	in-plane horizontal coordinate		
y	in-plane vertical coordinate		

References

1. Ding, D. *Calculation of Reinforced Concrete Floor*; Press of science and technology: Shanghai, China, 1954. (In Chinese)
2. Safar, A.; Lou, K.B. A study of the action of the beam and beamless (flush) floor slabs of the multistorey buildings. *Erciyes Üniversitesi Fen Bilim. Enstitüsü Fen Bilim. Derg.* **2007**, *23*, 127–135.
3. Newmark, N.M. Proposed design specifications for two-way floor slabs. *J. Proc.* **1950**, *46*, 597–607.
4. Nilson, A.H.; Walters, D.B. Deflection of two-way floor systems by the equivalent frame method. *J. Proc.* **1975**, *72*, 210–218.
5. Huang, Y.; Ma, K.; Zhang, H.; Xiao, J.; Jiang, S. Study and application of Vierendeel-sandwich-plate floor framing in multistoried and tall building. *J. Build. Struct.* **1997**, *18*, 55–64.
6. Pan, Y. Study of Load-Bearing Properties of PK Prestressed Composite Slab. Doctoral Dissertation, University of Hunan, Changsha, China, 2009. (In Chinese)
7. Pang, R. Research on the Mechanical Property and Seismic Design Method of New Type Precast RC Diaphragms. Master Dissertation, Southeast University, Nanjing, China, 2011. (In Chinese)
8. Naito, C.; Cao, L.; Peter, W. Precast concrete double-tee connections, part 1: Tension behavior. *PCI J.* **2009**, *54*, 49. [CrossRef]
9. Spadea, S.; Rossini, M.; Nanni, A. Design analysis and experimental behavior of precast concrete double-tee girders prestressed with carbon-fiber-reinforced polymer strands. *PCI J.* **2018**, *63*, 72–84. [CrossRef]
10. Jiang, Q.; Zhang, K.; Feng, Y.; Chong, X.; Wang, H.; Lei, Y.; Huang, J. Out-of-plane flexural behavior of full precast concrete hollow-core slabs with lateral joints. *Struct. Concr.* **2020**, *21*, 2433–2451. [CrossRef]
11. de Lima Araújo, D.; Sales, M.W.R.; Silva, R.P.M.; Antunes, C.D.F.M.; de Araújo Ferreira, M. Shear strength of prestressed 160 mm deep hollow core slabs. *Eng. Struct.* **2020**, *218*, 110723. [CrossRef]
12. Ghayeb, H.H.; Razak, H.A.; Sulong, N.R. Performance of dowel beam-to-column connections for precast concrete systems under seismic loads: A review. *Constr. Build. Mater.* **2020**, *237*, 117582. [CrossRef]
13. Chen, Y.; Shi, H.R.; Wang, C.L.; Wu, J.; Liao, Z.Q. Flexural mechanism and design method of novel precast concrete slabs with crossed bent-up rebar. *J. Build. Eng.* **2022**, *50*, 104216. [CrossRef]
14. Huang, Y.; Yang, J.; Zhong, C. Flexural performance of assembly integral floor structure voided with steel mesh boxes. *J. Build. Eng.* **2022**, *54*, 104693. [CrossRef]
15. Deng, B.Y.; Tan, D.; Li, L.Z.; Zhang, Z.; Cai, Z.W.; Yu, K.Q. Flexural behavior of precast ultra-lightweight ECC-concrete composite slab with lattice girders. *Eng. Struct.* **2023**, *279*, 115553. [CrossRef]
16. Zeng, X.; Feng, Y.; Ruan, S.; Xu, M.; Gong, L. Experimental and Numerical Study on Flexural Behavior of a Full-Scale Assembled Integral Two-Way Multi-Ribbed Composite Floor System. *Buildings* **2023**, *13*, 2517. [CrossRef]
17. Zhang, W.; Feng, Y.; Zeng, X.; Xu, M.; Gong, L.; Rui, L. Flexural Performances of Novel Wet Joints with Sleeve Connections in Precast Composite Floor System. *Buildings* **2024**, *14*, 822. [CrossRef]
18. Tilden, C.J. Kinetic effects of crowds. *Trans. Am. Soc. Civ. Eng.* **1913**, *76*, 2107–2126. [CrossRef]
19. Fuller, A.H. Dynamic effects of moving floor loads-stresses measured in the floor and balcony of a college gymnasium. *Am. Arch. Arch.* **1924**, *126*, 455–456.
20. Greimann, L.F.; Klaiber, F.W. Dynamic forces induced by spectators. *J. Struct. Div.* **1978**, *104*, 348–351. [CrossRef]
21. Ebrahimpour, A.; Sack, R.L. Modeling dynamic occupant loads. *J. Struct. Eng.* **1989**, *115*, 1476–1496. [CrossRef]
22. Ebrahimpour, A.; Sack, R.L. Design live loads for coherent crowd harmonic movements. *J. Struct. Eng.* **1992**, *118*, 1121–1136. [CrossRef]
23. Pernica, G. Dynamic load factors for pedestrian movements and rhythmic exercises. *Can. Acoust.* **1990**, *18*, 3–18.
24. Lenzen, K.H. Vibration of steel joist-concrete slab floors. *Eng. J.* **1966**, *3*, 133–136. [CrossRef]
25. Li, Q.; Fan, J.; Nie, J. Stochastic vibration of a long-span floor under crowd-induced rhythmic excitation. *J. Vib. Shock* **2012**, *31*, 9–15. (In Chinese)
26. Liu, J.; Zheng, X.; Li, J. Vibration serviceability of composite slab with prefabricated concrete straight ribbed panels. *J. Build. Struct.* **2019**, *40* (Suppl. S1), 41–46. (In Chinese)
27. Cui, W.X.; Chen, Z.F.; Yin, W.Y.; Liu, S.C. Serviceability of a new kind of precast hollow cross-grids floor system. *Appl. Mech. Mater.* **2017**, *873*, 158–163. [CrossRef]
28. *GB50010-2010*; National Standard of the People's Republic of China: Code for Design of Concrete Structures. China Architecture & Building Press: Beijing, China, 2015.
29. *GB/T 228.1-2010*; National Standard of the People's Republic of China: Metallic Materials—Tensile Testing—Part 1: Method of Test at Room Temperature. Standard Press of China: Beijing, China, 2015.
30. *JG/T 163-2013*; Coupler for Rebar Mechanical Splicing. Standards Press of China: Beijing, China, 2013.
31. Kalman, R.; Ho, B. Effective construction of linear state variable models from input output data. In Proceedings of the 3rd Allerton Conference, Monticello, VA, USA, 20–22 October 1965; pp. 449–459.
32. Van Overschee, P.; De Moor, B. *Subspace Identification for Linear Systems: Theory—Implementation—Applications*; Springer Science & Business Media: Berlin/Heidelberg, Germany, 2012.

33. Allen, G.R. *Ride Quality and International Standard ISO 2631 ("Guide for the Evaluation of Human Exposure to Whole Body Vibration")*; NASA, Technical Report No. NASA-TM-X-3295; NASA: Washington, DC, USA, 1975.
34. Chen, J.; Wang, H.; Wang, L. Experimental investigation on single person's jumping load model. *Earthq. Eng. Eng. Vib.* **2015**, *14*, 703–714. [CrossRef]
35. Zhang, S.; Xu, L. Human-induced vibration of cold-formed steel floor systems: Parametric studies. *Adv. Struct. Eng.* **2020**, *23*, 2030–2043. [CrossRef]
36. Pang, R.; Liu, J.; Zhou, F.; Dang, L.; Yang, Y.; Wang, W.; Duan, J. Vibration behavior of innovative discretely connected precast concrete sandwich floors. *J. Build. Eng.* **2024**, *82*, 108313. [CrossRef]
37. Li, Q.; Fan, J.; Nie, J.; Li, Q.; Chen, Y. Crowd-induced random vibration of footbridge and vibration control using multiple tuned mass dampers. *J. Sound Vib.* **2010**, *329*, 4068–4092. [CrossRef]
38. *JGJ 3-2010*; Technical Specification for Concrete Structures of Tall Buildings. China Architecture & Building Press: Beijing, China, 2010.

buildings

MDPI

Article

Study on the Composite Performance of Sandwich Wall Panels with SGCCs

Huanzhi Jiang [1], Libo Xie [1,*], Fengyuan Chang [1], Yu Cao [1] and Zhengxing Guo [2]

[1] School of Civil Engineering, Qingdao University of Technology, Qingdao 266520, China; jianghuanzhi@qut.edu.cn (H.J.); 18863658746@163.com (F.C.); caoyumarch@126.com (Y.C.)
[2] School of Civil Engineering, Southeast University, Nanjing 210096, China; guozx195608@126.com
* Correspondence: xielibo@qut.edu.cn

Abstract

This paper aims to explore the impact of different arrangements of new steel-glass FRP composite connectors (SGCCs) on the bending and composite performance of sandwich wall panels. Through monotonic loading bending tests on six full-size specimens, aspects such as their failure modes, load-deflection curves, load-strain relationships, slip between the thermal insulation layer and concrete, and composite action were analyzed. The results show that all sandwich wall panels experienced bending and ductile failure, and exhibit partial composite performance, with P4 having the best composite performance and P1 the worst. The degree of composite action is positively correlated with the flexural bearing capacity. The bending capacity mainly depends on the layout rather than the total number of SGCCs. Arranging connectors along the short side of the panel has a more significant impact, and the number of connectors at the panel's ends has a greater influence on the composite performance. Except for P1, the theoretical value of the composite degree of the other sandwich wall panels exceeds 70%, and P4 reaches 85%. The theoretical calculations are in good agreement with the experimental results. This study provides theoretical and data support for the rational configuration of connectors in sandwich wall panels and is of great significance for building engineering applications. Meanwhile, suggestions for configuring connectors in actual engineering are also given.

Keywords: sandwich wall panels; steel-glass FRP composite connector; bending capacity; composite performance; theoretical calculation

1. Introduction

Energy conservation and emission reduction in the construction industry are key measures to achieve the goals of carbon peak and carbon neutrality. External wall insulation accounts for a large proportion of building energy-saving systems. Sandwich wall panel, also known as precast concrete sandwich insulation wall panels, consists of inner and outer concrete wythes and a middle insulation layer [1]. The inner and outer concrete wythes are usually connected using connectors (such as metal connectors, concrete blocks, or FRP connectors) through the insulation layer to form a "sandwich" overall structure [2–4]. Due to its superior thermal and structural efficiency [5–7], sandwich wall panels have received widespread attention and application in the development of building industrialization.

According to the interaction between the inner and outer concrete wythes and the insulation layer, sandwich wall panels can be divided into three types: non-composite panels, fully composite panels, and partially composite panels [1]. These categories indicate the degree to which the two concrete wythes act in unison to resist loads. Non-composite means the connector is unable to transmit force or possess stiffness, and the inner and outer wythes bear loads independently. While fully composite behavior implies the presence of connections with infinite rigidity and enables all shear force transmission, the inner and outer wythes deform in full coordination and jointly bear the loads. Partially composite represents a state intermediate between fully composite and non-composite. In this state, the connectors can transfer a portion of the shear forces, realize the connection and stress transfer between the inner and outer wythes to a certain extent, and enable them to have a certain ability to work cooperatively. The bending strain distribution along the section height of each type of panel is shown in Figure 1. The degree of composite action mainly depends on the nature of connectors that connect the inner and outer concrete wythes.

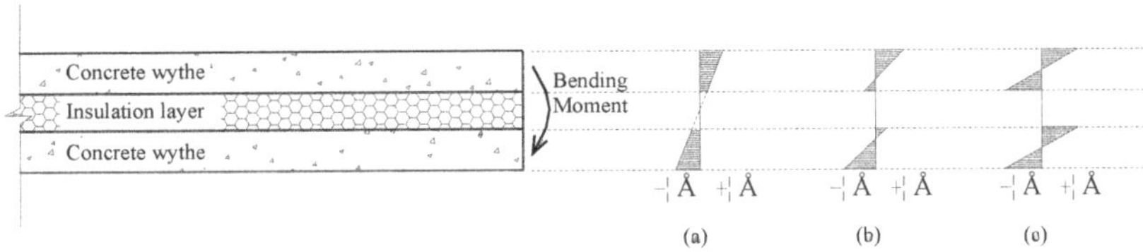

Figure 1. Strain distribution in sandwich wall panel under flexure: (**a**) Fully composite; (**b**) Partially composite; (**c**) Non-composite.

In high-rise buildings, sandwich wall panels provide thermal insulation shells for the building, mainly bearing vertical self-weight, horizontal wind loads, and seismic effects [5–7]. The load borne by the outer concrete wythe will be transmitted to the inner wythe through connectors. Connectors are the key components of sandwich wall panels.

Steel connectors are widely used in traditional sandwich wall panels, and their excellent mechanical properties can enable sandwich wall panels to achieve high composite action [8–10]. However, steel connectors have high thermal conductivity (45~60 W/(m·K)) and are prone to thermal bridging effects in sandwich wall panels, resulting in poor insulation performance [11], and no longer meeting current energy-saving requirements [12,13].

Fiber-reinforced polymer (FRP) has low thermal conductivity (0.2~0.3 W/(m·K)) and high tensile strength (usually 2~10 times that of steel bars). Using FRP connectors instead of traditional steel connectors can effectively solve the problem of the "thermal bridging effect" [14–17] and significantly reduce external wall energy consumption. However, the low shear strength of FRP connectors (usually 50~60 MPa and only 5~20% of the tensile strength of FRP connectors) can easily lead to interlayer shear failure of the inner and outer concrete wythes at the connection points [18,19], and their stiffness cannot ensure that the inner and outer concrete wythes of the sandwich wall panel work synergistically to achieve spatial composite action.

In practical applications of traditional sandwich wall panels, the performance of connectors has the above-mentioned limitations. Therefore, how to optimize the connectors of sandwich wall panels to meet the requirements of mechanical properties, thermal insulation performance, and durability simultaneously has become a crucial problem urgently to be solved at present, which is also the starting point of this study.

Considering the good ductility and high shear strength of the steel, as well as the excellent thermal efficiency and corrosion resistance of FRP, some scholars have attempted to combine the two to form a composite reinforcement and apply it in construction engineering.

Zheng Bailin took the initiative to propose a novel type of FRP-wrapped rebar [20], which had steel as the core with ±45° angle laminations of glass fiber impregnated with epoxy around it and which was then cured as a whole. Research showed that the mechanical deformation performance between the metal core and the FRP-wrapped layer was correlated, and the mechanical behavior of the FRP-wrapped rebar had a good bilinear character. After that, Wu Gang [21] conducted experimental and theoretical studies on the mechanical properties of steel-FRP composite bars (SFCB). The results showed that SFCB had a high elastic modulus, high tensile strength (about 50% higher than ordinary steel bars), stable post-yield stiffness, and excellent corrosion resistance. In view of these studies, the composite material composed of a mixture of metal and non-metal can also be applied to sandwich wall panels to improve the thermal and structural performance of buildings.

The team led by Guo Zhengxing from Southeast University has independently developed a new type of connector of W-shaped Steel Glass Fiber Reinforced Polymer (SGFRP) [22]. The SGFRP connector uses steel bars as the inner core, and glass fibers impregnated with epoxy resin are wrapped around the core in a circumferential manner. Researchers conducted a series of experimental studies on SGFRP connectors and sandwich wall panels and found that such W-shaped SGFRP connectors have high tensile and shear strength [23–25]. However, due to the difficulty in controlling the anchoring depth during construction, the connectors tend to anchor improperly and be pulled out prematurely when the sandwich wall panels bend, thereby reducing the bending resistance and composite performance of the sandwich wall panels.

To solve the anchoring and positioning problem of connectors, the research team has developed another type of Steel-Glass fiber-reinforced polymer Composite Connector (SGCC), as shown in Figure 2. The steel core skeleton of SGCC is formed in the following way: first, one end of the scored steel wire or ribbed steel bar is upset, and then it is inserted into a metal disc with a hole in the middle to make a disc expansion head. A certain thickness of GFRP resin layer is wrapped around the steel core skeleton, and then through the molding process, a steel-fiber hybrid connector is formed with a disc expansion head at one end, multiple continuous inverted conical platforms on the surface, and a circular positioning ring 30 mm away from the disc expansion head. The purpose of designing the disc expansion head and inverted conical platforms is to enhance the anchoring ability of the connector, while the purpose of the positioning ring is to facilitate the positioning of the connector when inserted into the insulation layer. The SGCCs are molded using a 315T hydraulic press equipment (Ningbo Hilead Hydraulic Co., Ltd., Ningbo, China), and the preparation process is as follows: apply a release agent to the mold→ filler preforming → molding → heating and solidifying → demolding → deburring. This process and the physical photo of SGCCs are shown in Figure 3. The SGCC can be set in two types: vertical insertion type and 45° oblique insertion type. As shown in Figure 4, the two types can be used in pairs to form a triangular structure to tie the three interlayers of the sandwich wall panels. The insertion direction of the oblique connector is perpendicular to the short side of the sandwich wall panels to transmit shear force.

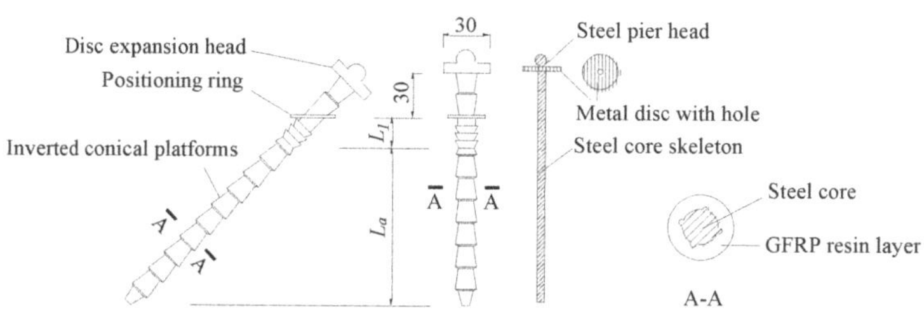

Figure 2. Constructional detail of SGCCs.

Previous studies have been conducted on the mechanical properties of SGCC and the flexural and composite performance of sandwich wall panels when uniformly arranged [26].

However, the influence of different arrangements of SGCCs on the bending strength and composite performance of sandwich wall panels is still unclear.

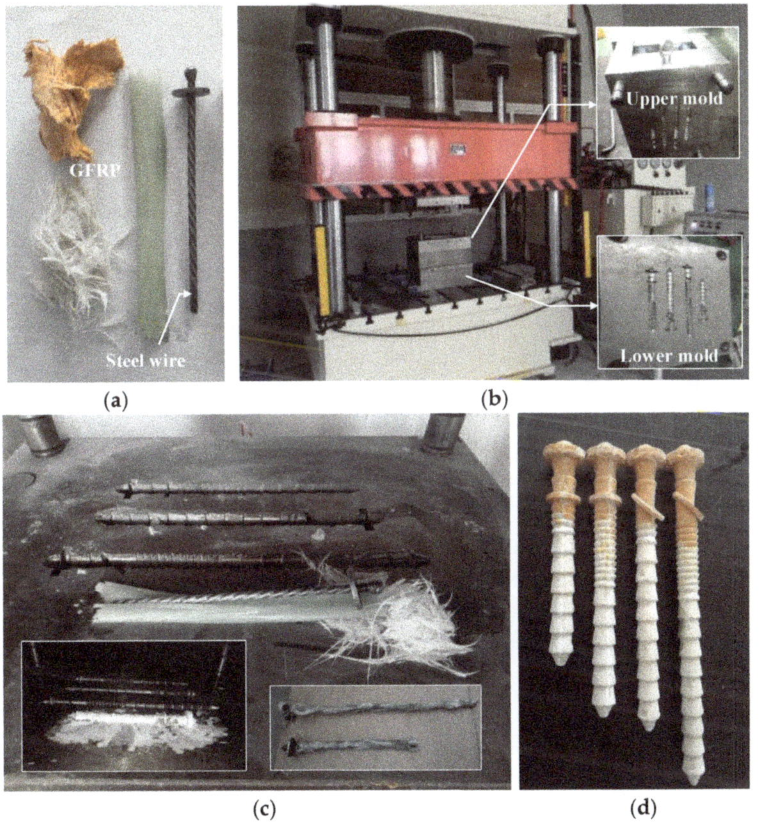

Figure 3. The preparation process of SGCCs: (**a**) Raw materials; (**b**) 315T hydraulic press equipment and molds; (**c**) Filling, molding, forming and demolding; (**d**) Physical photo of SGCCs.

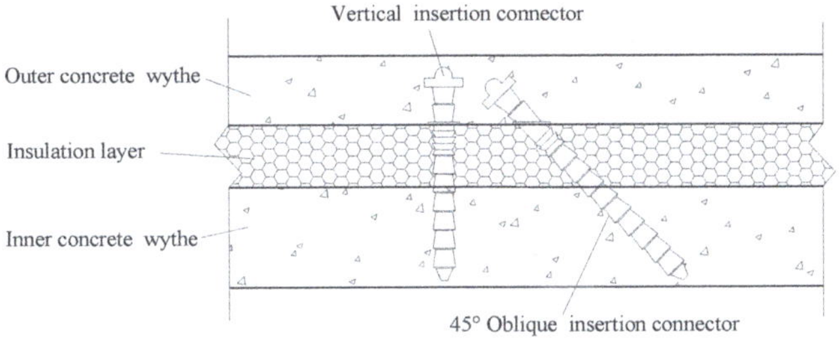

Figure 4. Components of sandwich wall panel with SGCCs.

Currently, there is limited research on the effects of different arrangements of connectors on the bending strength and composite performance of sandwich wall panels. Existing studies mainly focus on the performance of panels with uniformly spaced connectors [27–34], lacking a comprehensive and systematic comparative analysis of various connector arrangements. This study conducted monotonic loading bending tests on sandwich wall panels with different arrangements of SGCCs and evaluated their composite performance, filling the gap in research on the relationship between connector arrangement and panel performance in this field. It provides theoretical basis and data support for the reasonable arrangement of connectors in sandwich wall panels and helps to promote the further optimization and application of sandwich wall panels in construction engineering.

2. Experimental Program

2.1. Test Specimen Design

There are a total of six-panel specimens for the flexural test, including five sandwich wall panels and one concrete solid wall panel. The dimensions of each panel are 3200 mm × 1200 mm × 160 mm. The sandwich wall panels consist of a 50 mm thick outer concrete wythe, a 50 mm thick extruded polystyrene (XPS) insulation layer, and a 60 mm thick inner concrete wythe. The SGCC used in the sandwich wall panel specimen has a diameter of 10 mm and is composed of high-strength scored steel wire with a diameter of 5 mm and a GFRP wrapping layer with a thickness of 2.5 mm. The concrete used in the experiment is fine aggregate concrete with a strength of C40, and the compressive strength of the XPS insulation layer is 266 Kpa. All sandwich wall panels are equipped with HRB400 grade single-layer steel mesh with a diameter of 8 mm and a spacing of 200 mm in the inner and outer concrete wythes, and double-layer steel mesh is installed in the solid concrete wall panel. The dimensions and reinforcement details of each specimen are shown in Figure 5. According to the material performance test, the relevant strength indicators of the steel bars and connectors are measured in Table 1.

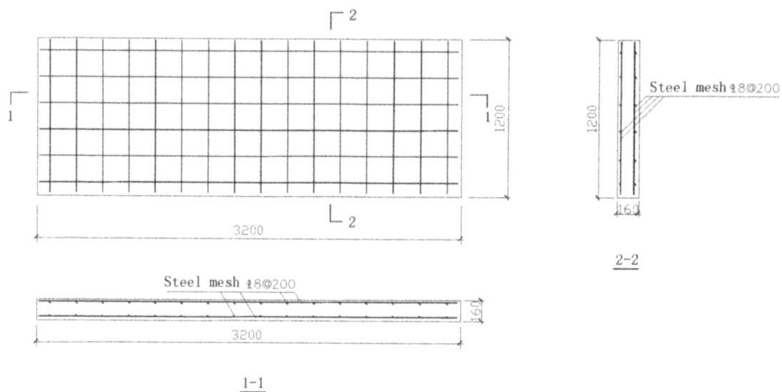

(a)

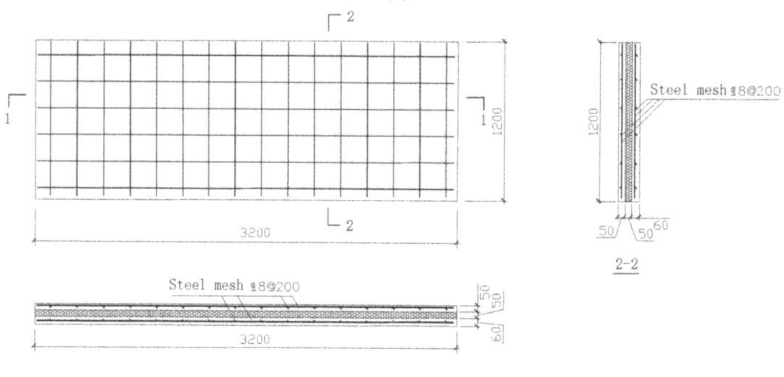

(b)

Figure 5. Reinforcement diagram of flexural test panels (all dimensions in mm): (**a**) Solid concrete wall panel; (**b**) Sandwich wall panel.

Table 1. Mechanical properties of materials.

Material Type	Diameter (mm)	Elastic Modulus (GPa)	Yield Strength (MPa)	Ultimate Strength (MPa)	Shearing Strength (MPa)
steel	8	200	495	656	--
SGCC	10	59	357	442	409

The five full-size sandwich wall panels use five different arrangements of SGCCs, as shown in Figure 6. The first panel P1(2-1-1-2) has two pairs of connectors at both ends, with one pair arranged in the middle. In the second panel P2(2-2-2-2), two rows of connectors are arranged along the short side, with two pairs at each end and middle section of the panel. Three pairs of connectors are arranged at both ends of the third panel P3(3-2-2-3), and two pairs of connectors are arranged in the middle section. The fourth panel P4(3-3-3-3) is based on the layout of P2(2-2-2-2), with the spacing between connectors reduced along the short side of the panel, and three pairs of connectors are arranged at both ends and the middle part of the panel. The fifth panel P5(2-2-2-2-2-2) is based on the layout of P2(2-2-2-2), with the spacing between connectors tightened along the long side of the panel. Two pairs of connectors are arranged at each end, quarter, and middle of the panel, that is, a total of six pairs of connectors are arranged in each row parallel to the long side direction. The purpose of arranging different numbers of connectors at different positions on the sandwich wall panel is to understand the influence of different connector layouts and quantities on the bending capacity and composite performance of sandwich wall panels, in order to find an optimal connector arrangement scheme. For the convenience of expression and to enhance the readability of the paper, each test panel will be abbreviated as P0, P1, P2, P3, P4, and P5 respectively in the following text.

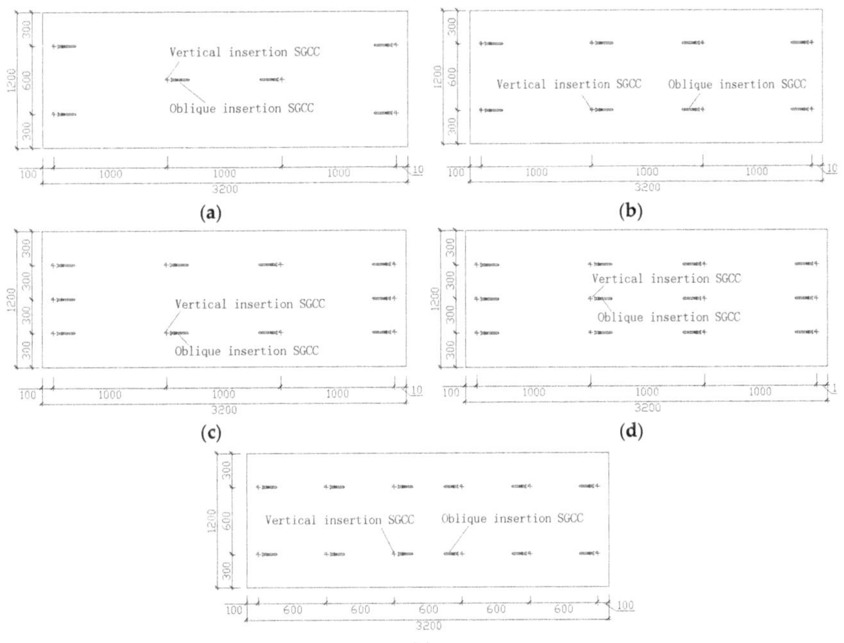

Figure 6. Plan view of the five test sandwich wall panels (all dimensions in mm): (**a**) P1 (2-1-1-2); (**b**) P2 (2-2-2-2); (**c**) P3 (3-2-2-3); (**d**) P4 (3-3-3-3); (**e**) P5 (2-2-2-2-2-2).

2.2. Preparation of Test Specimens

The production of sandwich wall panel specimens starts from the XPS insulation layer. Cut the XPS insulation layer and splice it together to match the width of the panel. Mark the position of the connectors on the surface of the XPS insulation layer according to the design, and then place them aside for future use.

The sandwich wall panel specimens are made using the forward striking process of inner concrete wythe, insulation layer and outer concrete wythe. After the steel formwork is prepared, the steel mesh is in place, concrete is poured and compacted by vibration to form the inner concrete wythe. Lay the prepared XPS insulation layer on the upper surface of the inner concrete wythe in order, insert SGCCs according to the marked points, perform self-positioning through the positioning ring on the connectors, and insert the lower end of the connectors into the inner concrete wythe. Then fill the gap of the insulation layer with canned expansion foam, and scrape off the excess foam. After ensuring that all connectors are inserted and positioned, place the upper steel mesh in place, pour the upper outer concrete wythe, and compact it by vibration. Finally, smooth and finish the surface. The forward striking process can firmly embed the enlarged head of SGCCs into the outer concrete wythe, providing a stronger anchoring effect. In addition, three cubic test blocks were prepared while the concrete was being poured. The production process of sandwich wall panels with SGCCs is shown in Figure 7.

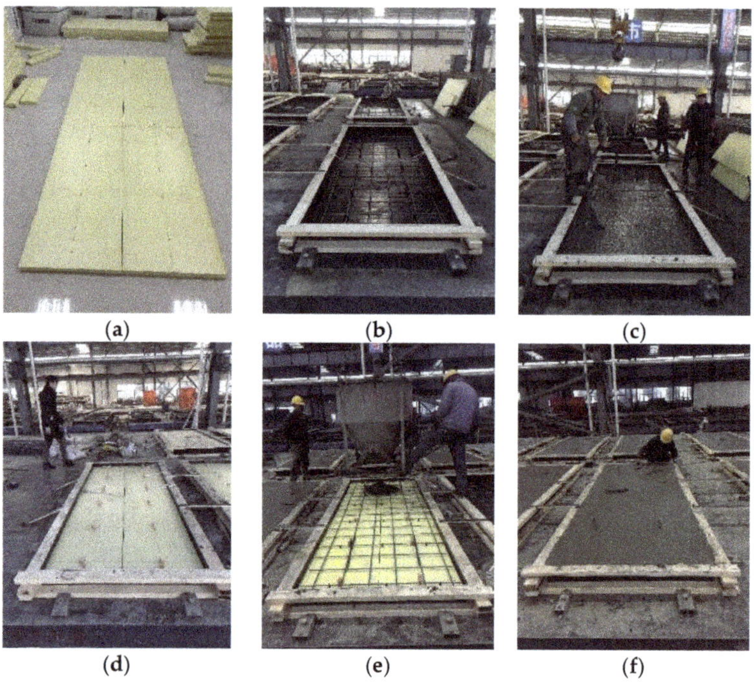

Figure 7. Fabrication and casting process of sandwich wall panels: (**a**) Cutting and splicing XPS insulation layer; (**b**) Laying steel mesh of inner concrete wythe; (**c**) Pouring and vibrating inner concrete wythe; (**d**) Arranging insulation layer and inserting SGCCs; (**e**) Laying the steel mesh of outer wythe and pouring concrete; (**f**) Smoothing and finishing the concrete surface.

2.3. Test Setup

The experiment adopts a four-point support and two-point symmetrical loading method. When placing the sandwich wall panel, the inner wythe is placed below and the outer wythe is placed above. The test setup is shown in Figures 8 and 9. The net span of all wall panels is 2900 mm, and a load distribution beam is used to distribute the load onto two I-shaped line load beams, forming a 600 mm pure bending section in the middle of the wall panel and an 1150 mm bending shear section on both sides. The panel specimens were tested and calibrated simultaneously using a manual digital pull-out instrument with a range of 30T and a force sensor with a range of 20T.

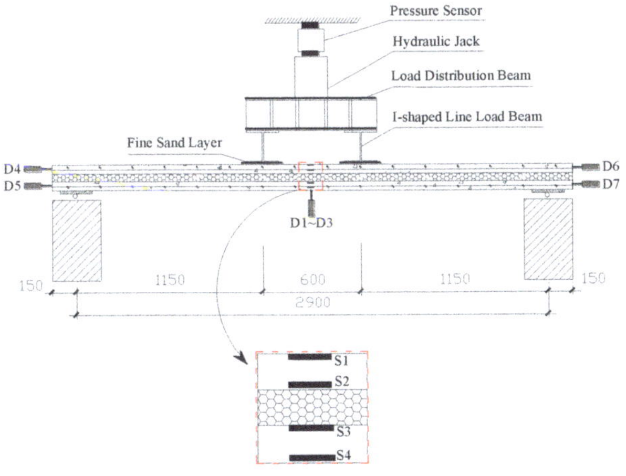

Figure 8. Design drawing of test setup (unit: mm).

Figure 9. Test setup at the scene.

2.4. Test Content and Process

A total of seven displacement gauges and four strain gauges were installed on each panel. Three displacement gauges (D1–D3) with a range of 300 mm were installed at the bottom of the mid-span of the panel to measure the deflection at the mid-span. Four displacement gauges with a range of 100 mm were installed in the middle of the inner and outer concrete wythes at both ends of the panel (D4 and D5 on the west side, D6 and D7 on the east side) to measure the relative slip between the inner and outer concrete layers, as shown in Figure 8. Four concrete strain gauges (S1–S4) with a length of 50 mm were pasted on the side of the mid-span of the panel to measure the strain along the section height of the panel. The positions of the strain gauges are shown in Figure 8. In addition, in order to understand the efficiency of SGCCs at different positions of the panel in the bending process, strain gauges with a length of 5 mm were pasted on the connectors at the end, quarter, and middle positions of panel P5.

Collect initial readings before placing the line load beams to deduct the deflection of the panel under its weight. Before formal loading, pre-load to 5 kN to ensure that all instruments and equipment are in close contact and working properly, and then unload to zero. The experimental loading is controlled by force, with each level being loaded in steps of 2 kN and held for 5 min until the load drops to 85% of the peak load and cannot rise again. This marks the complete failure of the panel and the loading is now terminated. During loading, the DH3816Net (V3.4.04) static strain testing system (DongHua Testing Technology Co., Ltd., Taizhou, China) is used to continuously collect and record data at 1-s intervals. At each level of the load stage, the crack morphology of the panel should also be recorded, the trend of the cracks should be traced with a marker pen, and the corresponding load value should be indicated. At the same time, the width of cracks and the slip between the insulation layer and the concrete wythes at both ends of the panels under each level of load is measured with a steel ruler and recorded.

2.5. Experimental Phenomena and Destructive Characteristics

All test panels experienced bending failure, and the failure process of each sandwich wall panel was very similar. The first full-length crack appeared in the pure bending section at the bottom of the inner concrete wythe and extended from the bottom to the side; As the load increased, new cracks continued to appear at the bottom of the inner concrete wythe, and the cracks on the panel side gradually extended upwards. Cracks also began to appear on the outer wythe side near the position of the line load beam, and slight slippage occurred between the insulation layer and the concrete layers. As the load further increased, the crack width also increased, and new cracks continued to appear on the outer concrete wythe side. The insulation layer broke, and there was obvious slippage between the insulation layer and the concrete layers. The deflection of the panel at the mid-span gradually increased. Finally, the surface of the outer concrete wythe cracked or the concrete was crushed, the crack width rapidly increased, and the deflection of the panel at the mid-span increased sharply. After the load dropped to 85% of the peak load and it could not continue to be loaded, the specimen was declared damaged and the test was stopped. After the experiment, the insulation material at the panel end was removed, and it was found that some SGCCs had fiber splitting and concrete anchorage failure. The entire failure process of all sandwich wall panels was slow, with significant deformation, all of which belonged to ductile failure. The crack distribution and failure mode of the test panels are shown in Figures 10 and 11, respectively.

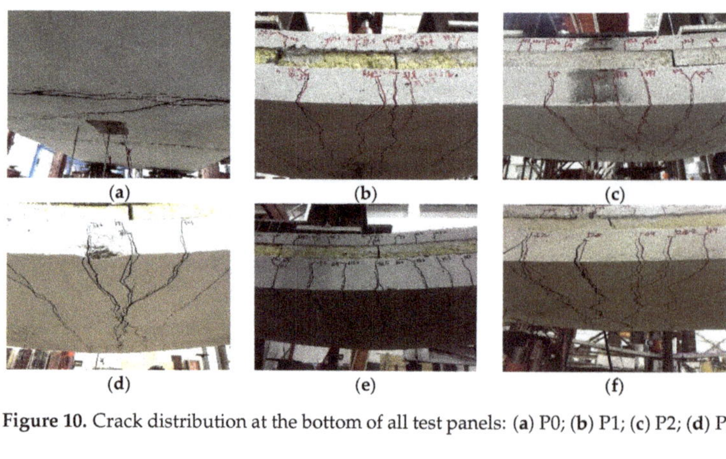

Figure 10. Crack distribution at the bottom of all test panels: (**a**) P0; (**b**) P1; (**c**) P2; (**d**) P3; (**e**) P4; (**f**) P5.

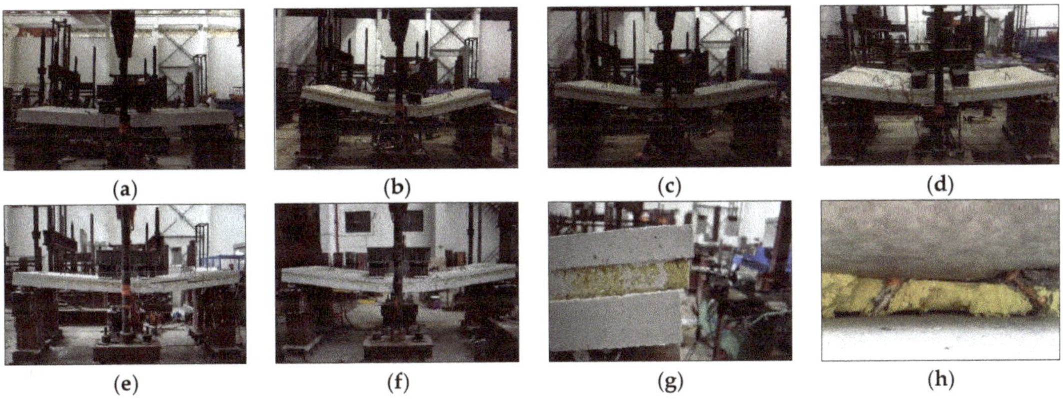

Figure 11. Failure modes of all test panels: (**a**) P0; (**b**) P1; (**c**) P2; (**d**) P3; (**e**) P4; (**f**) P5; (**g**) Relative interlayer slip at the panel end; (**h**) Failure mode of SGCCs at the panel end.

3. Results and Discussion

Based on the load-deflection curve, the load-strain relationship of connectors, strain variation along the section height, and load-relative slip relationship, the obtained experimental results were analyzed and discussed.

3.1. Load-Deflection Curve

The load-deflection curves of each flexural specimen are shown in Figure 12.

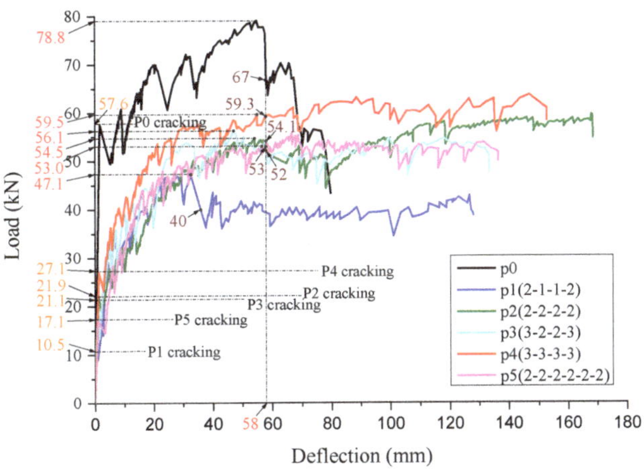

Figure 12. Load-deflection curves for test panels at mid-span.

From the load-deflection curve, it can be seen that during the loading process, the load of each specimen has varying degrees of fluctuations in both decrease and increase, and it has undergone a long process from loading to complete failure. The load-deflection curve of panel P1 tends to flatten after decreasing to 85% of the peak load, at which point the panel can still withstand a certain load, but with significant deformation. The load-deflection curves of panels P2 and P3 will slowly rise back after decreasing to 85% of the peak load, until reaching a higher load point, and then sharply decline, causing significant deformation of the panels. The load-deflection curve of panel P4 rises gently after the yielding of the panel, until the maximum load is reached and the load begins to decrease, resulting in significant deformation of the panel. The load-deflection curve of panel P5 gradually rises to the peak load and then decreases, and tends to flatten before reaching 85% of the peak load. At this point, the panel continues to bear the load, but undergoes significant deformation. In previous studies [31,32], the load reduction trend of test sandwich wall panels when reaching the ultimate state was relatively single. However, in this study, the load changes of each panel showed diverse fluctuation characteristics, which was different from previous studies. This is mainly attributed to the influence of different arrangements of connectors, which further highlights the complexity and uniqueness of the impact of reasonable arrangement of SGCCs on the mechanical properties of panels.

According to Article 9.3.2 of the Chinese standard "Code for Acceptance of Construction Quality of Concrete Structures" (GB50204-2015) [35], when the maximum crack width at the tensile main reinforcement of a bending member reaches 1.5 mm or the deflection reaches 1/50 of the span, the member is considered to have reached the ultimate bearing capacity state. In this experiment, considering the deformation of the panel, crack development, and the characteristics of the load-deflection curve, it is believed that even if the bearing capacity of the panel has not decreased to 85% of its peak value, as long as the mid-span deflection of the panel exceeds 58 mm ($l_0/50$), the panel has also reached its ultimate bearing capacity state.

The characteristic values on each load-deflection curve during the loading process are listed in Table 2, and all characteristic values are taken from the curve corresponding to the deflection at the mid-span of the panel before reaching 58 mm. In Table 2, P_{cr}, P_y, P_{max}, and Pu represent the loads corresponding to the cracking point, yield point, peak point, and ultimate point, respectively, while Δcr, Δy, Δmax, and Δu represent the corresponding displacements. The yield point is obtained through a geometric drawing method [36]. The ultimate point refers to the point where the bearing capacity of the specimen decreases to 85% of the peak load during the loading process. If it does not decrease to 85% of the peak load, the load value corresponding to the ultimate deflection of 58 mm is taken.

Table 2. Characteristic values of the load-deflection curves.

Specimens	Cracking Point		Yield Point		Peak Point		Ultimate Point	
	Pcr (kN)	Δcr (mm)	Py (kN)	Δy (mm)	Pp (kN)	Δp (mm)	Pu (kN)	Δu (mm)
P0	57.6	1.57	69.7	19.59	78.8	54.86	67	57.89
P1	10.5	0.58	39.7	15.36	47.1	32.54	40.0	35.98
P2	21.9	1.24	50.5	37.97	54.5	54.11	52.0	58
P3	21.1	1.01	50.7	25.71	56.1	47.03	54.1	58
P4	27.1	1.27	57.4	30.48	59.5	55.02	59.3	58
P5	17.1	0.96	45.7	19.66	53.0	57.62	53	58

The cracking load, peak load, and ultimate load values of each test panel are also clearly marked in Figure 12 for easy reference and comparison.

From the load-deflection curve, it can be observed that before cracking, each panel exhibits an approximately linear increase in deflection with increasing load. However, after the panel cracks, the load-deflection curve demonstrates nonlinear behavior, indicating a more complex relationship between load and deflection. The long time from yielding to failure of the panel indicates that its ductility is excellent. According to the data in Table 2, the initial stiffness (ratio of cracking load to cracking deflection) of the six test panels was calculated to be 36.7, 18.1, 17.7, 20.9, 21.3, and 17.8, respectively. The initial stiffness of all sandwich wall panels is generally higher than that in the existing studies [32,37]. This might be because the SGCCs used in this study are more conducive to improving the overall collaborative performance of the panels in structural design, thus enabling the panels to resist deformation more effectively when subjected to initial loads. In addition, the fact that the initial stiffness of panels P1, P2, and P5 is close, while the initial stiffness of panels P3 and P4 is also close and relatively high indicates that the initial stiffness of these panels is positively correlated with the number of connectors at the end of the panels.

In the load-deflection curve graph, solid concrete wall panel P0 is considered as a fully composite panel to provide a reference for the composite performance of sandwich wall panels. It can be seen that, compared to other sandwich wall panels, the load-deflection curve of panel P4 is significantly higher and closest to the P0 curve. It exhibits higher bending capacity and composite performance. The load-deflection curve of P3 is slightly lower than that of P4, and the composite performance of P3 is inferior to that of P4. The load-deflection curves of P2 and P5 are very similar, while the curve of panel P1 is located at the very bottom, farthest from the P0 curve, exhibiting the lowest composite performance.

Comparing the load-deflection curves of P2, P4, and P5, it can be inferred that adding the same number of connectors along the short side of the panel has a more pronounced effect on its bending capacity and composite performance than adding them along the long side. In other words, reducing the spacing between connectors along the short side of the panel has a more significant effect on its bending capacity and composite performance than reducing the spacing between connectors along the long side. Meanwhile, it should be noted that panels P4 and P5 have the same number of connectors, but P4 exhibits a 10.6% higher bending capacity than P5, indicating that their bending and composite performance characteristics are not identical. On the other hand, although P5 has 33% more connectors than P2, their bending capacities differ by less than 2%. This indicates that the bending resistance and composite performance of sandwich wall panels are not proportional to the total number of connectors, but are mainly influenced by the layout of connectors.

In actual engineering projects, for the sandwich wall panels of high-rise buildings or those bearing relatively large loads on the exterior walls of buildings, it is recommended to preferentially adopt a connector arrangement method similar to that of P4. That is, appropriately densify the connector spacing in the short-side direction of the panel to enhance the flexural and composite performance of the panel. Meanwhile, attention should be paid to the quantity and arrangement of the connectors at the end of the panel. Since the connectors at the end have a significant impact on the composite performance, a sufficient quantity and reasonable anchorage depth should be ensured. For example, the composite performance of the P3 and P4 panels, which are equipped with more connectors at the end, is superior to that of other panels. In addition, when determining the quantity and layout of the connectors, it is necessary to combine the structural design calculation results of specific projects and comprehensively consider factors such as the height, number of floors, wind load, and seismic fortification intensity of the building, and conduct optimized design to avoid blindly increasing the quantity of connectors while neglecting the rationality of the layout, so as to ensure the safety and stability of the sandwich wall panels during long-term use.

3.2. Strain of SGCCs

The attachment position of strain gauges on SGCCs in P5 and the load-strain relationship are shown in Figure 13. In the figure, Z1, X1, Z2, X2, Z3, and X3 respectively represent strain gauges on the vertical insertion and oblique insertion connectors at the end, quarter, and middle positions of the panel.

It can be observed that the strain in the connectors cannot be ignored before the panel reaches its ultimate load. For the connectors at the end and quarter of the panel, they mainly bear tensile strain, and the strain value changes approximately linearly with the increase in load. For the connectors in the middle of the panel, they mainly bear compressive strain in the early stage and gradually transform into tensile strain in the later stage. At the same time, the tensile stress of connectors at the end is significantly greater than that at the quarter position, and the tensile stress of the oblique insertion connectors is greater than that of the vertical insertion connectors. The above phenomena indicate that under bending, the interlayer shear force between the inner and outer concrete wythes of the sandwich wall panel is mainly transmitted through the oblique insertion connectors. The connectors at the panel end exert the greatest resistance to bending, indicating that the bending stiffness of the panel is mainly determined by the stiffness provided by the connectors at the ends of the panel. This further confirms the conclusion in Section 3.1.

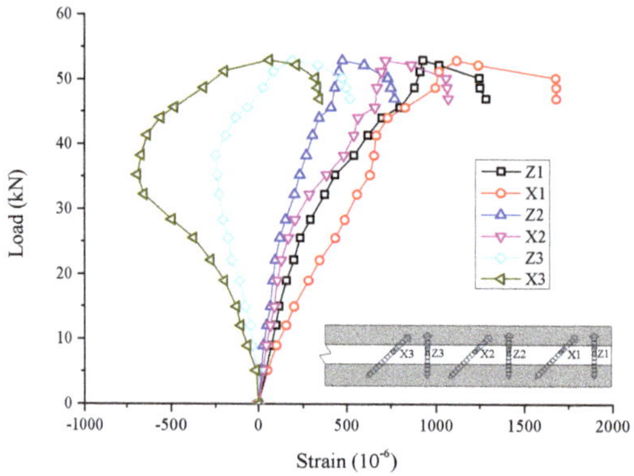

Figure 13. Strain of SGCCs at different positions in panel P5.

3.3. Strain Variation Along the Section Height

Figure 14 shows the strain variation along the section height of six test wall panels at mid-span at different load stages.

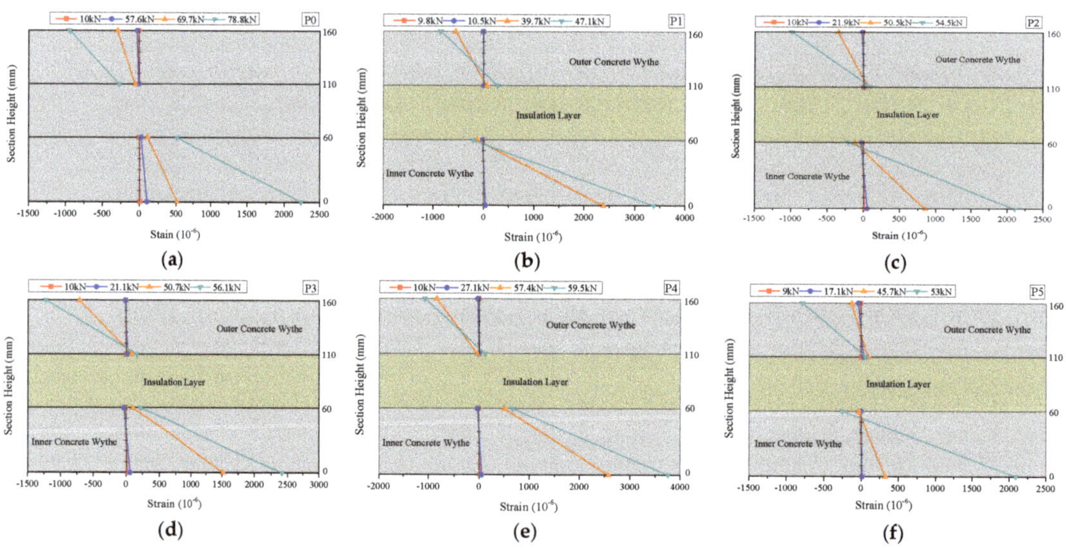

Figure 14. Strain variation along section height of the test panels at mid-span at different load stages: (**a**) P0; (**b**) P1; (**c**) P2; (**d**) P3; (**e**) P4; (**f**) P5.

In the initial loading stage, the strain of two concrete wythes was small, and the strain variation of all panels along the section height was close to continuous and basically linearly distributed, which conforms to the assumption of a flat section. The composite performance of the inner and outer concrete wythes was good. After the panel cracked, the discontinuity of strain between the inner and outer concrete wythes began to emerge and gradually became significant as the load increased because the composite action would degrade with the cracking of wythes, the sliding between the insulation layer and concrete wythes, the yielding of the steel mesh, and the damage of the connectors.

It can be observed that sandwich wall panels exhibit partial composite behavior. The strain variation level of panel P4 is closest to that of panel P0, indicating that the composite performance of panel P4 is the best. Compared with other panels, panel P1 features a more conspicuous strain discontinuity between the layers. This panel exhibits weaker composite performance and tends to exhibit non-composite behavior, as both the inner and outer concrete wythes approximately form their neutral axes. Compared with panel P2, the strain variation of panel P3 is closer to that of P4, indicating that adding the same number of connectors at the end of the panel is more significant in improving the composite performance of sandwich wall panels than in the middle of the panel. The strain variation of panel P5 is similar to that of P2, and the discontinuity is more pronounced under the ultimate loading state. In summary, the degree of composite action of sandwich wall panels is not directly proportional to the total number of connectors but mainly depends on the distribution and spacing of connectors. This also confirms the conclusion described in the previous section.

3.4. Load-Relative Slip Relationship

In the initial bending load stage, the interlayer shear force of sandwich wall panels was mainly resisted by the connectors in the bending shear section and the interfacial bonding force between the XPS insulation layer and concrete layers. After bending cracks occurred, as the load increased, the XPS insulation layer gradually separated from the concrete layer until the insulation layer fractured, as shown in Figure 15.

Figure 15. Interface debonding between XPS insulation and concrete wythes and XPS insulation layer fracture.

The initial bonding failure mainly occurred at the interface between the insulation layer and the bottom concrete layer near the quarter-span position of the panel, and then sliding began at the interface. As the bending deflection increased, the insulation layer experienced brittle fracture, followed by debonding between the insulation layer and the upper concrete layer. At this stage, most of the shear force was transmitted by the connectors, and the sandwich wall panels exhibited a certain degree of composite action.

As the load continued to increase, significant relative slippage began to occur between the upper and lower concrete layers and the insulation layer.

Figure 16 shows the relationship curve between the load and the total relative slip at the end of the inner and outer concrete wythes before the sandwich wall panel reaches the ultimate bearing capacity state.

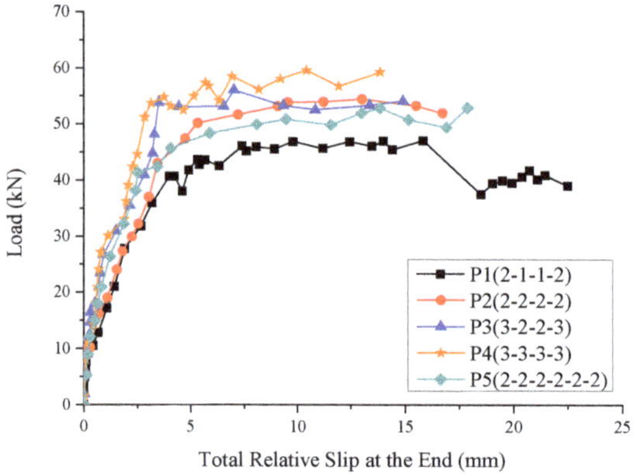

Figure 16. Load-total relative slip relationships of test sandwich wall panels.

The total relative slip at the end of each panel represents the sum of the relative displacement between the inner and outer concrete wythes and the insulation layer at the left and right ends of the panel. The total relative slip at the end of the panel can, to some extent, reflect the degree of partial composite action. When the total relative slip at the end is small, it indicates that the collaborative performance between the layers of the sandwich wall panel is good and can effectively share the load, which means that the degree of partial composite action is high. On the contrary, when the total relative slip at the end is large, it indicates poor collaborative performance between the layers and a lower degree of partial composite action.

From Figure 16, it can be seen that under ultimate load, the total relative slip at the end of each test panel is ranked from small to large as P4 < P3 < P2 < P5 < P1, indicating that the degree of composite action of each panel decreases in order. Furthermore, it is indicated that the composite performance of sandwich wall panels mainly depends on the number and layout of connectors at the ends of the panels.

In practical engineering applications, to ensure the stability and safety of sandwich wall panels during long-term use, the quantity and layout of connectors at the ends of the panels should be reasonably determined according to specific engineering requirements and structural designs. For regions with relatively large wind loads or high seismic fortification intensities, the number of connectors at the ends should be appropriately increased, and the anchorage depth should be ensured to meet the design requirements, so as to enhance the overall shear resistance and composite performance of the panels. At the same time, during the construction process, the installation quality of the connectors should be strictly controlled to ensure their accurate positions and reliable anchorage and to avoid connector failures caused by improper construction, which would affect the mechanical properties and service functions of the sandwich wall panels.

4. Theoretical Analysis

The degree of composite action between the inner and outer concrete wythes directly affects the ultimate flexural bearing capacity of the sandwich wall panel. Conversely, the ultimate load can also be used to evaluate the degree of composite action of sandwich wall panels. Referring to reference [37], we can evaluate the degree of composite action of sandwich wall panels from the perspective of strength using Equation (1). Assuming that the ultimate bending capacity corresponding to the ultimate load of the solid concrete wall panel P0 represents the ultimate bending capacity of the fully composite panel, while the ultimate bending capacity of non-composite sandwich wall panels is calculated using Equations (2)–(4) according to the theoretical method in reference [23].

$$\kappa = \frac{P_e - P_{non}}{P_{full} - P_{non}} \times 100(\%) \tag{1}$$

$$P_{non} = \frac{8M_{Nu}}{l} \tag{2}$$

$$M_{Nu} = F_s\left(h_{01} - {}^{x}\!/_{2}\right) + F_s\left(h_{02} - {}^{x}\!/_{2}\right) \tag{3}$$

$$F_s = F_c,\ F_s = A_s f_y,\ F_c = f_c b x,\ x = 0.8x_c \tag{4}$$

κ is employed to signify the degree of composite action achieved by the sandwich wall panel. This parameter precisely reflects the extent to which the inner and outer concrete wythes cooperate in a composite fashion within the panel, offering a quantitative measure of their combined effectiveness.

P_{full} serves to represent the experimental ultimate load value under fully composite action. This value is obtained through experimental procedures and showcases the maximum load capacity that the panel can bear when it functions as a fully composite structure.

P_{non} is designated to denote the theoretically calculated ultimate load value under non-composite action. Such a calculation is rooted in established theoretical frameworks and relevant computational principles, aiming to estimate the peak load the panel could endure in non-composite behavior.

P_e represents the experimental ultimate load value of the test specimen. It captures the actual maximum load that the specific specimen is capable of withstanding during the testing process, providing real-world data for analysis.

M_{Nu} is used to represent the ultimate bending moment of the sandwich wall panel under non-composite action.

F_s stands for the force in tensile steel bars in the inner and outer concrete wythes of the non-composite panels.

F_c denotes the compressive force on the cross-section of the inner and outer concrete wythes of the non-composite panels.

f_y represents the yield strength of the tensile steel bars.

f_c signifies the design value of the axial compressive strength of the concrete.

l refers to the net span of the panel.

A_s represents the cross-sectional area of the tensile steel bars.

b is the cross-sectional width of the panel.

h_{01} and h_{02} are utilized to represent the effective heights of the cross-section of the inner and outer concrete wythes of the sandwich wall panel.

x_c represents the height of the neutral axis of the inner and outer concrete wythes of the sandwich wall panel.

x stands for the height of the concrete compression zone of the equivalent rectangular.

Upon conducting calculations based on the above equations and parameters, the estimated degree of composite action of the test sandwich wall panels can be found in Table 3.

Table 3. Degree of composite action of sandwich wall panels.

Specimens	P_e (kN)	P_{full} (kN)	P_{non} (kN)	κ, %
P1	40.0	67	15.7	47.4
P2	52.0			70.8
P3	54.1			74.7
P4	59.3			85.0
P5	53			72.7

It can be clearly seen from the calculation results that, except for panel P1, the composite action degrees of other sandwich wall panels are all higher than 70%, which indicates that most of the panels exhibit good composite performance under the test conditions. Among them, the composite action degree of panel P4 reaches 85%, which is at a relatively high level. This shows that the arrangement and structural design of its connectors enable the synergistic work effect between the inner and outer concrete wythes to be remarkable so that they can resist bending loads more effectively and thus achieve a relatively high degree of composite action. The composite action degrees of panels P2 and P5 are similar, which indicates that although the total number of connectors of P5 is 33% more than that of P2, their similar connector layouts result in similar flexural capacities and composite performances. In addition, the composite action degree of P5 is 12.3% lower than that of P4 containing the same number of connectors, which is consistent with the fact that the flexural capacity of P5 is 10.6% lower than that of P4 in the experiment. It means that the flexural capacity and composite performance of the panels are not proportional to the total number of connectors, but are mainly influenced by the layout of the connectors. However, the composite action degree of panel P1 is the lowest, only 47.4%, which indicates that such a connector layout fails to fully exert the synergistic effect between the inner and outer wythes, thus affecting the overall composite performance.

This result is highly consistent with the conclusion drawn from the test phenomena and data in Section 3.4, further verifying our understanding of the composite performance of sandwich wall panels. Namely, the sandwich wall panels all exhibit partial composite performance during the actual loading process, and there is an obvious positive correlation between the composite action degree and the flexural bearing capacity. This implies that optimizing the layout scheme of connectors and improving the composite action degree of the panels, is expected to significantly enhance their flexural bearing capacity, thus providing more reliable structural performance guarantees for the application of sandwich wall panels in construction projects. Meanwhile, the good agreement between the theoretical calculation results and the test results also proves the accuracy and reliability of the theoretical analysis methods and calculation formulas adopted in this paper, providing powerful theoretical support for further in-depth research and optimization of the structural design of sandwich wall panels in the future.

5. Conclusions

This article discussed the composite performance and bending resistance of five sandwich wall panels with different numbers and distributions of SGCCs based on the analysis of the bending test results. Furthermore, the degree of composite action of sandwich wall panels was estimated. The conclusion is as follows:

1. During the test process, all test panels underwent bending failure and generally exhibited ductile failure characteristics. The failure process was relatively slow and accompanied by large deformations. Differences existed in the flexural bearing capacity

and composite performance among sandwich wall panels with different arrangements of connectors. Among all the sandwich wall panels, P4 showed the highest flexural bearing capacity and composite performance, followed in sequence by P3, P2, and P5, while the performance of panel P1 was the lowest.

2. By comparing the influence of different connector arrangement methods on the performance of sandwich wall panels, it was found that when the same number of connectors was added, increasing them along the short side of the wall panel had a greater impact on the flexural bearing capacity and composite degree of the panel than along the long side. That is, reducing the spacing of connectors along the short side of the panel had a more significant effect on improving the flexural performance and composite performance of the panel. At the same time, it indicated that the flexural capacity and composite performance of the sandwich wall panel were not proportional to the total number of connectors but mainly depended on the layout of the connectors.
3. Under the bending action of the sandwich wall panel, the SGCCs played a crucial role. The connectors at the end and at the 1/4 position mainly bore tensile stress, and the tensile stress of the end connectors was greater than that at the 1/4 position. The tensile stress of the oblique insertion connectors was greater than that of the vertical insertion connectors. Consequently, it meant that the shear force between the inner and outer concrete wythes was mainly transmitted through the tensile force borne by the oblique insertion connectors, and the flexural stiffness of the panel was mainly determined by the stiffness provided by the connectors at the end of the panel.
4. When the same number of connectors was added, increasing them at the end of the panel had a more significant impact on the composite performance than adding them in the middle of the panel.
5. All sandwich wall panels exhibited partial composite performance. Except for panel P1, the composite action degree of the other sandwich wall panels was higher than 70%. Among them, the composite action degree of panel P4 reached 85%, and that of panels P2 and P5 was similar. The theoretical evaluation result was consistent with the conclusion drawn from the test phenomena and related relationships. The composite action degree was positively correlated with the flexural bearing capacity.

In actual construction, the connectors should be reasonably configured according to factors such as the type of building, the wind load and the seismic fortification intensity of the region where the building is located. This includes preferentially adopting an arrangement method similar to that of P4, densifying the connectors along the short side of the panel, paying attention to the quantity and layout of the connectors at the ends of the panel, and ensuring the installation quality of the connectors during the construction process. The aim is to better apply the experimental research results to the actual engineering construction and ensure the mechanical properties and service functions of the sandwich wall panels.

In summary, this research has clearly elucidated the influence of different arrangements of SGCCs on the flexural strength and composite performance of sandwich wall panels through tests and theoretical analysis. It can provide a scientific basis for the rational arrangement of connectors in sandwich wall panels in practical engineering and help to further optimize the application of such panels in building envelopes.

Author Contributions: The manuscript was written with contributions all authors. H.J. and L.X.: Conceptualization, investigation, methodology, data curation, software, validation, formal analysis, writing—original draft preparation, writing—review and editing, and visualization. F.C. and Y.C.: Writing—review and editing, visualization. Z.G.: Conceptualization and resources. All authors have read and agreed to the published version of the manuscript.

Funding: This research was funded by National Natural Science Foundation of China, grant number 52278154, the Jiangsu Province Natural Science Foundation, grant number BK20231429, and the Fundamental Research Funds for the Central Universities (2242024RCB0008), as well as support from the program of Zhishan Young Scholar of Southeast University.

Data Availability Statement: The data presented in this study are available on request from the corresponding author. The data are not publicly available due to we want to ensure data security and compliance with regulations while protecting the privacy and rights of study participants and associates.

Conflicts of Interest: The authors declare no conflicts of interest.

References

1. Einea, A.; Salmon, D.C.; Fogarasi, G.J. State-of the-art of precast concrete sandwich panels. *PCI J.* **1991**, *36*, 78–98. [CrossRef]
2. Frankl, B.A. Structural Behavior of Insulated Precast Prestressed Concrete Sandwich Panels Reinforced with CFRP Grid. Ph.D. Thesis, North Carolina State University, Raleigh, NC, USA, 2008.
3. Bunn, W.G. CFRP Grid/Rigid Foam Shear Transfer Mechanism for Precast, Pre-Stressed Concrete Sandwich Wall Panels. Ph.D. Thesis, North Carolina State University, Raleigh, NC, USA, 2011.
4. Soriano, J.G. GFRP Shear Grid for Precast, Prestressed Concrete Sandwich Wall Panels. Ph.D. Thesis, North Carolina State University, Raleigh, NC, USA, 2013.
5. Wang, Y.; Yi, G.; Qin, C.; Zhang, G. Review on the precast concrete sandwich panels. *Constr. Des. Proj.* **2023**, *7*, 212–216. [CrossRef]
6. Zhao, Z.; Zhang, J.; Yang, S. Study on Integrated Decorative Lightweight Sandwich Insulation Panel. *China Concr.* **2022**, *12*, 52–58.
7. Xiong, F.; Bian, Y.; Liu, Y.; Ran, M.; Ge, Q. Review on structural performance of precast concrete sandwich insulation wall panels study. *Build. Struct.* **2022**, *52*, 26–34. [CrossRef]
8. Benayoune, A.; Abdul Samad, A.A.; Trikha, D.N.; Abang Ali, A.A.; Ellinna, S.H.M. Flexural behavior of pre-cast concrete sandwich composite panel—Experimental and theoretical investigations. *Constr. Build. Mater.* **2008**, *22*, 580–592. [CrossRef]
9. Mohamad, N.; Hassan, N. The structural performance of precast lightweight foam concrete sandwich panel with single and double shear truss connectors subjected to axial load. *Adv. Mater. Res.* **2013**, *634–638*, 2746–2751. [CrossRef]
10. Bush, T.D.; Stine, G.L. Flexural behavior of composite precast concrete sandwich panels with continuous truss connectors. *PCI J.* **1994**, *39*, 112–121. [CrossRef]
11. Busher, S.; Govindarajulu, K.V. Thermal performance and reduced carbon footprint of precast concrete sandwich wall panels (PCSWPs) with composite shear connector—An experimental assessment. *Energy Sources Part A Recovery Util. Environ. Eff.* **2023**, *45*, 12202–12214. [CrossRef]
12. McCall, C.W. Thermal properties of sandwich panels. *Concr. Int.* **1985**, *7*, 35–41.
13. Lee, B.J.; Pessiki, S. Thermal performance evaluation of precast concrete three-wythe sandwich wall panels. *Energy Build* **2006**, *38*, 1006–1014. [CrossRef]
14. Gu, M. Thoughts on Sandwich Wall Panel Design at Home and Abroad. *Hous. Real Estate* **2015**, *27*, 53–59.
15. Zhang, J.; Zhang, X.; Chen, P. Experimental investigation on flexural performance of precast sandwich insulation panels with GFRP screw connectors. *J. Shenyang Jianzhu Univ. (Nat. Sci.)* **2024**, *40*, 203–211. [CrossRef]
16. Maximos, H.N.; Pong, W.A.; Tadros, M.K. *Behavior and Design of Composite Precast Prestressed Concrete Sandwich Panels with NU-Tie*; University of Nebraska—Lincoln: Lincoln, NE, USA, 2007.
17. Xue, W.; Yang, J. A Fiber-Reinforced Polymer Connector for Precast Concrete Sandwich Insulation Walls. Patent CN 202298982 U, 30 March 2018.
18. Zhu, H.; Qian, Y. Mechanics Performance of FRP Tendons Used in Engineering Structure. *J. Archit. Civ. Eng.* **2006**, *23*, 26–31.
19. Wan, Z.; Chen, G. The research status of insulated connector in precast insulation all panels. *Compos. Sci. Eng.* **2015**, *11*, 81–84.
20. Zheng, B.; Li, W.; Zhang, W. Mechanics behavior of FRP wrapped rebar reinforced concrete(I) Microstructure design and analyses. *Acta Mater. Compos. Sin.* **2004**, *21*, 33–37. [CrossRef]
21. Wu, G.; Luo, Y.; Wu, Z. Experimental and theoretical studies on the mechanical properties of steel-FRP composite bars. *China Civ. Eng. J.* **2010**, *43*, 53–61. [CrossRef]
22. Liu, J.; Guo, Z.; Jiang, H.; Zhi, Q. A New Precast Concrete Sandwich Panel with W-Shaped Steel Glass Fiber Reinforced Polymer Shear Connectors. Patent CN205591421 U, 27 April 2016.
23. Jiang, H.; Guo, Z.; Liu, J. Composite behavior of sandwich panels with W-shaped SGFRP connectors. *KSCE J. Civ. Eng.* **2018**, *22*, 1889–1899. [CrossRef]
24. Jiang, H.; Guo, Z.; Liu, J. The Shear Behavior of Precast Concrete Sandwich Panels with W-shaped SGFRP Shear Connectors. *KSCE J. Civ. Eng.* **2018**, *22*, 3961–3971. [CrossRef]

25. Zhi, Q.; Guo, Z. Experimental evaluation of precast concrete sandwich wall panels with steel-glass fiber-reinforced polymer shear connectors. *Adv. Struct. Eng.* **2016**, *20*, 1476–1492. [CrossRef]
26. Jiang, H. Research on Behavior of Precast Concrete Sandwich Panels with Steel-GFRP Composite Connectors. Ph.D. Thesis, Southeast University, Nanjing, China, 2019.
27. Tang, T.; Huang, J. Study on flexural performance of the precast concrete sandwich panel enabled by FRP plate type connector. *Guangdong Archit. Civ. Eng.* **2023**, *30*, 42–45. [CrossRef]
28. Zhang, M.; Feng, W.; Chen, K.; Li, B. Flexural behavior of a new precast insulation mortar sandwich panel. *Appl. Sci.* **2024**, *14*, 2071. [CrossRef]
29. Jithin, P.U.; Joseph, A. Thermal and structural behavior of precast concrete sandwich panels. In Proceedings of the SECON'22: Structural Engineering and Construction Management 284, Angamaly, India, 1–3 June 2022; Springer: Cham, Switzerland, 2023; pp. 511–522. [CrossRef]
30. Xie, J.; Chen, F.; Zhao, J.; Lu, P.; Liu, F.; Li, L. Flexural behaviour of full-scale precast recycled concrete sandwich panels with BFRP connectors. *J. Build. Eng.* **2022**, *56*, 104816. [CrossRef]
31. Chen, Y.; Kang, C.; Wu, Y.; Qian, Z. Bending performance of precast ceramsite-concrete-insulated sandwich panel with stainless steel shear connectors. *Buildings* **2022**, *12*, 1640. [CrossRef]
32. Bian, Y.; Xiong, F.; Liu, Y.; Ge, Q. Flexural performance of novel ECC-RC composite sandwich panels. *Eng. Struct.* **2023**, *292*, 116547. [CrossRef]
33. Ahmadi, A.D.; Torabi, A.; Totonchi, A.; Safari, D. Experimental and numerical investigation on the behaviour of precast concrete sandwich panels with different shear connectors. *Iran. J. Sci. Technol. Trans. Civ. Eng.* **2024**, *48*, 3087–3102. [CrossRef]
34. West, R.P.; Kinnane, O. Composite behaviour of thin precast concrete sandwich panels. In *Emerging Trends of Advanced Composite Materials in Structural Applications*; Springer: Singapore, 2021; pp. 1–29. [CrossRef]
35. *GB50204-2015*; Code for Quality Acceptance of Concrete Structure Construction. China Architecture & Building Press: Beijing, China, 2015.
36. Guo, Z.; Shi, X. *Reinfored Concrete Theory and Analyse*; Tsinghua University Press: Beijing, China, 2003; pp. 163–337.
37. Kim, J.; You, Y.-C. Composite behavior of a novel insulated concrete sandwich wall panel reinforced with GFRP shear grids: Effects of insulation types. *Materials* **2015**, *8*, 899–913. [CrossRef]

buildings

MDPI

Article

Optimization of Shear Resistance in Horizontal Joints of Prefabricated Shear Walls through Post-Cast Epoxy Resin Concrete Applications

Peiqi Chen [1,2], Shilong Zhao [1,2], Pengzhan Xu [3,*], Xiaojie Zhou [1,2] and Yueqiang Li [1,2]

1 Tianjin Key Laboratory of Civil Buildings Protection and Reinforcement, Tianjin 300384, China; cpq@tcu.edu.cn (P.C.); 15713558280@163.com (S.Z.); zhouxj@tcu.edu.cn (X.Z.); septlib0928@163.com (Y.L.)
2 School of Civil Engineering, Tianjin Chengjian University, Tianjin 300384, China
3 Tangshi Jianhua Building Materials (Tianjin) Co., Ltd., Tianjin 301500, China
* Correspondence: sunny756@126.com

Abstract: The horizontal joint is a critical component of the prefabricated shear wall structure, responsible for supporting both horizontal shear forces and vertical loads along with the wall, thereby influencing the overall structural performance. This study employs direct shear testing and finite element analysis to investigate the horizontal joint in walls with ring reinforcement. It examines the impact of various factors on joint shear performance, including the type of joint material, joint configuration, buckling length of ring reinforcement, strength of precast concrete, reinforcement ratio of ring reinforcement and dowel bars, and the effect of horizontal binding force. The findings indicate that the shear bearing capacity and stiffness of joints incorporating post-cast epoxy resin concrete and keyways are comparable or superior to those of integrally cast specimens. A larger buckling length in ring reinforcement may reduce shear strength, suggesting an optimal buckling length at approximately one-third of the joint width. As the strength of precast concrete increases, ductility decreases while bearing capacity increases, initially at an increasing rate that subsequently declines. Optimal results are achieved when the strength of precast concrete closely matches that of the post-cast epoxy concrete. Enhancing the reinforcement ratio of ring reinforcement improves shear capacity, but excessively high ratios significantly reduce ductility. It is recommended that the diameter of ring reinforcement be maintained between 10 mm and 12 mm, with a reinforcement ratio between 0.79% and 1.13%. Increasing horizontal restraint enhances stiffness and shear capacity but reduces ductility; thus, the axial compression ratio should not exceed 0.5.

Keywords: epoxy resin concrete; prefabricated shear wall; horizontal joint; shear resistance; experiment; finite element analysis

1. Introduction

In prefabricated shear wall structures, the quality of connections between the prefabricated walls is critical for ensuring the safety and reliability of the overall structure [1]. An appropriately designed connection structure facilitates a rational and orderly force transfer mechanism, thereby enhancing the overall performance and seismic resilience of the structure. Existing research indicates that the primary methods for improving the performance of prefabricated shear wall joints include [2–5] the following: the implementation of reliable steel bar connection techniques, the proper treatment of joint surfaces between new and existing concrete, the utilization of high-performance post-cast joint materials such as fiber-reinforced concrete, polymer concrete, and high-strength concrete, as well as the optimization of joint reinforcement ratios.

In terms of reinforcement connections, common connection technologies for steel bars include the sleeve grouting method, the bolt anchoring method, and the slurry anchor connection method [6–8]. Engineering practices indicate that these steel bar connection methods can fulfill the quality requirements for steel bar connections. However, they also present challenges, such as high costs, inconvenient construction processes, low efficiency, and difficulties in ensuring the quality of sleeve grouting. The ring reinforcement buckle connection technology, introduced by Yu, Z.-W., et al. [9,10], represents a significant simplification in the reinforcement connection process. This technology not only ensures superior quality in wall connections but also equates the load-bearing capacity and seismic performance of prefabricated concrete shear walls with those of cast-in-place walls. Yu et al. [11] extended this connection technology to precast concrete columns, which exhibited commendable transverse strength and ductility. The ring bar joint technology [9–11] is characterized by its straightforward construction process. It involves fastening a U-shaped hoop, which protrudes from the upper and lower prefabricated walls, to an appropriate length, threading the longitudinal bar through the joint area, tying the longitudinal bar and U-shaped hoop together, and finally pouring concrete to form a horizontal joint, thereby achieving the connection of two prefabricated walls (Figure 1).

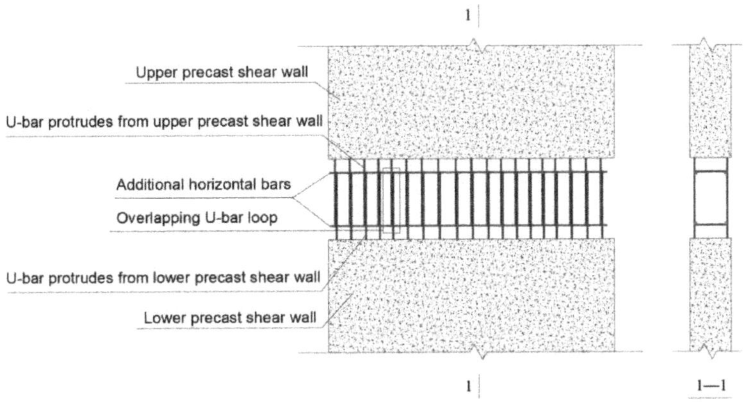

Figure 1. Technical diagram of the buckle connection of ring reinforcement.

When it comes to epoxy resin concrete, epoxy concrete exhibits strong interfacial bond strength and possesses favorable thermal stability, chemical resistance, and mechanical properties. Consequently, this material is frequently employed in reinforcement projects, including concrete crack repair and road construction [12–16]. Natarajan et al. [17] investigated the effects of partially replacing cement with epoxy resin, suggesting that this substitution enhanced both the compressive and flexural strength of the concrete. Qian et al. [18] conducted axial tensile tests on both unreinforced and reinforced epoxy concrete, revealing that the tensile and cracking properties of epoxy concrete surpassed those of ordinary concrete, while also demonstrating excellent bonding characteristics with steel reinforcement bars. Additionally, El Mandouh et al. [19] assessed the shear performance of 18 simply supported super-reinforced epoxy resin concrete beams, discovering that these beams exhibited less deformation, higher cracking loads, greater ultimate shear capacity, and enhanced ductility compared to ordinary concrete beams. Chen [20] conducted both experimental and numerical analyses on post-cast epoxy resin concrete shear walls, concluding that the seismic performance of these walls surpassed that of conventional cast-in-place reinforced concrete walls. Consequently, this study proposes the adoption of epoxy resin concrete as the post-cast material to further explore its impact on the shear performance of the horizontal joints in prefabricated shear walls.

Regarding interface treatments between new and old concrete surfaces, Julio et al. [21] conducted an experimental investigation into various treatment technologies. Their findings indicated that sandblasting yielded the highest bond strength. However, the efficiency of field construction was relatively low. Rizkalla et al. [22] examined different horizontal joint configurations in prefabricated shear walls, concluding that shear walls featuring keyways exhibited superior mechanical properties compared to those with conventional horizontal joints. Gopal et al. [23] demonstrated through shear testing that the use of epoxy resin adhesive in keyways significantly enhances the shear bearing capacity of the bonding surface when compared to dry joints. These findings also provides a reference for the research direction of this study.

In terms of finite element analysis, Fan et al. [24] introduced a novel reinforcement lap method that involves the assembly of nodes with X-shaped reinforcement. They employed ABAQUS software to simulate the U-hoop lap in conjunction with X-shaped reinforcement. In this simulation, the interface between the new and old concrete within the precast concrete structure was represented through surface contact, while the tangential interactions were modeled based on a Coulomb friction model. The analysis indicated that the incorporation of additional X-shaped reinforcement alters the force transmission path at the joint, reduces the stress in the cross-sectional area of the joint, and enhances the reliability of the vulnerable sections of the assembled shear wall. Building upon extensive research, the European Union code [25] offers recommendations regarding the friction coefficient values for the joint surfaces of new and old concrete. Jin et al. [26] utilized ABAQUS finite element simulation to investigate the mechanical properties of joints in epoxy concrete truss structures, demonstrating that the numerical model could accurately simulate and predict the strength and failure behavior of the specimens. These research methods provide the basis for the finite element analysis in this study.

In summary, to enhance the shear resistance of the horizontal joint in prefabricated shear walls, seven symmetrical direct shear specimens were designed and fabricated utilizing toroidal joint technology. The design parameters included the joint material, the length of the toroidal joint, and the joint configuration (i.e., with or without a keyway). Direct shear testing and finite element simulation were employed to analyze the effect of these design parameters on the performance of the horizontal joint in prefabricated shear walls, with the aim of identifying the optimal construction scheme for the horizontal joint.

2. Experimental Overview

2.1. Specimen Design and Production

The assembled specimen is categorized into two components: prefabrication and post-casting. The post-casting component specifically pertains to the horizontal joint of the shear wall, which employs ring bar joint technology (Figure 2). In accordance with the "Technical standard for precast reinforced concrete shear wall structure assembled by anchoring closed loop reinforcement " (JGJ/T 430-2018) [27], a U-shaped rebar retention length of 120 mm ensures that the anchoring performance of the shear wall matches that of a cast-in-place structure. Therefore, U-shaped rebar retention length was set to 120 mm for this experiment. As part of the design parameters, 6 prefabricated direct shear specimens were developed, alongside one cast-in-place specimen designated as a comparison specimen. The fundamental information and identification numbers for each specimen are presented in Table 1, while the dimensions and reinforcement details of typical specimens are depicted in Figure 2. For instance, specimen ZHY-1-80 is exemplified in Figure 2a, where the No. 3

rib (6C8) serves as the horizontal joint ring rib, with an annular reinforcement ratio of 0.50%. Additionally, the No. 5 bar (4C8) functions as the horizontal joint insert bar, exhibiting an insert bar reinforcement ratio of 0.83%.

Table 1. Basic information and structural details of test specimens.

Specimen Number	Horizontal Joint Construction		
	After-Cast Material	**Whether There Is a Keyway**	**Ring Buckle Length (mm)**
ZHY-1-80	Epoxy concrete	Yes	80
ZHY-1-60		Yes	60
ZHY-1-100		Yes	100
ZXJ-2-60	Ordinary concrete (C40)	No	60
ZXJ-1-60		Yes	80
ZXJ-2-100		No	100
ZZJ	-	-	60

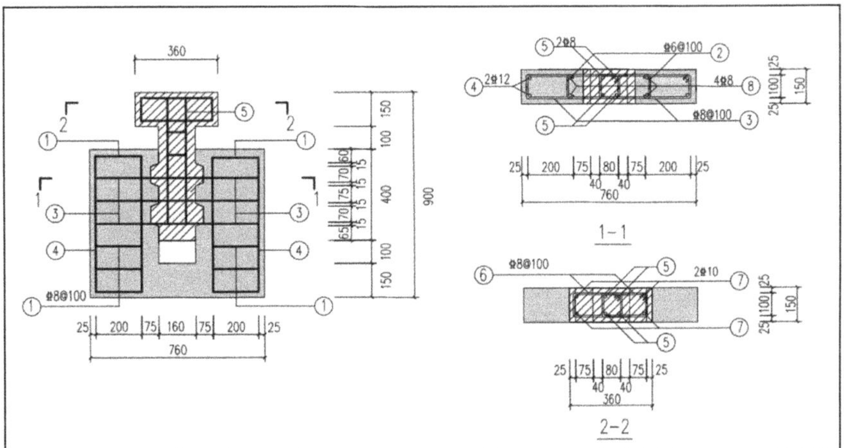

(a) ZHY-1-80

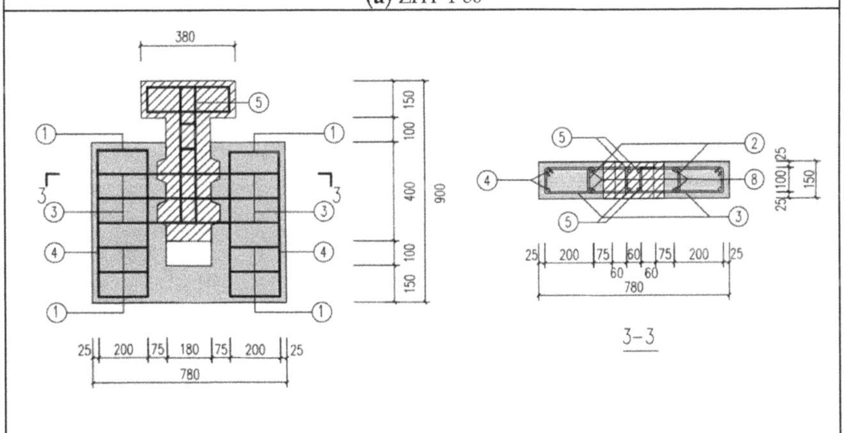

(b) ZHY-1-60

Figure 2. *Cont.*

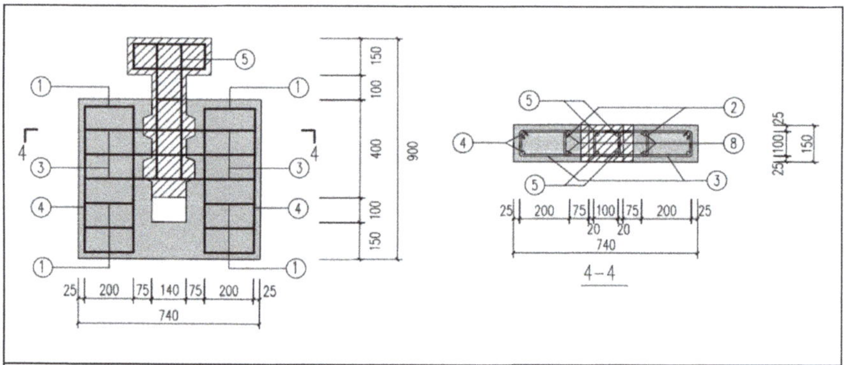

(c) ZHY-1-100

Figure 2. Size and reinforcement diagram of typical specimen.

2.2. Basic Mechanical Properties of Materials

In the proposed test, self-mixed C40 concrete was utilized as the standard for ordinary concrete. The average measured cubic compressive strength of precast concrete was found to be 42.1 MPa, while that of post-cast concrete was 40.6 MPa. Additionally, the average compressive strength of self-mixed epoxy resin concrete was recorded at 55.9 MPa. Furthermore, the average measured yield strength of HPB300 steel bar with a diameter of 6 mm was 268.3 MPa. In contrast, the average yield strengths of the HRB400 steel bars with diameters of 8 mm and 12 mm were measured at 361.1 MPa and 360.0 MPa, respectively.

2.3. Loading Scheme

The test device is shown in Figure 3. The vertical loading equipment utilizes a 30,000 kN electro-hydraulic servo dynamic and static universal testing machine, which can be used for static testing of different materials with high precision and repeatability; it can carry out tensile, compression, bending, shear, and spalling tests of various materials. The test piece is placed on the test bench, and the loading end of the testing machine, the test piece, and the positioning line of the test bench are kept on the same vertical line so as to ensure that the old and new concrete joint surface under load is in a pure shear state.

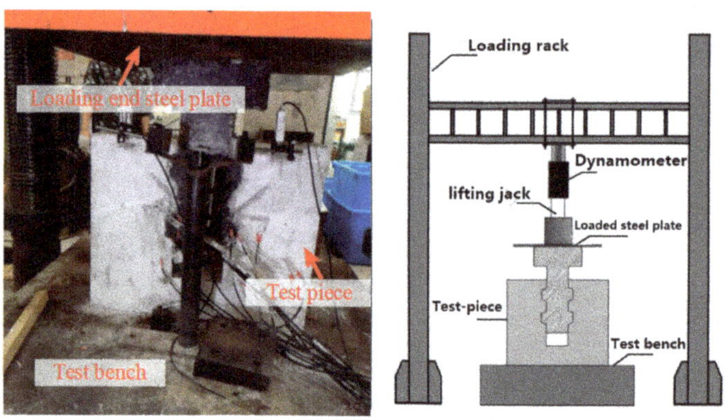

(a) Test equipment (b) Schematic test device

Figure 3. Test device diagram.

Load displacement control is employed for vertical loading. When the vertical load is less than the estimated cracking load of 200 kN, load control is utilized. Within the range of 0 kN to 100 kN, loading increments are set at 20 kN, while in the range of 100 kN to 200 kN, increments are adjusted to 10 kN. Once the vertical load exceeds the cracking load, the system switches to displacement control load, with a loading rate of 0.5 mm/min. The test concludes when the displacement of the specimen exceeds 10 times the cracking displacement and the residual load stabilizes. The configuration of the test loading system is depicted in Figure 4.

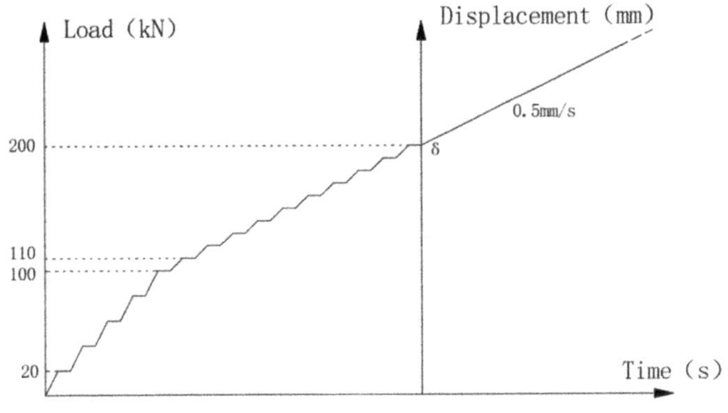

Figure 4. Loading system.

2.4. Displacement Measurement Scheme

Two displacement meters, designated as W-1 and W-3, were symmetrically positioned on the upper section of the prefabricated component of the specimen. Additionally, two displacement meters, W-2 and W-4, were installed on both the upper and lower surfaces of the post-cast joint to measure the slip at the joint interface between the prefabricated and post-cast sections of the specimen (Figure 5). Specifically, the slip at the joint surface is determined by calculating the difference between the mean values obtained from displacement meters W-2 and W-4.

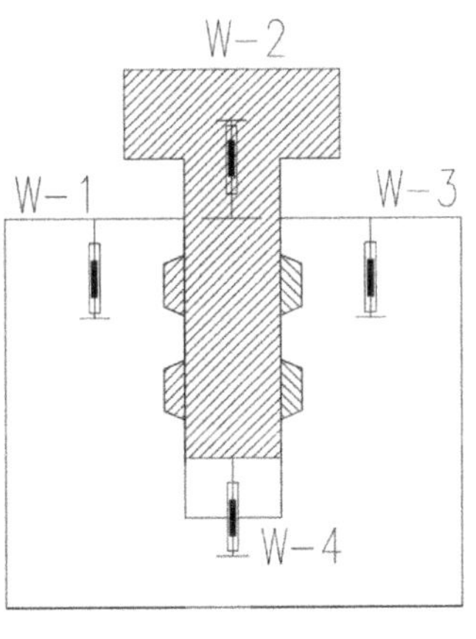

Figure 5. Displacement meter layout.

3. Analysis of Experimental Phenomena and Results

3.1. Effect of Buckle Length on Joint Shear Performance

3.1.1. Specimens with Post-Poured Epoxy Concrete and Bonding Surface with Keyway

The specimens ZHY-1-60, ZHY-1-80, and ZHY-1-100 are constructed from epoxy resin concrete, serving as joint materials. Each bonding surface is designed with a keyway. However, the lengths of the buckles vary among the specimens. The failure characteristics of each specimen are illustrated in Figure 6, while the load-displacement curves for each specimen are presented in Figure 7.

(**a**) ZHY-1-60 (**b**) ZHY-1-80 (**c**) ZHY-1-100

Figure 6. Failure condition of each specimen.

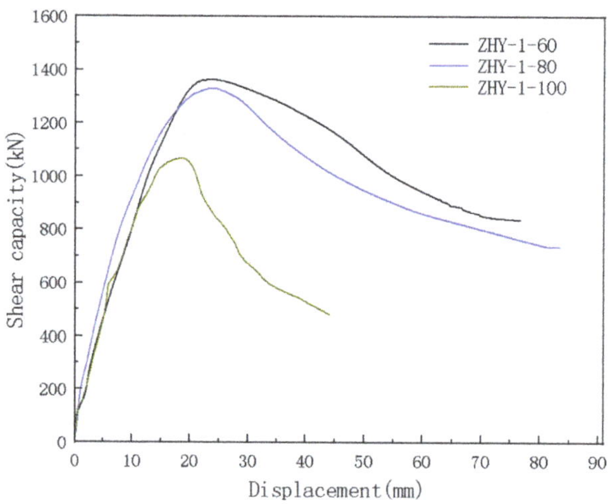

Figure 7. Load-displacement curves of ZHY specimen.

The typical failure process observed in this group of specimens is characterized by the following sequence: Initially, upon the commencement of loading, small cracks emerge at the ends of the key teeth located on the lower side of the specimens. As the load increases, these cracks progressively extend along the bonding surface, subsequently leading to the formation of cracks at the root of the keyway. With further increases in load and displacement, the crack at the root of the keyway propagates into the post-poured epoxy concrete. At the peak load, the cracks on the joint surface gradually extend from the bottom to the base, resulting in the cracks at the root of the keyway becoming fully developed and the key teeth experiencing severance. Following this, the spalling of precast concrete adjacent to the keyway becomes increasingly pronounced, while the cracks in the post-cast epoxy concrete also develop, albeit less visibly. When the displacement exceeds 10 times the initial cracking displacement and the residual load stabilizes, the specimen's bearing capacity is compromised, leading to the conclusion of the test.

In the case of specimen ZHY-1-60, when the load decreased to 64.2% of the peak load, the displacement surpassed 10 times the cracking displacement, at which point the residual load stabilized, leading to the conclusion of the test. For specimen ZHY-1-80, the spalling of concrete adjacent to the keyway following the peak load, as well as the cracking of both the main and secondary diagonals, was more pronounced compared to specimen ZHY-1-60. The test for this specimen concluded when the load diminished to 57.4% of the peak load. Regarding specimen ZHY-1-100, spalling was noted when the load reached 73.6% of the peak load. At the peak load, significant damage was observed at the neck of the keyway, with portions of the epoxy concrete exhibiting spalling and the cracks on the joint surface becoming penetrating. The test was terminated when the load fell to 44.5% of the peak load.

As seen in Figure 7, there is no significant difference in the shear stiffness of the specimens as the buckle length increases. However, a decrease in shear capacity is observed. The shear bearing capacities of the two specimens with buckle lengths of 60 mm and 80 mm are comparable. Notably, the reduction in bearing capacity for specimen ZHY-1-60 is less pronounced after reaching the peak value, suggesting superior ductility.

3.1.2. After Pouring Ordinary Concrete Specimens with Joint Surfaces without Keyways

Specimens ZXJ-2-60 and ZXJ-2-100 are constructed using standard concrete as the joint material. The bonding surfaces lack a keyway. However, the annular joint lengths differ between the specimens. The failure characteristics of each specimen are illustrated in Figure 8, while the load-displacement curves are presented in Figure 9.

(**a**) ZXJ-2-60

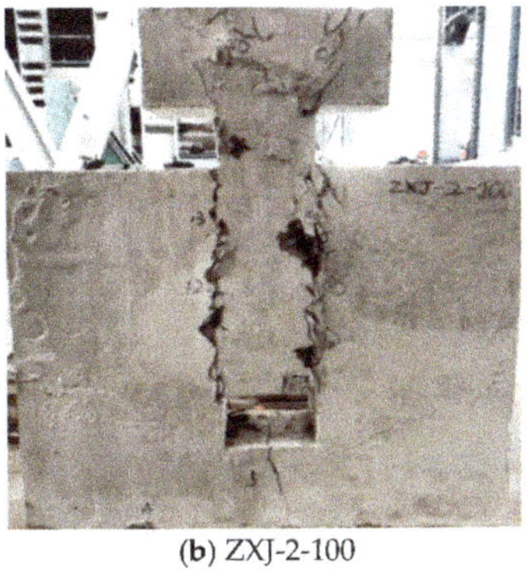

(**b**) ZXJ-2-100

Figure 8. Failure of each specimen.

The typical failure process observed in this group of specimens is characterized by the following sequence: Upon loading, multiple small cracks emerge on both sides of the interface between the new and old concrete. As displacement increases, these cracks progressively connect vertically and widen. Concurrently, spalling of the concrete at the joint surface occurs. At the peak load, the majority of the concrete at the joint surface exhibits significant spalling. The test concludes when the residual load stabilizes.

As shown in Figure 9, both the peak load and shear stiffness of specimen ZXJ-2-60 exceed those of test specimen ZXJ-2-100. This observation suggests that the shear capacity and shear stiffness of specimens cast with ordinary concrete, which lack a keyway on the joint surface, diminish as the binding length of the annulus increases. Furthermore, the absence of a descending section in the load-displacement curves for both specimens indicates that the shear failure of the bonding surface exhibits pronounced brittleness in the absence of a keyway.

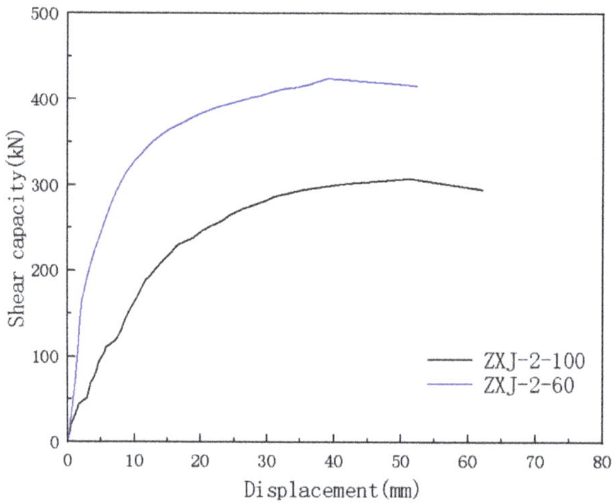

Figure 9. Load-displacement curves of ZXJ-2-60 and ZXJ-2-100 specimen.

3.1.3. Mechanism Analysis

In summary, the optimal performance in terms of shear capacity, stiffness, and ductility is achieved when the buckle length of the ring reinforcement is minimized, such as at 60 mm. This observation can be explained through the structural composition of the joint, as depicted in Figure 10. The joint primarily consists of three segments: the reinforcement skeleton (width b) created by the buckle of the ring reinforcement in the center, the concrete segment (width b_1) extending from the edge of the reinforcement skeleton to the joint interface, and the keyway (width b_2). The shear bearing capacity of the joint surface is derived from four main components: the friction force at the joint surface, the bite force of the keyway, the pin bolt force exerted by the reinforcement, and the shear capacity of the concrete within width b_1. The critical role of the concrete within width b_1 becomes evident under shear stress; it interacts with the pin bolt force from the ring reinforcement to enhance the joint's shear capacity. Thus, the larger the b_1 dimension, the more significant its contribution to the overall shear resistance. However, it is essential to maintain a balance, as excessively increasing b_1 can diminish the integrity of the reinforcement skeleton (width b), thereby adversely affecting the joint's shear performance.

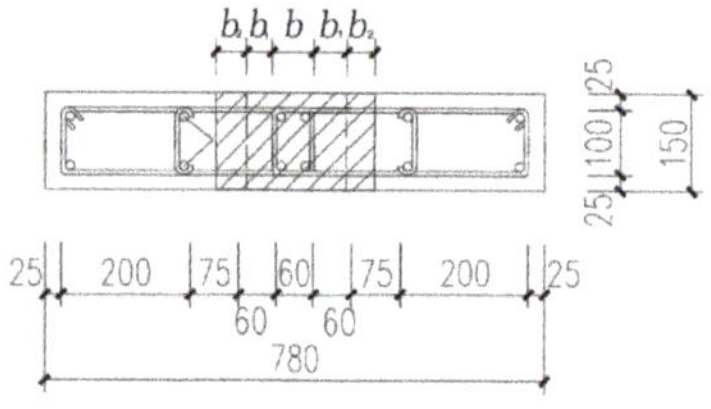

Figure 10. Schematic diagram of ZHY-1-60 joint structure.

When the buckle length of the ring reinforcement is maintained at 60 mm, the width b of the reinforcement skeleton occupies one-third of the total joint width. Consequently, the dimension b_1, from the edge of the reinforcement skeleton to the joint interface, also measures 60 mm, mirroring the buckle length. With the retaining length of the U-shaped reinforcement held constant, larger buckle lengths reduce the size of b_1. Therefore, in comparison to specimens with buckle lengths of 80 mm and 100 mm, the specimen with a 60 mm buckle length offers the most substantial contribution to shear resistance and enhances the shear ductility of the joint surface.

3.2. Effect of Post-Poured Concrete Type on Joint Shear Performance

Specimens ZHY-1-60 and ZXJ-1-60 are characterized by a joint length of 60 mm and a bonding surface featuring a keyway. However, they differ in the materials used for the joints following the pouring process. The failure modes observed in each specimen are illustrated in Figure 11. Additionally, the load-displacement curves for each specimen are presented in Figure 12.

(**a**) ZHY-1-60 (**b**) ZXJ-1-60

Figure 11. Damage condition of each specimen.

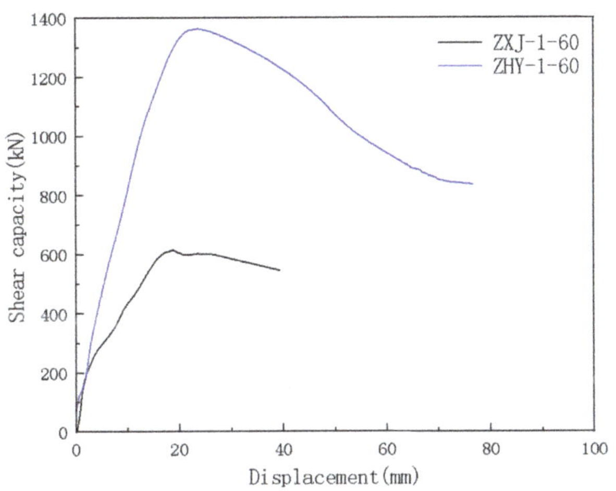

Figure 12. Load-displacement curves of ZXJ-1-60 and ZHY-1-60 specimen.

The failure process of the post-cast epoxy concrete specimen ZHY-1-60 exhibits similarities to that of the post-cast ordinary concrete specimen ZXJ-1-60, as previously discussed. In the case of specimen ZXJ-1-60, spalling of the concrete was observed at the interface of the new and old concrete surfaces when the load reached 78.2% of the peak load. Following the attainment of the peak load, all cracks were identified at the interface, resulting in the crushing of the keyway concrete. Although neither specimen exhibited brittle failure during the testing process, the damage observed at the joint of the post-poured ordinary concrete specimens was notably more severe.

In Figure 12, the peak load of specimen ZHY-1-60 exhibits a substantial increase of 120% in comparison to specimen ZXJ1-60. However, the rate of decline in the curve and the overall trend remain relatively consistent between the two specimens. This observation suggests that the utilization of epoxy resin concrete as the material for joint pouring enhances the shear performance of the joint surface.

3.3. *Effect of Keyway on Joint Shear Performance*

Specimens ZXJ-1-60 and ZXJ-2-60 are characterized by a joint length of 60 mm and are constructed from ordinary concrete following the pouring process. However, these specimens differ in the design of the keyways incorporated into the joint surface. The failure conditions for each specimen are illustrated in Figure 13, while the corresponding load-displacement curves are presented in Figure 14.

(**a**) ZXJ-1-60 (**b**) ZXJ-2-60

Figure 13. Comparison of failure conditions of each specimen.

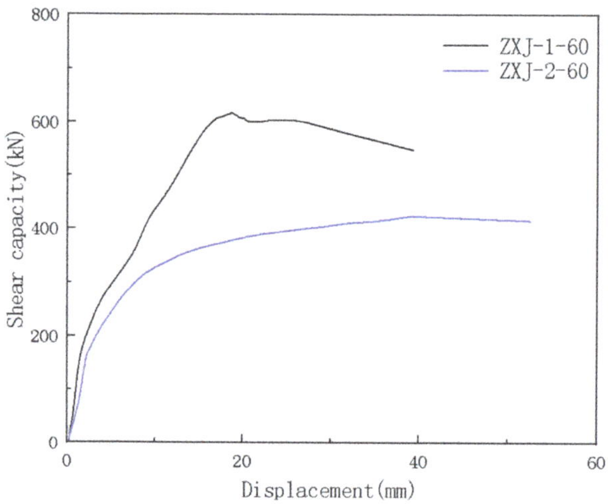

Figure 14. Load-displacement curve of an ordinary concrete specimen with a 60mm length of U-shaped steel bar.

As previously noted, during the initial loading of the ZXJ-2-60 specimen, which lacks a keyway on the bonding surface, penetrating cracks emerged at the interface between the old and new concrete, accompanied by spalling of the concrete at this junction. In contrast, during the loading process of the ZXJ-1-60 specimen, which features a keyway on the bonding surface, significant damage was observed at both the bonding surface of the old and new concrete and the keyway itself as the load and displacement increased. This observation suggests that the presence of the keyway is critical to enhancing the shear resistance of the bonding surface.

As shown in Figure 14, a comparison between specimen ZXJ-2-60 and specimen ZXJ-1-60, which incorporates a keyway, reveals a significant increase in both peak load and shear stiffness for the latter. Specifically, the peak load for specimen ZXJ-1-60 increased by 46.1%. The load-displacement curve for specimen ZXJ-2-60 does not exhibit a different decreasing section, suggesting that brittle shear failure occurs at the bonding surface in the absence of a keyway. Conversely, the presence of a keyway enhances the interlocking force between the prefabricated component and the cast-in-place element of the specimen, thereby improving the shear resistance of the bonding surface

3.4. Comparison and Analysis of Monolithic Cast Specimen and Assembled Specimen

The damage condition of the entire pouring specimen ZZJ is illustrated in Figure 15. A comparison of the load-displacement curves for all specimens is presented in Figure 16, while the characteristic point data corresponding to these curves are provided in Table 2.

Figure 15. Failure condition of specimen ZZJ.

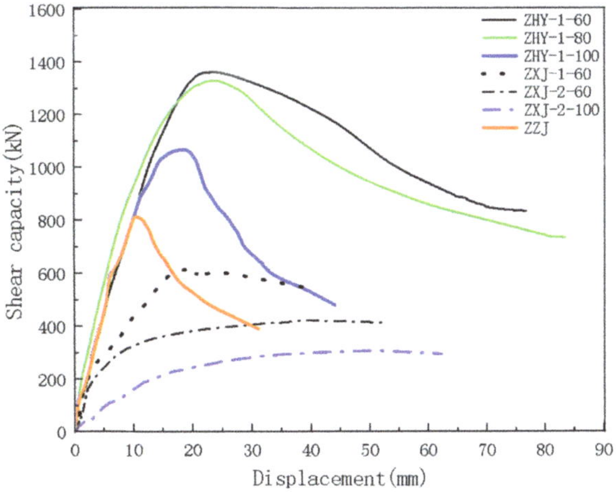

Figure 16. Comparison of load-displacement curves of all specimens.

Table 2. Summary of primary load-displacement curve metrics for each specimen.

Specimen Number	Cracking Load F_C (kN)	Cracking Displacement (mm)	Peak Load F_m (kN)	Peak Displacement (mm)
ZHY-1-60	245	2.35	1384.3	23.6
ZHY-1-80	280	2.57	1295.5	21.1
ZHY-1-100	243	3.51	1073.4	18.9
ZXJ-1-60	103	1.02	621.8	18.8
ZXJ-2-60	106	1.63	425.7	39.0
ZXJ-2-100	77	4.14	307.7	42.1
ZZJ	180	1.72	813.3	10.3

Due to the in situ casting of specimen ZZJ, which exhibited commendable integrity, the failure during loading was not localized to the two shear planes. Upon the initiation of loading, cracks emerged in the lower right corner of the specimen. As the load increased, vertical cracks developed in proximity to the shear plane, resulting in the crushing and spalling of the concrete at the neck of the specimen. When the load decreased to 44.7% of the peak load, the displacement surpassed 10 times the cracking displacement, and the residual load stabilized, marking the conclusion of the test.

In Figure 16 and Table 2, the following results can be summarized:

(1) The shear stiffness of the test specimen featuring a keyway and poured epoxy resin concrete at the joint is comparable to that of the entire poured test specimen. However, the shear strength of the test specimen exceeds that of the complete poured test specimen. Specifically, the shear strength of the specimen with a buckle length of 60 mm exhibits an increase of 70%, while the specimen with a buckle length of 80 mm shows an increase of 59%. Furthermore, specimens with a bonding length of 100 mm demonstrate an improvement of 32%. These results suggest that the incorporation of keyways on the bonding surface, along with the use of epoxy resin concrete as a joint material, are effective strategies for enhancing the shear resistance of the interface between new and old concrete. Additionally, it is recommended that the bonding length of the ring reinforcement not exceed a certain limit, ideally being approximately one-third of the total width of the horizontal joint.

(2) The shear stiffness and shear bearing capacity of the post-poured ordinary concrete specimens at the joint are lower than those of the fully poured specimens. Specifically, the shear bearing capacity of the specimens featuring a keyway decreases by 23%, while the specimens without a keyway and with a rib length of 60 mm exhibit a decrease of 47%. Furthermore, the specimens with a rib length of 100 mm demonstrate a reduction of 62%. This indicates that an increase in buckle length adversely affects the shear performance of the joint surface, assuming a constant retaining length of the annulus. Additionally, the load-displacement curve for the test specimen without a keyway does not exhibit a different decreasing section; rather, it shows a prolonged peak load holding time and significant shear displacement. This suggests that the shear resistance of the joint surface is predominantly attributed to the pin force provided by the reinforcement at the joint surface. Upon the loss of this pin force, the bearing capacity experiences an immediate decline from the peak value, thereby illustrating the brittle property of shear failure.

4. Development and Verification of Finite Element Model

4.1. Material Constitutive Model

The finite element model was developed and analyzed using ABAQUS software. A plastic damage model was employed for the concrete material. The uniaxial stress-strain curve recommended in "Code for Design of Concrete Structures" (GB 50010-2010) [28] was utilized as the constitutive relationship for ordinary concrete. For epoxy resin concrete, the constitutive relationship is represented by the complete stress-strain curve equation, which was fitted in this study based on prior experimental results (Equation (1)). The stress-strain behavior of the rebar is modeled using a double broken-line approach:

$$y = \begin{cases} ax + (4.9 - 4.23a)x^2 + (-4.67 + 6.67a)x^3 + (-0.27 - 4.74a)x^4 + (1.07 + 1.27a)x^5, \ 0 \le x \le 1 \\ \frac{x}{b(x-1)^2 + x}, \ x > 1 \end{cases} \tag{1}$$

where a and b are the undetermined parameters, which can be determined as follows:

$$\begin{cases} 0 < a < 1.0 \\ 1.0 < b < 10.0 \end{cases}$$

where a is set as 0.7 and b is set as 7.0 in this study.

4.2. Model Establishment

A finite element model of the typical specimen ZHY-1-60 has been established as shown in Figure 17. The solid element type C3D8R is employed for both epoxy concrete and ordinary concrete, while 2D truss element type T2D2 is utilized for the reinforcement bars. Structured grid technology is implemented to discretize the mesh. Referring to the numerical modeling of perforated composite materials by Khan et al. [29] and others, achieving more accurate results necessitates a reduction in the mesh size around the keyway. Consequently, the mesh size at the keyway is determined to be 15 × 35 mm, whereas the remainder of the mesh is set at 30 × 35 mm, and the mesh size for the reinforcement is standardized at 30 mm. A surface-to-surface contact model is adopted to simulate the interfacial contact between precast concrete and post-cast concrete. The normal interaction is governed by a "hard contact", while the tangential interaction is modeled using the Coulomb friction model, with a friction coefficient of 0.6 [15].

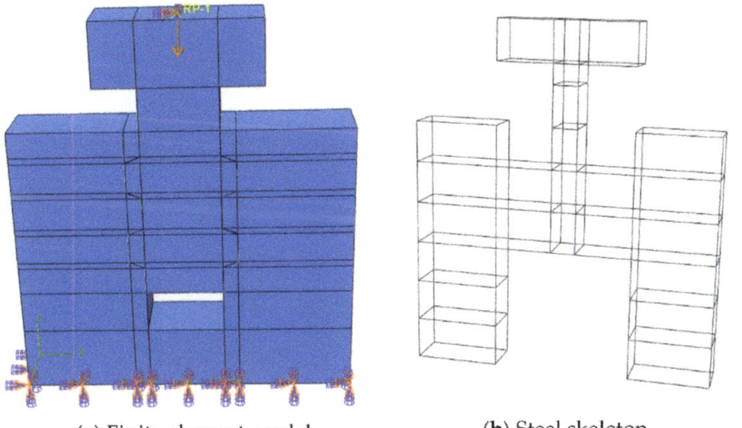

(**a**) Finite element model (**b**) Steel skeleton

Figure 17. Finite element model of specimen ZHY-1-60.

The boundary conditions of the model align with those established by the test. Specifically, the bottom of the component is designated as a fixed constraint, while the center point of the top surface of the loading beam serves as the reference point for controlling vertical displacement during loading. Additionally, a coupling method is employed to mitigate stress concentration throughout the loading process.

4.3. Verification of Model

Figure 18 presents a comparative analysis of the experimental results and finite element simulations regarding the failure morphology of the ZHY-1-60 specimen. As shown in Figure 18b, the distribution of concrete compression damage factors is employed to approximate the model's failure. Both the experimental data and the finite element simulations indicate that the damage is predominantly concentrated in the keyway. As displacement increases, the damage at the root of the keyway progressively extends along the bonding surface until the damaged surface becomes interconnected, resulting in the severance of the key teeth.

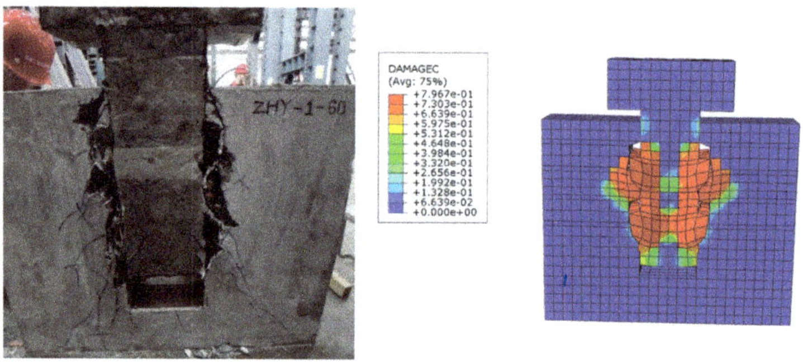

(**a**) Test results (**b**) Finite element simulation results

Figure 18. Comparison of test and simulation results of failure morphology of specimens.

Figure 19 depicts a comparison between the test results and the finite element simulation results for the load-displacement curve of specimen ZHY-1-60. The overall trend of the curve exhibits a similar pattern, and the discrepancies between the two sets of results fall within an acceptable range.

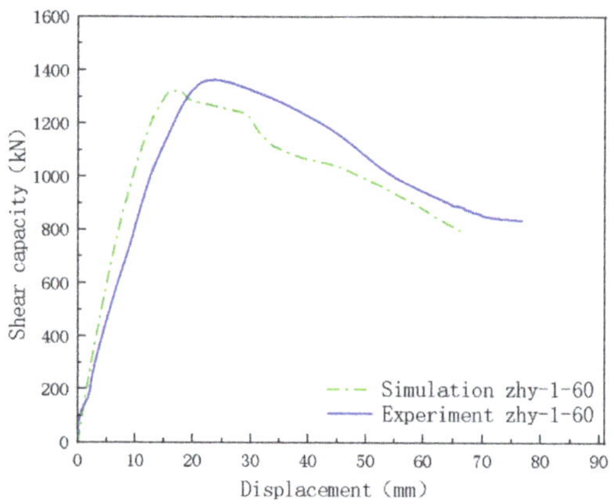

Figure 19. Comparison of experimental and finite element simulation results of load-displacement curves of specimens.

In summary, the modeling approach presented in this paper demonstrates the capability to effectively simulate the shear behavior of direct shear specimens.

5. Finite Element Analysis of Factors Affecting Horizontal Joint Shear Performance

A finite element variable parameter analysis was conducted using specimen ZHY-1-60. Following the modification of four parameters (e.g., the grade of precast concrete, the reinforcement ratio of the horizontal joint annulus, the reinforcement ratio of the horizontal joint insertion, and the horizontal binding force), a total of 17 finite element models were developed across four groups. The parameters of the test specimens are outlined in Table 3.

Table 3. Parameters and configurations of finite element models.

Specimen Number	Strength Grade of Concrete	Ratio of Joint Ring Reinforcement (Diameter)	Reinforcement Ratio of Horizontal Joint Insertion (Diameter)	Horizontal Binding Force (kN)
ZHY-1-60	C40	0.50% (8 mm)	0.74% (8 mm)	0
ZJC1	C30	0.50% (8 mm)	0.74% (8 mm)	0
ZJC2	C50	0.50% (8 mm)	0.74% (8 mm)	0
ZJC3	C60	0.50% (8 mm)	0.74% (8 mm)	0
ZJC4	C70	0.50% (8 mm)	0.74% (8 mm)	0
ZJH1	C40	0.28% (6 mm)	0.74% (8 mm)	0
ZJH2	C40	0.79% (10 mm)	0.74% (8 mm)	0
ZJH3	C40	1.13% (12 mm)	0.74% (8 mm)	0
ZJH4	C40	1.54% (14 mm)	0.74% (8 mm)	0
ZJS1	C40	0.50% (8 mm)	0.42% (6 mm)	0
ZJS2	C40	0.50% (8 mm)	1.20% (10 mm)	0
ZJS3	C40	0.50% (8 mm)	1.70% (12 mm)	0
ZJS4	C40	0.50% (8 mm)	2.30% (14 mm)	0
ZJF1	C40	0.50% (8 mm)	0.74% (8 mm)	100
ZJF2	C40	0.50% (8 mm)	0.74% (8 mm)	200
ZJF3	C40	0.50% (8 mm)	0.74% (8 mm)	300
ZJF4	C40	0.50% (8 mm)	0.74% (8 mm)	400
ZJF5	C40	0.50% (8 mm)	0.74% (8 mm)	500

5.1. Effect of Precast Partial Concrete Strength

Finite element models ZJC1, ZHY-1-60, ZJC2, ZJC3, and ZJC4 were developed utilizing precast concrete with strength grades of C30, C40, C50, C60, and C70, respectively (Table 3). This study examines the effect of the strength of precast concrete on the shear performance of horizontal joints. Notably, the strength grade of the precast concrete in model ZHY-1-60 is approximately equivalent to C40. Figure 20 illustrates the stress distribution across the concrete and steel reinforcement of each specimen, while Figure 21 depicts the relationship between the strength of precast concrete and the shear capacity of the joints.

As illustrated in the concrete stress distribution cloud map in Figure 20, the keyway serves as a critical shear element of the bonding surface, exhibiting an increase in stress value corresponding to the enhancement of concrete strength. This phenomenon underscores the keyway's effectiveness in shear resistance. However, when the strength of the precast concrete reaches C70, the stress value decreases. Analysis of the stress distribution nephogram for reinforcement, along with the distribution of maximum stress points, reveals substantial stress at the joint interface between new and existing concrete. This indicates that the ring reinforcement plays a vital role in shear resistance by exerting a pin bolt force. This force is subsequently transmitted to the surrounding concrete through the embedded effect of the ring reinforcement, enhancing the shear resistance of the overall structure and ensuring effective bonding between the new and old concrete. Additionally, the stress concentration in the upper part of the horizontal joint dowel bar and the lower part of the vertical reinforcement on both sides of the horizontal joint is significant, forming a force transmission pathway from the loading end to the dowel bar, then to the ring reinforcement, and finally to the vertical reinforcement of the prefabricated wall. Thus, to ensure adequate shear resistance at horizontal joints, the reinforcement quantity of dowel bars, ring bars, and longitudinal reinforcement in prefabricated walls adjacent to horizontal joints should not be minimal.

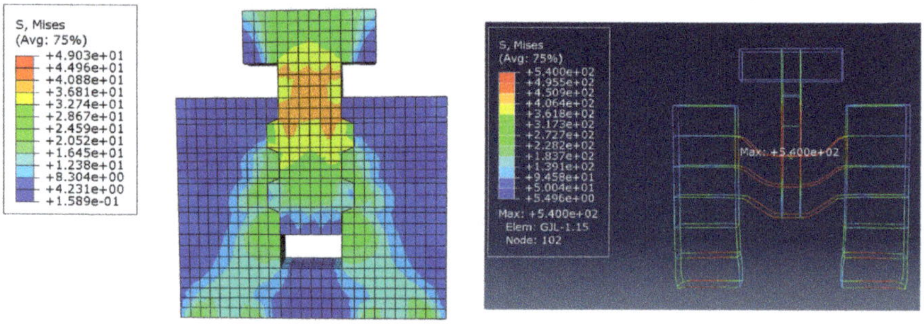

(**a**) Specimen ZJC1

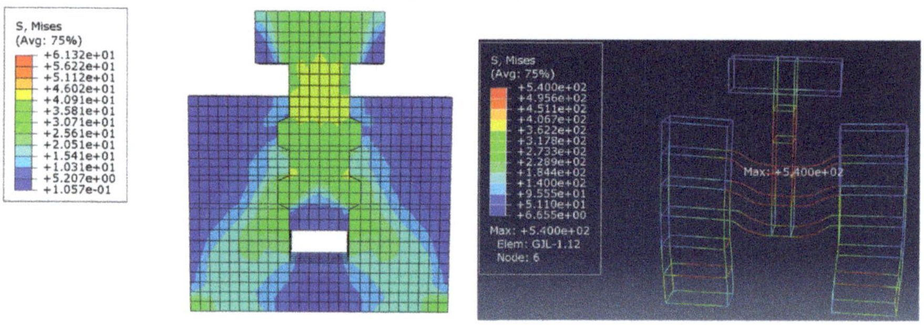

(**b**) Specimen ZHY-1-60

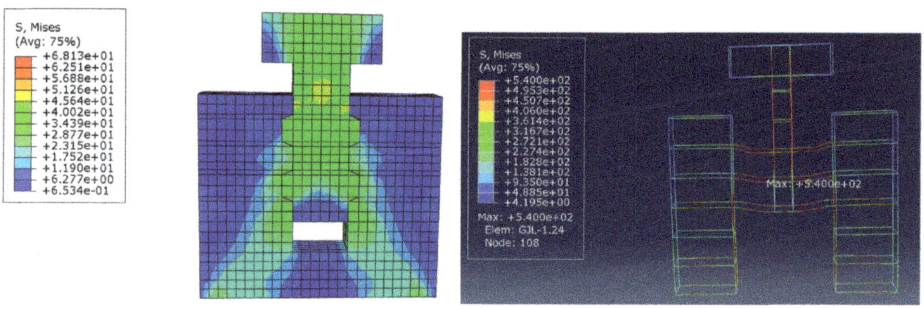

(**c**) Specimen ZJC2

Figure 20. *Cont.*

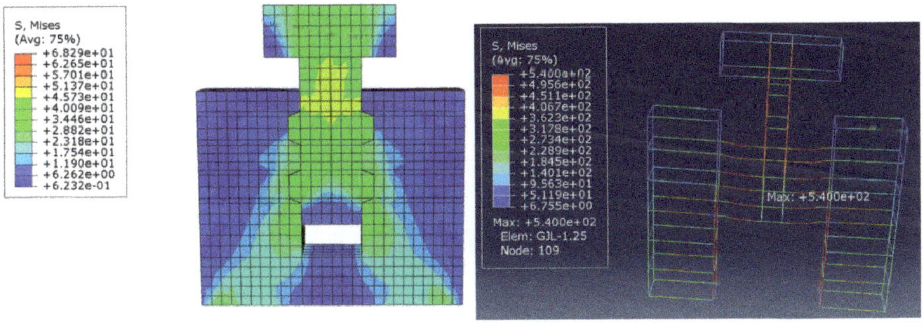

(**d**) Specimen ZJC3

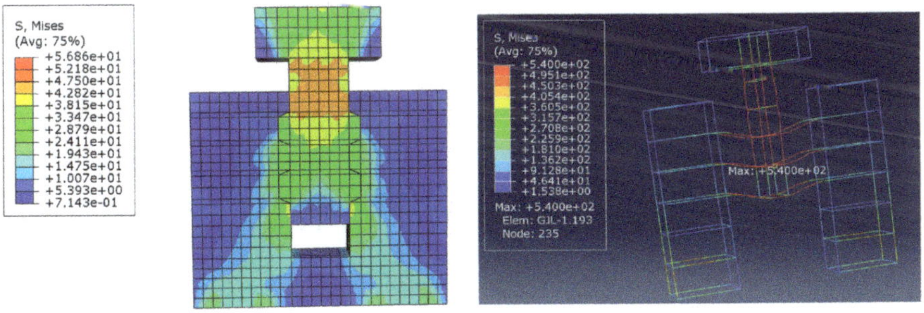

(**e**) Specimen ZJC4

Figure 20. Stress distribution of concrete and steel bars under peak load.

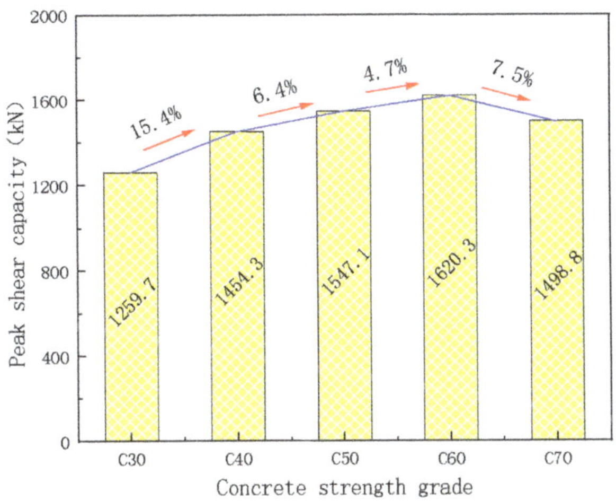

Figure 21. Influence of precast partial concrete strength on joint shear capacity.

In Figure 21, an increase in the strength grade of precast concrete from C30 to C40 results in a 15.4% enhancement in interfacial shear resistance. Subsequently, when the strength grade is elevated from C40 to C50, the shear resistance improves by 6.4%. Further increases from C50 to C60 yield a 4.7% rise in shear resistance, while an increase from C60 to C70 leads to a decrease of 7.5% in shear resistance. These results suggest that the specimen's performance is optimized when the strength of the post-cast epoxy concrete closely matches that of the precast concrete, particularly within the C40 to C60 range. This phenomenon can be attributed to the equal interaction capacity of the new and old concrete at the joint surface, which includes shear bond characteristics, thereby resulting in the most favorable overall performance.

5.2. Effect of Horizontal Joint Annulus Reinforcement Ratio

Finite element models ZJH1, ZHY-1-60, ZJH2, ZJH3, and ZJH4 (Table 3) were developed utilizing horizontal joint annulus diameters of 6 mm, 8 mm, 10 mm, 12 mm, and 14 mm, respectively. This study explored the effect of the reinforcement ratio (diameter) on the shear performance of horizontal joints. The load-displacement curves for each specimen are presented in Figure 22, while Figure 23 illustrates a comparison of the shear capacities of the specimens.

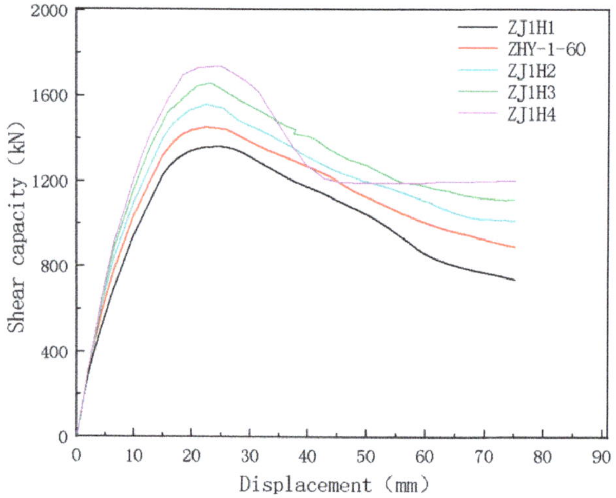

Figure 22. Load-displacement curves of specimens with different annulus diameters.

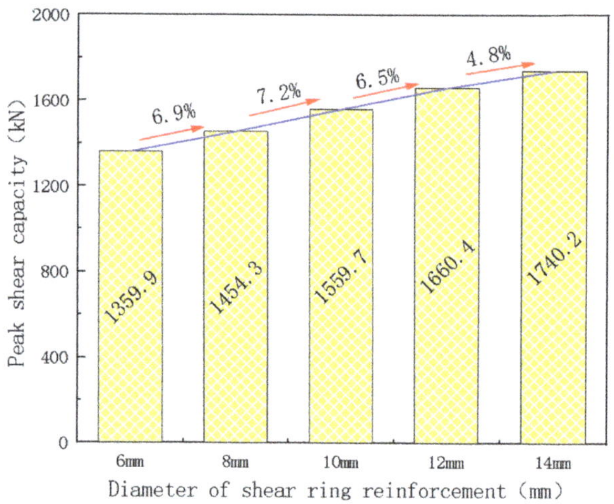

Figure 23. Comparison of the peak load of each specimen under different annulus diameters.

In Figure 22, an increase in the reinforcement ratio (diameter) of the annulus significantly enhances the shear capacity of the joint surface. Nonetheless, when the reinforcement ratio reaches 1.54% (diameter 14 mm), there is a notable decrease in the specimen's ductility. This reduction may be attributed to the excessive reinforcement ratio and diameter, which over-constrain the concrete at the keyway and lead to stress concentration in the local area, particularly at the keyway. This stress concentration tends to diminish the overall deformation energy of the concrete and increases the risk of brittle failure. Despite the relatively high shear bearing capacity at this reinforcement level, the ductility performance is markedly compromised. Based on the findings of this study and relevant standards, it is recommended that the diameter of the ring reinforcement be limited to 12 mm.

As shown in Figure 23, an increase in the diameter of the annulus from 6 mm to 14 mm resulted in corresponding increases in the shear strength of each specimen, specifically by 6.9%, 7.2%, 6.5%, and 4.8%, respectively. This trend indicates an initial increase in shear strength followed by a subsequent decrease. Furthermore, the ductility of the joint diminishes when the reinforcement ratio of the ring reinforcement is excessively high. Consequently, for practical design considerations, it is advisable to select a ring diameter ranging from 10 mm to 12 mm, which corresponds to a reinforcement ratio of 0.79% to 1.13%, as this range is more appropriate.

5.3. Effect of Horizontal Joint Insertion Reinforcement Ratio

The insert bar is defined as a long steel bar that traverses the U-shaped closed sleeve. The presence of the insert bar facilitates the buckle connection of the ring reinforcement within the prefabricated wall, thereby forming a concealed beam with a rectangular cross-section. This configuration contributes to the enhancement of structural integrity and reliability. Finite element models, designated as ZJS1, ZHY-1-60, ZJS2, ZJS3, and ZJS4 (Table 3), were developed using insert bar diameters of 6 mm, 8 mm, 10 mm, 12 mm, and 14 mm, respectively. This study investigates the effect of the reinforcement ratio (diameter) of the insert bar on the shear resistance of horizontal joints. The load-displacement curves for each test specimen are presented in Figure 24, while Figure 25 illustrates a comparative analysis of the shear capacities of the various specimens.

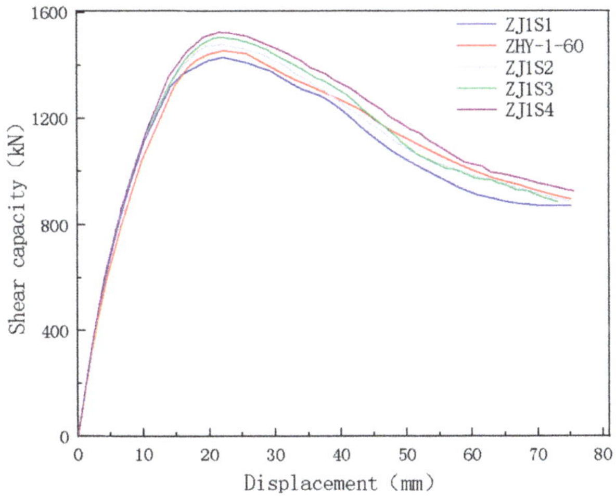

Figure 24. Load displacement curve of each specimen under different dowel bar diameters.

2.4. Test Method and Sensor Distribution

The dynamic responses of the assembled integral two-way multi-ribbed composite floor, including the vibration mode and acceleration time-history curve under a pedestrian load, were tested in the civil engineering laboratory of Southeast University, Nanjing, China. The vibration modes and frequencies were tested using the pulsating method and analyzed by the subspace identification (SSI) method, with all data collected at midnight to minimize disturbances from human activity. The acceleration-history curves were obtained under pedestrian loads for single-person and six-person loads, respectively. For the single-person load, the involved volunteer was 61 kg. For the six-person load, the average weight was controlled at 63 kg.

The volunteers followed a standardized walking routine from left to right with a step frequency of 2 Hz. To maintain these frequencies, an application was installed on their phones, which provided a reminder tone every 0.5 s. Upon reaching the opposite side of the floor, the volunteers waited for several seconds and then turned around and walked back, ensuring the total walking duration was more than 40 s. This procedure was followed for single-person and group walking.

To capture the acceleration-history curves of the floor, 16 acceleration meters (DASP, China Orient Institute of Noise & Vibration, Beijing, China) were evenly distributed across the specimen, as presented in Figure 5. All the data were automatically logged by a dynamic logger (DASP, China Orient Institute of Noise & Vibration, Beijing, China) at a sampling rate of 256 Hz.

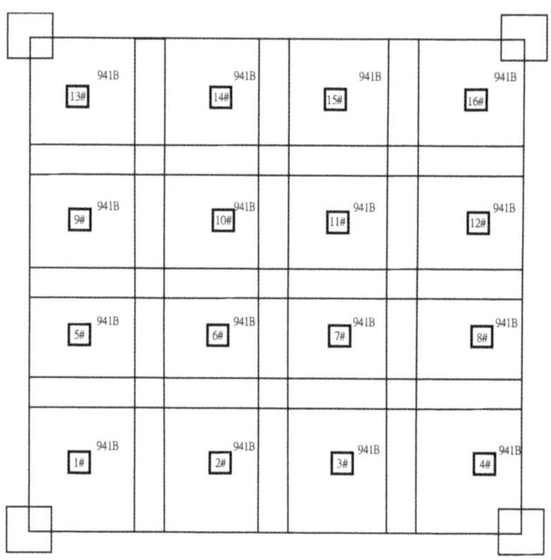

Figure 5. Distribution of acceleration meters.

The mechanism of SSI is explained by the following equations. For linear multi-DOF systems, the equation of motion could be presented as Equation (1). In the space state equation form, it could be transformed to Equations (2) and (3). Discretized to time, Equations (4) and (5) could be obtained. The main aim of the method is to obtain the A matrix based on the measurement values.

$$M\ddot{X} + C\dot{X} + KX = u(t) \tag{1}$$

$$\dot{Z} = HZ + Fu \tag{2}$$

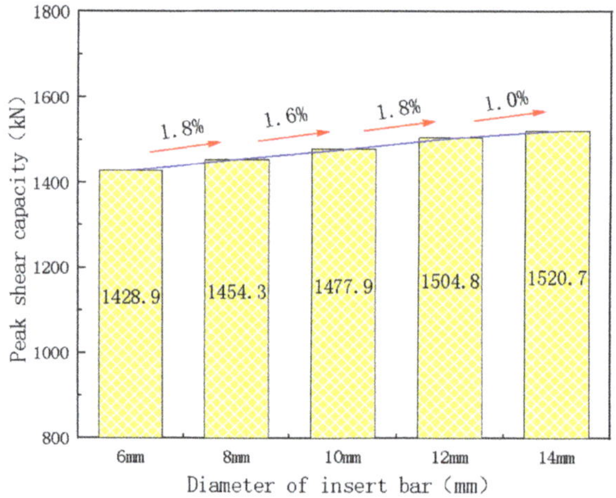

Figure 25. Peak load of each specimen under different insert bar diameters.

As shown in Figures 24 and 25, an increase in the reinforcement ratio (diameter) of the insert bars correlates with an enhancement in the shear capacity of the joint surface for each specimen. However, the extent of this improvement is limited. Specifically, when the diameter of the insert bar is increased from 6 mm to 14 mm, the shear capacity each specimen exhibits increases by 1.8%, 1.6%, 1.8%, and 1.0%, respectively. This indicates that joint reinforcement has a minimal effect on the shear capacity of the joint surface. Consequently, when selecting the diameter of the insert bar in practical design applications, it is sufficient to meet the fundamental structural requirements or to utilize the same diameter as that of the prefabricated wall. There is no necessity to increase the diameter of the insert bar to enhance the bearing capacity of the joint surface.

5.4. Effect of Horizontal Binding Force

The horizontal joint of a precast shear wall experiences vertical compressive stress during its operational phase, and the magnitude of this stress significantly impacts the performance of the horizontal joint. To simulate vertical compressive stress, horizontal binding forces of 0 kN, 100 kN, 200 kN, 300 kN, 400 kN, and 500 kN were applied to the side of the model, corresponding to axial compression ratios of 0.09, 0.17, 0.26, 0.35, and 0.44, respectively. Finite element models designated as ZHY-1-60, ZJF1, ZJF2, ZJF3, ZJF4, and ZJF5 were developed to examine the effect of vertical compressive stress on the shear performance of horizontal joints. Figure 26 illustrates the load-displacement curves for each specimen under varying binding forces, while Figure 27 presents the peak load diagrams for each specimen at different binding force levels.

As shown in Figures 26 and 27, an increase in lateral horizontal constraint correlates with an enhancement in the initial elastic modulus and shear capacity of the specimens. Conversely, the peak displacement and ductility exhibit a decline. Specifically, when the horizontal restraint force is elevated from 0 kN to 500 kN, the shear capacity of each specimen increases by 2.4%, 3.7%, 5.2%, 4.0%, and 0.9%, respectively. Notably, in comparison to specimen ZJF4, which has a horizontal constraint of 400 kN, the bearing capacity of specimen ZJF5, subjected to a horizontal constraint of 500 kN, shows minimal improvement. Furthermore, the shear capacity of specimen ZJF5 decreases significantly after reaching the peak load. This suggests that while increased horizontal constraint offers limited enhancements in bearing capacity, it may also lead to rapid degradation post-peak load. Engineering standards and requirements recommend controlling the maximum axial compression ratio to 0.5.

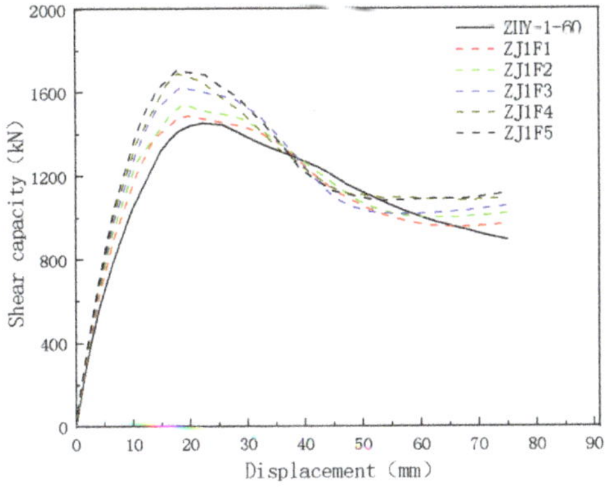

Figure 26. Load displacement curve of each specimen under different horizontal binding forces.

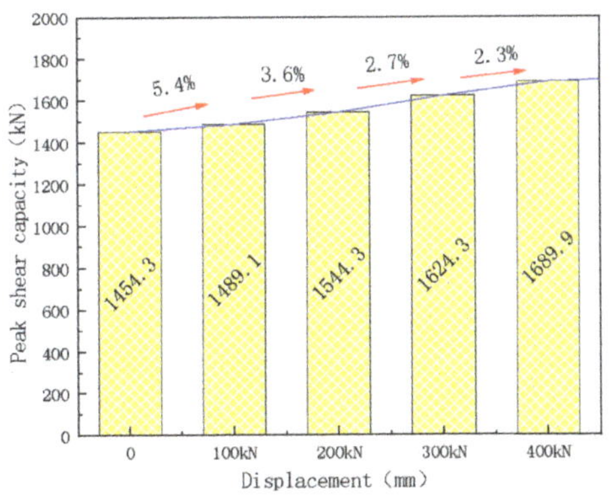

Figure 27. Peak load under different levels of restraint.

6. Conclusions

In this study, epoxy resin concrete was utilized as the post-cast material. The horizontal joint of a shear wall reinforced with ring buckles was subjected to direct shear tests and finite element analysis. The conclusions derived from this research offer valuable insights for the application of epoxy resin concrete in structural engineering, which can be summarized as follows:

(1) When epoxy resin concrete is employed as the post-cast material in prefabricated components, it exhibits excellent bonding performance with the ordinary concrete of the prefabricated sections. Additionally, incorporating a keyway on the joint surface significantly enhances the interlock between the old and new concrete, thereby improving the shear resistance of the joint surface. Therefore, the shear bearing capacity and stiffness of horizontal joints with epoxy resin concrete and keyways can match or even exceed those of fully cast specimens.

(2) An increase in the buckle length of the ring reinforcement negatively affects the shear bearing capacity of the joint surface. When ordinary concrete is used as the post-cast material alongside a large buckle length, the shear stiffness may also be compromised. This study recommends that the buckle length of the ring reinforcement be approximately one-third of the total width of the horizontal joint to optimize performance.

(3) As the strength of the precast concrete increases, there is a corresponding increase in bearing capacity, but a decrease in ductility. The rate of increase in bearing capacity initially rises and then diminishes. Therefore, it is recommended that the strengths of the precast concrete and the post-cast epoxy concrete be closely matched to achieve the optimal strength design.
(4) The shear capacity of the wall improves with higher reinforcement ratios in both the horizontal joint ring reinforcement and the inserted reinforcement. The effect of the ring reinforcement is more significant than that of the inserted reinforcement. However, when the reinforcement ratio surpasses a certain threshold, the wall's ductility decreases markedly. It is advised that the diameter of the ring reinforcement be between 10 mm and 12 mm, with a reinforcement ratio ranging from 0.79% to 1.13%.
(5) Increasing the lateral horizontal constraint force on the specimen enhances both stiffness and shear capacity. However, it also reduces ductility, particularly when the constraint force reaches 500 kN, leading to a significant decrease in ductility. Therefore, it is recommended that the axial compression ratio of the wall not exceed 0.5.

7. Outlook

Future research will provide a comprehensive technical reference for the application of epoxy resin concrete in prefabricated buildings. Planned activities include the following:

(1) Focusing on key parameters, such as joint surface roughness, keyway slotting angle, and the number of keyways, to conduct more detailed experimental and numerical simulation studies and develop construction recommendations for shear walls composed of post-cast epoxy resin concrete.
(2) Based on the findings, proposing a set of formulas to calculate the bearing capacity of the horizontal joint in walls using ring reinforcement buckle connection technology, thereby providing theoretical support for practical engineering applications.
(3) Investigating the durability of joints in prefabricated shear walls with post-cast epoxy resin concrete and optimizing construction technology.

Author Contributions: Software, S.Z. and Y.L.; Investigation, P.X. and X.Z.; Resources, P.X.; Data curation, X.Z.; Writing—original draft, S.Z.; Writing—review & editing, P.C.; Project administration, P.C. All authors have read and agreed to the published version of the manuscript.

Funding: This research was funded by the National Natural Science Foundation of China, grant number 51678389.

Data Availability Statement: The original contributions presented in the study are included in the article, further inquiries can be directed to the corresponding author.

Conflicts of Interest: Author Pengzhan Xu was employed by the company Tangshi Jianhua Building Materials (Tianjin) Co., Ltd. The remaining authors declare that the research was conducted in the absence of any commercial or financial relationships that could be construed as a potential conflict of interest.

References

1. Cholewicki, A. Loadbearing capacity and deformability of vertical joints in structural walls of large panel buildings. *Build. Sci.* **1971**, *6*, 163–184. [CrossRef]
2. Hemamalini, S.; Vidjeapriya, R.; Jaya, K.P. Performance of precast shear wall connections under monotonic and cyclic loading: A state-of-the-art review. *Iran. J. Sci. Technol. Trans. Civ. Eng.* **2021**, *45*, 1307–1328. [CrossRef]
3. Hu, R.; Fang, Z.; Shi, C.; Benmokrane, B.; Su, J. A review on seismic behavior of ultra-high performance concrete members. *Adv. Struct. Eng.* **2021**, *24*, 1054–1069. [CrossRef]

4. Basereh, S.; Okumus, P.; Aaleti, S. Seismic retrofit of reinforced concrete shear walls to ensure reparability. In Proceedings of the Structures Congress 2020, St. Louis, MO, USA, 5–8 April 2020; American Society of Civil Engineers: Reston, VA, USA, 2020; pp. 498–509.
5. Sharma, S.; Aaleti, S.; Okumus, P. Experimental testing of RC shear wall seismic retrofit using selective weakening, self-centering and Ultra High performance concrete. *Resilient Cities Struct.* **2023**, *2*, 76–90. [CrossRef]
6. Han, W.L.; Zhao, Z.Z.; Qian, J.R. Global experimental response of a three-story, full-scale precast concrete shear wall structure with reinforcing bars spliced by grouted couplers. *PCI J.* **2019**, *64*, 65–80. [CrossRef]
7. Xiong, F.; Malla, P.; Cai, G.; Larbi, A.S.; Zhong, Y.; Tufail, R.F.; Chen, W.; Huang, W. Numerical analysis of precast concrete shear walls with horizontal bolted joints under seismic loads. *J. Earthq. Eng.* **2022**, *26*, 4737–4760. [CrossRef]
8. Pall, A.S.; Marsh, C.; Fazio, P. Friction joints for seismic control of large panel structures. *J. Prestress. Concr. Inst.* **1980**, *25*, 38–61. [CrossRef]
9. Yu, Z.-W.; Peng, X.-D.; Guo, W.; Peng, M.-P. Seismic performance of precast concrete shear wall with U-type reinforcements ferrule connection. *Zhejiang Daxue Xuebao (Gongxue Ban)/J. Zhejiang Univ. (Eng. Sci.)* **2015**, *49*, 975–984.
10. Yu, Z.; Peng, X.; Guo, W.; Peng, M. New node connection mode and seismic performance of precast concrete shear wall structure. *Xi'an Jianzhu Keji Daxue Xuebao/J. Xi'an Univ. Archit. Technol.* **2015**, *47*, 160–164+191.
11. Yu, Z.; Lv, X.; Yu, Y.; Ding, F.; Peng, X. Seismic performance of precast concrete columns with improved u-type reinforcement ferrule connections. *Int. J. Concr. Struct. Mater.* **2019**, *13*, 54. [CrossRef]
12. Issa, C.A.; Debs, P. Experimental study of epoxy repairing of cracks in concrete. *Constr. Build. Mater.* **2007**, *21*, 157–163. [CrossRef]
13. Rahman, M.M.; Akhtarul Islam, M. Application of epoxy resins in building materials: Progress and prospects. *Polym. Bull.* **2022**, *79*, 1949–1975. [CrossRef]
14. Yu, T. The Experimental Study of Repairing Concrete Cracks with Epoxy Resin Grouting. Master's Thesis, Beijing University of Technology, Beijing, China, 2016. (In Chinese).
15. Xiang, Q.; Xiao, F. Applications of epoxy materials in pavement engineering. *Constr. Build. Mater.* **2020**, *235*, 117529. [CrossRef]
16. Zhang, W.; Guan, X.; Ren, J.; Gu, X. Experimental study on chloride permeability of concrete surface-treated with epoxy resin. *J. Build. Mater.* **2008**, *11*, 340–344.
17. Natarajan, S.; Neelakanda Pillai, N.; Murugan, S. Experimental investigations on the properties of epoxy-resin-bonded cement concrete containing sea sand for use in unreinforced concrete applications. *Materials* **2019**, *12*, 645. [CrossRef]
18. Qian, Y.; Li, Z.; Jin, Y.; Wang, R. Experimental Study on Axial Tension Members of a New Epoxy Resin Concrete. *Sci. Adv. Mater.* **2021**, *13*, 2005–2015. [CrossRef]
19. El-Mandouh, M.A.; El-Maula, A.S.A. Shear strength of epoxy-modified reinforced concrete beams. *Innov. Infrastruct. Solut.* **2021**, *6*, 105. [CrossRef]
20. Chen, P.; Liu, Z.; Zhou, X.; Xu, S.; Wang, J. Research on Work Performance of Monolithic Precast Concrete Shear Walls with Post-Cast Epoxy Resin Concrete. *Buildings* **2024**, *14*, 1675. [CrossRef]
21. Julio, E.N.B.S.; Branco, F.A.B.; Silva, V.D. Concrete-to-concrete bond strength. influence of the roughness of the substrate surface. *Constr. Build. Mater.* **2004**, *18*, 675–681. [CrossRef]
22. Rizkalla, S.H.; Serrette, R.L.; Heuvel, J.S.; Attiogbe, E.K. Multiple shear key connections for precast shear wall panels. *PCI J.* **1989**, *34*, 104–120. [CrossRef]
23. Gopal, B.A.; Hejazi, F.; Hafezolghorani, M.; Voo, Y.L. Shear strength of dry and epoxy joints for ultra-high-performance fiber-reinforced concrete. *ACI Struct. J.* **2020**, *117*, 279–288. [CrossRef]
24. Fan, H.; Zhang, Y.; Zhang, Y.; Jiang, H.; Du, L.; Liu, Q. Mechanical Properties and Energy Conservation Analysis for The New Connection of The Fabricated Shear Wall. *IOP Conf. Ser. Earth Environ. Sci.* **2021**, *804*, 022094. [CrossRef]
25. *EN1992-1-2*; Euro-Code2: Design of Concrete Structures-Part1: General Rules and Rules for Buildings. European Committee for Standardization: Bruxelles, Belgium, 2021.
26. Jin, Y.; Zhao, Y.; Xie, X.; Zhang, X.; Cai, J. Simulation Analysis of Joint Connection of New Epoxy Resin Concrete Truss Structure. *Int. J. Perform. Eng.* **2021**, *17*, 143.
27. *JGJ/T 430-2018*; Technical standard for precast reinforced concrete shear wall structure assembled by anchoring closed loop reinforcement. China Architecture & Build Press: Beijing, China, 2018.
28. *GB 50010-2010*; Code for Design of Concrete Structures. China Architecture & Building Press: Beijing, China, 2015.
29. Khan RM, A.; Shafighfard, T.; Ali, H.Q.; Mieloszyk, M.; Yildiz, M. Strength prediction and experimental damage investigations of plain woven CFRPs with interacting holes using multi-instrument measurements. *Polym. Compos.* **2023**, *44*, 3594–3609. [CrossRef]

buildings

MDPI

Article

Study on the Bending Performance of Prefabricated H-Shaped Steel Beams with Different Bolt Hole Types

Xin Zhang [1,*], Shenlu Yu [1], Shuaike Feng [1], Dawei Fan [2], Fang Zhang [1] and Han Cao [1]

[1] School of Civil Engineering, Shandong Jianzhu University, Jinan 250101, China; 15668357886@163.com (S.Y.)
[2] Qihang (Shandong) Investment and Construction Group Co., Ltd., Jinan 250013, China
* Correspondence: 15969698504@163.com

Abstract: This paper investigates the structural performance of a new prefabricated H-shaped steel beam assembled using high-strength bolts under three-point bending. The study evaluates four bolt hole types in five layout schemes through experimental tests. The results show that specimens with standard round holes in both the H-shaped steel and connecting plates exhibited 11% to 30% higher flexural bearing capacity compared to other hole types. Additionally, ANSYS simulations closely matched the experimental results, with a 6% difference. The research results provide important references for the design of prefabricated H-shaped steel beams with different bolt hole types, offering a practical foundation for enhancing the flexural performance of steel beam designs.

Keywords: high-strength bolts; prefabricated; H-shaped steel; bolt hole types; three-point bending load; flexural strength; numerical simulation

1. Introduction

The development of prefabricated steel structures is a key driver for advancing sustainability and intelligence in construction. These structures offer advantages such as controlled resource consumption, shortened construction periods, and excellent structural performance [1–5]. Figure 1 illustrates the assembly of a prefabricated H-shaped steel beam [6]. Without the need for on-site welding, this beam is formed by bolting together dual HM150 steel sections and connecting plates, creating the main unit of the prefabricated H-shaped steel beam. The main unit is further connected with stiffening rib plates through bolts, forming an integrated structure that can be assembled on-site to the required length according to project specifications. These prefabricated components feature numerous holes on the connecting plates, as well as on the web and flanges of the H-shaped steel, which can affect the failure modes and flexural strength of the components. Moreover, current design guidelines, such as those from the American Iron and Steel Institute (AISI) [7], do not provide any calculation procedures to determine the flexural capacity of such steel beams with flange perforations and variations in bolt hole types.

In recent decades, researchers such as Swanson [8], Teh [9], Cavène [10], and PENG [11] have conducted extensive experimental and theoretical studies on the impact of hole type and preloading on the shear performance of bolts, and the effect of slip resistance on the mechanical properties of bolted connections. In engineering, reamed high-strength bolt connections are often used to improve the energy dissipation, structural ductility, and load-bearing performance of prefabricated components. Based on these studies, the reaming coefficient for high-strength bolts was improved and explicitly specified in the relevant codes [12]. With the widespread application of prefabricated steel structures in foundation

https://doi.org/10.3390/buildings14092988

pit support, the use and performance of prefabricated dual H-shaped main steel units have gained increasing attention.

To date, studies on the mechanical performance of steel components with openings have primarily focused on factory-produced steel sections. Studies have examined issues such as the distortion and overall buckling performance of cold-formed thin-walled web-perforated composite beams [13], as well as the impact of web opening size, location, and shape on the load-bearing capacity and stiffness of high-strength welded I-beams and H-beams [10,14–18]. Additionally, some researchers [19,20] conducted finite element analyses on cold-formed stainless steel channel sections with web openings or circular web openings under end or internal single-flange loading conditions and proposed design guidelines. Recent studies on the fire resistance and impact behaviour of H-shaped steel members [21,22] have provided significant insights into how temperature variations influence the structural integrity of H-shaped steel and how geometric factors affect its dynamic response under lateral impact loading.

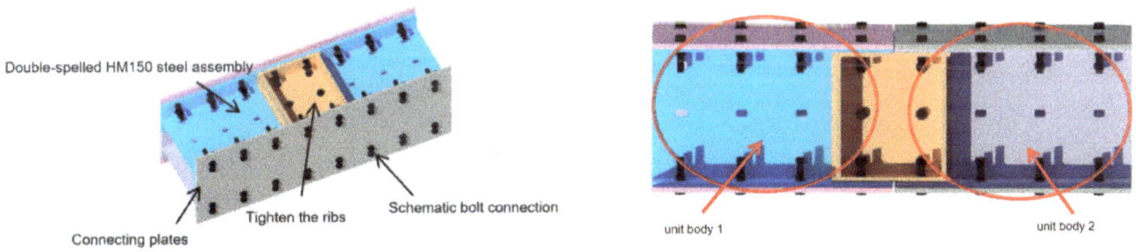

Figure 1. Construction and assembly joints of the prefabricated H-shaped steel main unit. Note: The colours in the diagram distinguish different assembly components.

Moreover, recent studies by Karalar et al. on the fatigue performance of H-shaped steel beams in integrated structural systems [23–26] provide a comprehensive analysis of the fatigue behaviour of H-shaped steel piles under various conditions. These findings offer an important theoretical foundation and practical guidance for the design and application of H-shaped steel piles. In addition, to enhance the application of H-shaped steel beams in frame structures, Xiao et al. conducted research on connection types and damage identification, expanding the theoretical understanding of the structural performance of H-shaped steel beams in terms of shear deformation, semi-rigid connections, and structural parameter identification [27–29]. This work contributes to a deeper understanding of how H-shaped steel behaves under different loading conditions, offering valuable recommendations for optimising design in civil engineering projects.

Previous research has primarily focused on the mechanical performance of H-shaped steel after perforation, with limited studies on the effects of hole size. Therefore, this experiment compared and analysed the flexural and flexural–shear bearing capacity, and the degree of flexural stiffness reduction in specimens with different bolt hole schemes, resulting in an optimised bolt hole design. This study focused on the influence of different bolt hole types on the failure modes, ultimate load-bearing capacity, load–deflection curves, and stress distribution of prefabricated H-beam components. Additionally, extensive parametric analyses were performed on prefabricated H-beam components with various bolt hole configurations using ANSYS 2024 finite element analysis software and validated by experimental results. The analysis verifies the optimised bolt hole design scheme. The findings of this research provide valuable insights into the bending design and application of prefabricated H-beam components with various bolt hole types, offering a useful reference for engineering applications.

2. Experimental Investigation

In the structural laboratory of Shandong Jianzhu University, five groups of three-point bending static tests were conducted on prefabricated H-beams with different bolt hole types. The failure modes, ultimate load-bearing capacity, load–deflection curves, and strain distribution of these steel beams were analysed. The results of the finite element simulations were also validated against the experimental results obtained from the physical tests.

2.1. Specimen Description

The main unit of the prefabricated H-shaped steel beam component consisted of a double-combined HM150 (148 × 100 × 6 × 9 mm) steel section and connecting plates. On both sides of the H-shaped steel component, 9 mm thick connecting plates were installed, which were attached to the H-shaped steel flanges using grade 8.8 M8 bolts, as shown in Figure 1. The direction of the force applied to the steel beam component is along the y-axis, and the Cartesian three-dimensional coordinate system referenced throughout this study adheres to the standard shown in Figure 2. The main unit of the prefabricated H-shaped steel beam component was categorised into different specimens based on the hole sizes of the connecting plates and double-combined HM150 steel sections.

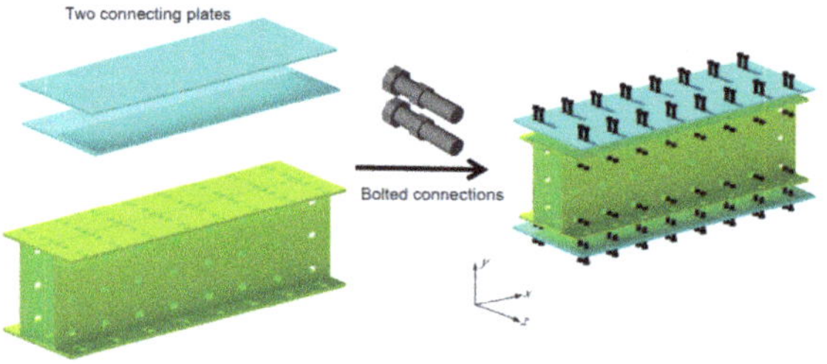

Figure 2. Diagram of the construction of the prefabricated H-shaped steel beam main unit and bolt connection. Note: Blue represents the connection plate, green represents the H-beam, and dark gray represents the bolts.

2.2. Experimental Specimens

The geometric parameters of the specimen cross-sections are shown in Figure 3. The height of the T-section formed by the bolt holes in the HM150 steel member is 130 mm, with bolt hole spacing of 40 mm along the height of the main unit of the prefabricated H-shaped steel beam component. The selected specimens consist of prefabricated H-shaped steel beam components with a length of 1.995 m. Along the length of the main unit, the edge distance of the bolt holes is 47.5 mm, and the spacing between the bolt holes is 100 mm. The connection plates are made of 9 mm thick steel, and the stiffening ribs are made of 6 mm thick steel, positioned 80 mm from both ends and at the mid-span of the component. Along the length of the main unit of the steel beam, the bolt hole edge distance is 47.5 mm, and the bolt hole spacing is 100 mm. Five specimens were fabricated with assembly joints located at the centre of the bolt holes in the plates. The specimen numbering scheme was based on the hole pattern in the connection plates and HM150 steel section. For example, in the designation "BL15 × 9-HC9", "B1" represents the connection plate, "L" indicates a long slot hole, and "9 × 15" specifies the size of the long slot hole. The "H" stands for HM150 steel, "C" denotes a circular hole, and "9" indicates the diameter of the standard circular hole. The construction of the specimens is illustrated in Figure 3, and the design parameters of the assembly joints for each specimen are listed in Table 1.

The specimen fabrication process is illustrated in Figure 4. Holes were created in the H-shaped steel and connection plates using laser cutting machines. After the cutting process, the rust on the steel surface around the cutting areas was removed and the surfaces were wiped with acetone to eliminate grease. The specimens were then assembled using high-strength bolts. To prevent variations in the bearing capacity of the specimens caused by discrepancies in the bolt pre-tightening force, a torque wrench was used during the installation process with a specified installation torque of 16 NM.

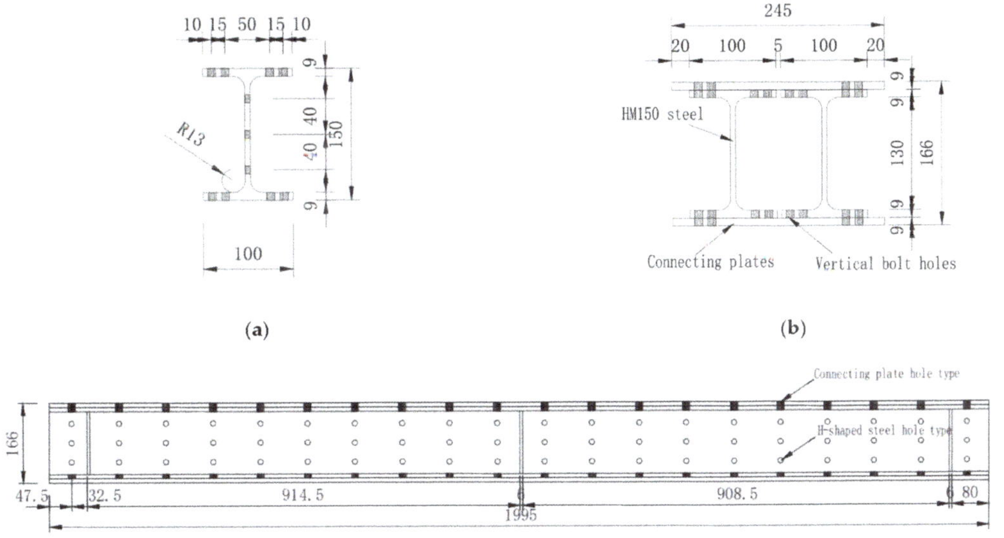

(a) (b) (c)

Figure 3. Schematic diagram of the cross-sectional parameters of the prefabricated H-shaped steel beam component specimen. (**a**) HM150 Steel Section; (**b**) Prefabricated H-shaped Steel Beam Component; (**c**) Side elevation view (using BC9-HC9 as an example).

Table 1. Design parameters of the specimen assembly joints.

Specimen Name	Bolt Hole Type of Connection Plate	Bolt Hole Type of HM150 Steel
BC9-HC9	9 mm Diameter Standard Circular Hole	9 mm Diameter Standard Circular Hole
BC9-HC11	9 mm Diameter Standard Circular Hole	11 mm Diameter Large Circular Hole
BC9-HL12 × 9	9 mm Diameter Standard Circular Hole	9 mm × 12 mm Short Slot Hole
BC9-HL15 × 9	9 mm Diameter Standard Circular Hole	9 mm × 15 mm Long Slot Hole
BL15 × 9-HL15 × 9	9 mm × 15 mm Long Slot Hole	9 mm × 15 mm Long Slot Hole

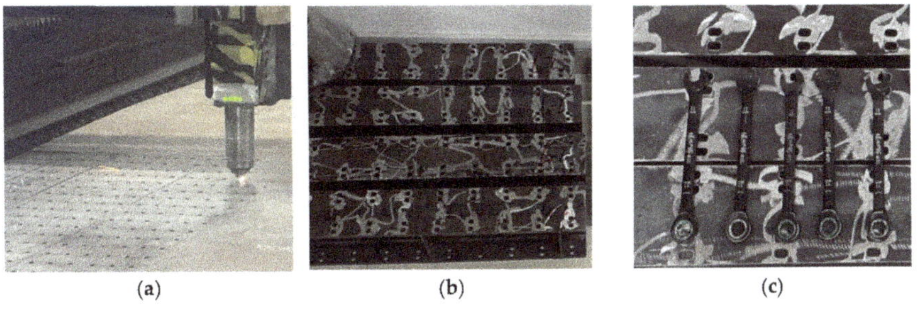

(a) (b) (c)

Figure 4. Specimen preparation. (**a**) Hole Drilling; (**b**) Grinding and Rust Removal; (**c**) Installation Using Torque Wrench.

2.3. Material Properties

Specimens were fabricated using Q235-grade cold-rolled steel plates. Tensile coupon tests were performed to determine the material properties of the specimens. According to the recommendations of GB/T 228.1 [30], material property samples were obtained from the connection plates, stiffening connection plates, and H-shaped steel sections. The selected samples and rolled materials were sourced from the same steel batch. Three samples were obtained from each location, resulting in three groups of nine samples with different material properties. The dimensions of the material property samples, testing apparatus, and test data are presented in Figure 5 and Table 2.

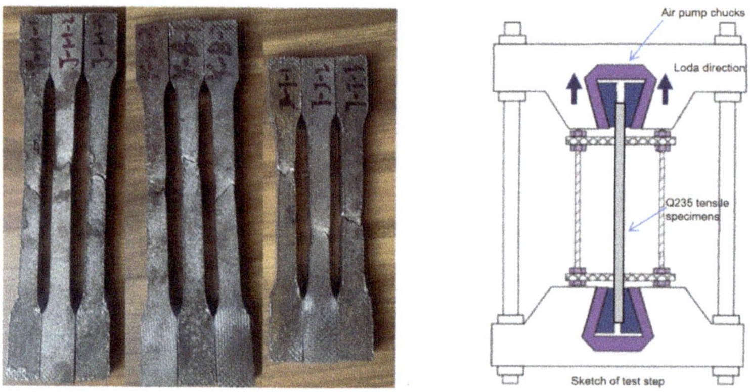

(**a**) (**b**)

Figure 5. Tensile test of material property specimens. (**a**) Tensile Specimen for Material Properties; (**b**) Schematic Diagram of the Tensile Test for Material Properties Specimens.

Table 2. Processing dimensions of material property specimens and material property test results (average value of three coupons).

Specimen ID	Sampling Location	Thickness /mm	Gauge Length (L0) /mm	Yield Strength /MPa	Ultimate Strength /MPa	Elastic Modulus /MPa	Elongation /%
H	H-shaped Steel	9	100	260	390	255	20.2
B	Connection Plate	9	100	255	390	248	23.7
L	Stiffening Rib	6	80	245	430	256	19.6

2.4. Testing Apparatus and Measurement System

A three-point bending test was conducted using a high-precision hydraulic jack to apply the load; both the testing procedures and data processing adhered to the GB/T 1041 standard [31], as illustrated in Figure 6. The specimen was supported by a hinge on the left side and a fixed support on the right side, with a displacement transducer positioned at the mid-span. At the beginning of the experiment, a uniform loading rate of 20 kN per stage was applied at the mid-span of the composite beam. During the tests, a high-precision hydraulic jack simultaneously recorded the load and relative displacement at the bottom of the composite steel beam. When the displacement at the mid-span increased at a significantly faster rate than the load, the loading rate was adjusted to 10 kN.

It is evident from summarising previous research [32] that the loading process was terminated upon the occurrence of any of the following phenomena: fracture of the high-strength bolts, tearing of the connected cover plates or end plates, steel components entering a state of noticeably large deformation, local buckling at the loading point of the H-shaped steel section, or any other unexpected occurrence. Seven transverse strain gauges were installed along the vertical height of the beam at the mid-span to measure the strain variations during the loading process. The data acquisition system automatically captured and recorded strain data throughout the entire experiment.

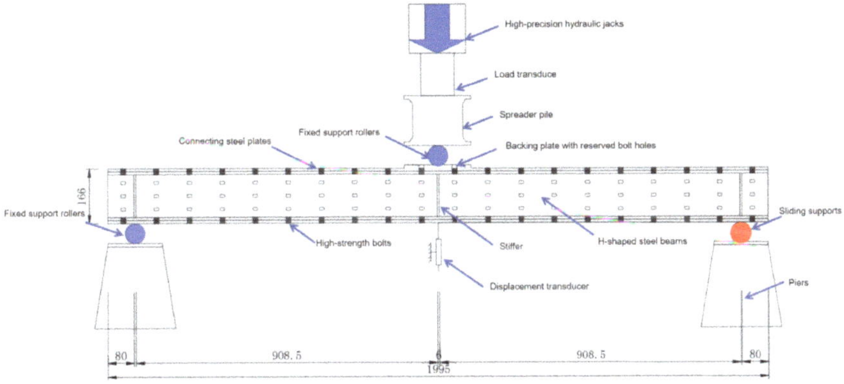

(**a**) Loading rig

(**b**) Strain gauge attachment position

(**c**) Right view of loading rig

(**d**) Full view of loading rig

Figure 6. Test loading rig.

2.5. Experimental Results

2.5.1. Experimental Observations

Table 3 summarises the specific failure modes and ultimate bending moments of the specimens. Taking specimen BC9-HL12 × 9 as an example, no abnormal phenomena were observed before the vertical load at the mid-span reached 240 kN. The real-time displacement–load relationship exhibited good linearity, indicating that the specimen remained in the elastic phase with no significant slip between the components. As the load increased to 250 kN, bolt slip sounds were heard, and slip between the H-shaped steel and connection plates began. And at this point, the mid-span displacement no longer maintains a linear relationship with the applied load. The slip in the upper section was slower than that in the lower section, likely because the tensile region at the bottom of the mid-span entered the plastic stage first, where the prying forces of the bolts constrained the relative slip, leading to a nonlinear slip response in the section.

Table 3. Experimental results analysis of specimens under three-point bending.

Specimen Name	Mp /(kN·m)	Reduction Factor of Ultimate Moment Capacity for the BC9-HC9 Component /%	Reduction Factor of Ultimate Moment Capacity for the BC9-HC9 Component /%	Failure Mode	Failure Stiffness /(kN·m²)
BC9-HC9	238.80	--	78.12	FF-B	152,460
BC9-HC11	199	16.67	76.5	FF-B	129,740
BC9-HL12 × 9	211.94	11.25	77.35	FF-B	136,658
BC9-HL15 × 9	175.12	26.67	65.43	FF-L	133,487
BL15 × 9-HL15 × 9	169.15	29.17	46.31	FF-L	182,171

As the load continued to increase to 420 kN, significant slip occurred at the bottom of the specimen, and the slip at the upper section increased to some extent. Noticeable plastic deformation occurred at the loading point, making further loading difficult, with even small increments in the load causing large vertical displacements. When the load reached 426 kN, the steel plate around the bolt hole on the lower flange of the mid-span H-shaped steel section fractured.

2.5.2. Failure Modes

The primary failure phenomenon observed during the three-point bending test was buckling deformation at the mid-span of the H-shaped steel section (Figure 7). Based on the condition of the lower flange at the point of failure, the failure modes were further categorised as flange plate failure (labelled as FF-L) and flange plate fracture (labelled as FF-B). This type of failure indicates that the maximum stress in the specimens occurred around the bolt holes, where the stress concentrations were induced by the bolt hole openings. The localised stress at these bolt holes exceeded the bending capacity of the H-shaped steel section, resulting in redistribution of the plastic stress. Eventually, the entire section reaches its bending strength, leading to specimen failure.

All cases of flange plate failure occurred in specimens with long slot holes of 9 × 15 mm in the H-shaped steel section, whereas specimens with shorter long slot holes and other circular hole configurations experienced flange plate fracture. This phenomenon can be attributed to the increasing mid-span deflection during loading, which caused a relative slip between the components on both sides of the mid-span (Figure 8). The bolts disrupted the load transfer mechanism, which relied solely on friction, resulting in a loss of pre-tensioning force. Ultimately, the long slot hole specimens failed to reach flange plate fracture.

(a) (b)

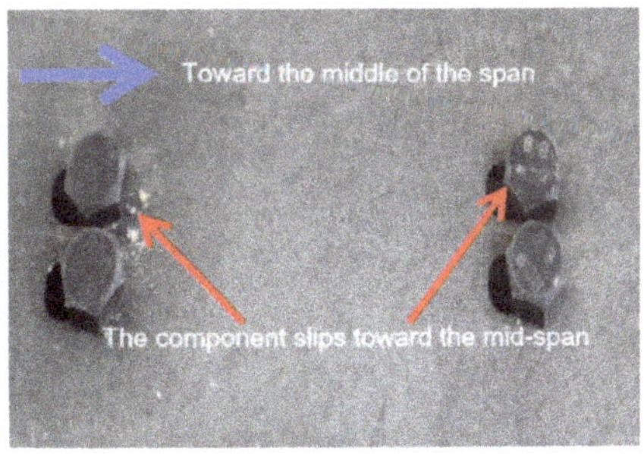

(c) (d)

Figure 7. Failure mode of specimens (FF-L); (**a**) BL15 × 9-HL15 × 9; (**b**) BC9-HL15 × 9; (**c**) H-beam lower flange necking failure; (**d**) relative slippage between components.

(**a**) BC9-HL12 × 9

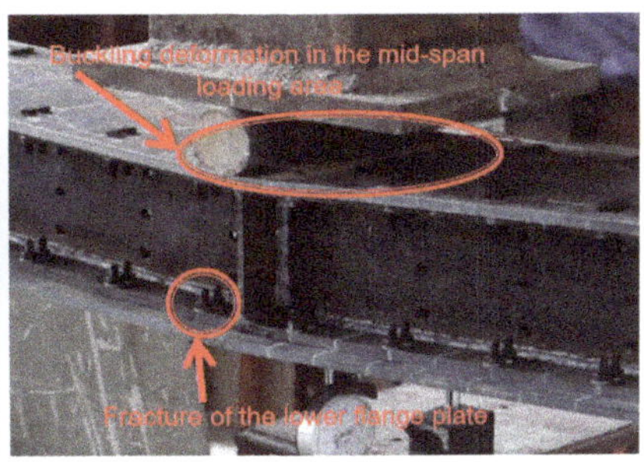

(**b**) BC9-HC11

Figure 8. *Cont.*

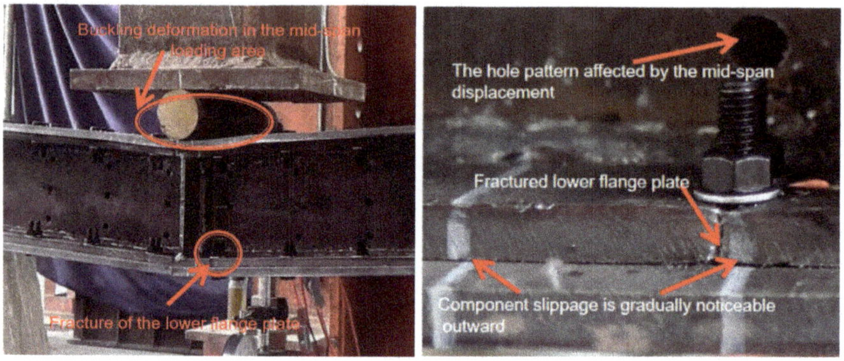

(**c**) BC9-HC9 (**d**) BC9-HC9

Figure 8. Failure mode of specimens (FF-B): (**a**) BC9-HL12 × 9; (**b**) BC9-HC11; (**c**) BC9-HC9; (**d**) BC9-HC9. (**d**) The lower flange of the H-beam is fractured and damaged.

3. Mechanical Behaviour of Prefabricated Double-H-Shaped Steel Components

3.1. Mid-Span Load–Deflection Relationship

The bending moment versus deflection curves for the prefabricated double H-shaped steel specimens at the mid-span are shown in Figure 9. Mid-span deflection data were obtained from displacement transducers installed at the mid-span during the tests. The overall deformation of the perforated specimens can be divided into three stages.

1. Stage 1: Linear increase in the moment–deflection curve: As the load was incrementally applied, the mid-span deflection of the specimens increased linearly with the load. This linear relationship characterises the initial behaviour of the moment–deflection curves of all specimens. During this stage, the incremental deflection at the mid-span was relatively small, indicating that a pre-tightening force of the bolts was applied during specimen preparation. The bolt heads and nuts compress the H-shaped steel flanges and connection plates, generating static friction, and the applied pre-tension increases the frictional resistance between the components, ensuring a tight connection between the structural elements, thus resisting relative slip under bending moments. At the end of this stage, the H-shaped steel components remained in the elastic phase, and no significant relative slip was observed on either side of the mid-span.
2. Stage 2: Nonlinear increase in the moment–deflection curve: In this stage, as the load continued to increase, the stress in the bolts could no longer accommodate the deformation, and the force distributed to the tensioned steel plates continuously increased. When the distributed force exceeded the shear force generated by the pre-tightening of the bolts, slippage occurred. The root cause is that the frictional resistance provided by the pre-tension is insufficient to counteract the applied load. At this point, most of the load was borne by the bolts on both sides of the mid-span. Because of the smaller moment of inertia of these bolts, the slope of the load–displacement curve decreased. Therefore, during the yielding phase of the specimens, the rate of the deflection increase at the mid-span was significantly higher than that of the load increase. This phase was also accompanied by the yielding and hardening of the double-H-shaped steel components. A distinct inflection point (Point A) can be observed in the moment–deflection curves of specimens FF-L and FF-B, as shown in Figure 9c. The elastic limit bending moment and deflection at this point were recorded as Me and Dy, respectively. During this stage, the bolts on both the upper and lower flanges began to slip toward

the mid-span, and the holes in the web of the H-shaped steel exhibited downward tensile deformation as the deflection increased. This indicated that the local plastic deformation at the bolt holes began to propagate from the flange surface to the web.

3. Stage 3: Ultimate stage of the specimens: After the yielding and hardening phases, the load reached the ultimate bending moment Mp, and the specimens completed their plastic deformation, with the mid-span displacement reaching the ultimate deflection Su. The specimens that experienced FF-B failure exhibited higher ultimate bending moments, Mp, and ultimate deflections, Su, than those that experienced FF-L failure. This demonstrates the influence of the bolt hole geometry on the bending capacity and ductility of double-H-shaped steel components. Furthermore, differences in the bending capacity and mid-span deflection capability were observed among the specimens with similar failure modes but different configuration parameters, as shown in Figure 9a–c. These differences are discussed in detail below.

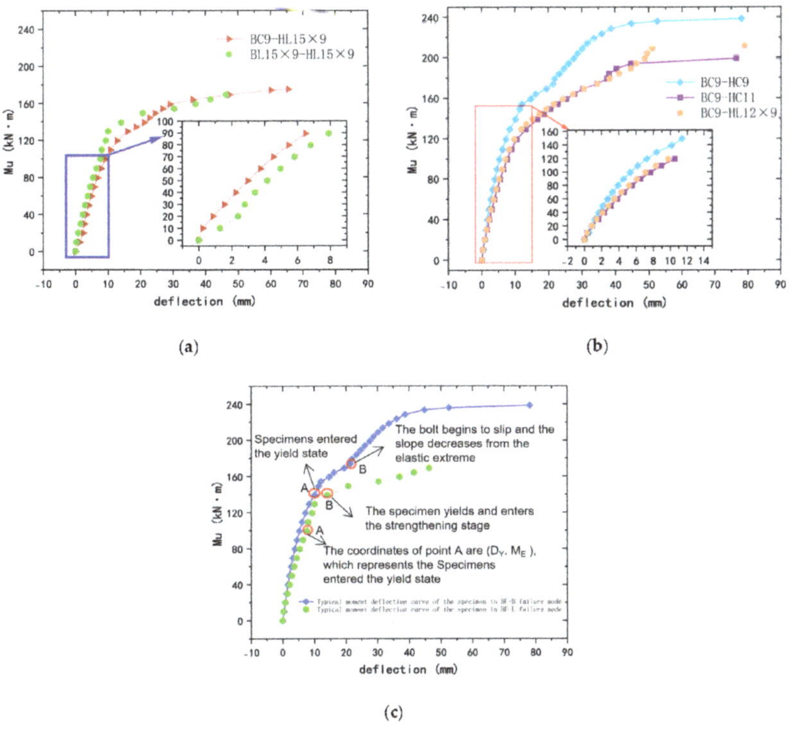

Figure 9. Mid-span moment–deflection curves of the specimens. Note: (**a**) shows the curve for the FF-l failure mode; (**b**) shows the curve for the FF-B failure mode; (**c**) presents the comparative curves for both failure modes.

3.2. Bending Performance

As shown in Table 3, the ultimate bending moment of the specimens with long slot holes and large circular holes was 11–30% lower than that of the standard circular hole specimen (BC9-HC9). The analysis results demonstrate that larger bolt holes in the main unit of the double H-shaped steel components led to a reduction in the bending capacity of the specimens. The specimen with standard circular holes in the connection plates (BC9-HC9) exhibited a significantly higher failure moment than the specimen with long slot holes in both the connection plates and HM150 steel under the same design parameters.

Comparing specimens BL15 × 9-HL15 × 9, BC9-HL15 × 9, BC9-HC11, and BC9-HL12 × 9, it can be observed that specimens with standard circular holes in the connection plates have higher bending capacity and stiffness than those with long slot holes in the connection plates under the same parameters. Furthermore, a comparison between BC9-HL12 × 9 and BC9-HC11 revealed that specimens with long slot holes in the double H-shaped steel exhibited greater mid-span deflection than those with circular holes in the H-shaped steel, despite having a bending capacity difference of less than 5%.

Based on the experimental results, the elastic limit bending moment Me of the specimens was determined (because the load was applied incrementally, an exact calculation of the elastic limit bending moment was not possible; therefore, the last applied load before the specimen entered the yielding phase was used for the calculation). The corresponding mid-span node displacement in the y-direction was measured for each specimen. Combining these results with the calculation of the mid-span deflection using the principle of superposition for pure bending deformation in the mechanics of materials, the simplified equivalent bending stiffness can be expressed as shown in Equation (1).

$$\omega = \frac{M_z l}{4EI} \Rightarrow EI = \frac{M_z l}{4\omega} \tag{1}$$

In the equation, *Mz* represents the bending moment of the specimen in the elastic phase (N·mm); I represents the moment of inertia of the specimen in the direction of the applied force (mm4); l represents the span length of the specimen, taken as 1995 mm; ω represents the y-direction displacement at the mid-span node of the specimen (mm). The equivalent bending stiffness for the remaining specimens can be found in Table 4.

Table 4. Bending stiffness of three-point bending specimens.

Specimen Name	Bending Moment in the Elastic Phase /(kN·m)	The Maximum y-Direction Displacement Value of the Bending Moment in the Elastic Phase /mm	Flexural Stiffness /(kN·m^2)
BC9-HC9	149.25	11.4	65,297
BC9-HC11	119.4	10.64	56,180
BC9-HL12 × 9	119.4	9.36	63,352
BC9-HL15 × 9	89.55	5.70	78,356
BL15 × 9-HL15 × 9	89.55	6.05	73,823

Table 3 reveals that the failure stiffness of the specimens decreases as the opening size in the H-shaped steel increases, particularly when the connection plate has standard circular holes. When the opening area in the H-shaped steel exceeds the 12 × 9 threshold, the failure mode shifts from local failure to overall failure. However, when the hole size in the connection plates also increases, the reduction in load-bearing capacity is more significant than the impact on failure stiffness, preventing a change in the failure mode.

As listed in Table 4, during the elastic deformation phase of the specimens, the bending stiffness was greater for specimens with standard circular holes in the connection plates, given the same elastic bending moment and consistent web hole parameters in the H-shaped steel. The analysis suggests that a reduced bolt hole diameter in the connection plates increases the contact area between the bolts and connection plates. This allows the bolt pre-tightening force to more effectively limit relative slip between the components, thereby better controlling the overall bending deformation of the specimen. For specimen BC9-HL12 × 9, which exhibits greater bending stiffness than specimen BC9-HC11, the analysis indicates that the larger 11 mm circular holes create a gap between the bolts and the connection plates at various angles, leading to a less effective constraint compared to short slot holes, which do not adequately limit displacement in the x-direction.

In summary, variations in bolt hole configurations can lead to a reduction in the bending capacity and stiffness of the specimens to some extent. Among the specimens with bolted holes, those with standard circular holes in both the connection plates and H-shaped steel exhibited the highest bending capacity and stiffness. For the remaining specimens, considering construction errors on site and ease of installation, the BC9-HL12 × 9 configuration offers a certain cost-effectiveness.

3.3. Strain Analysis

Figure 10 shows the strain measurements for the upper flange, lower flange, and one side of the web for H-shaped steel beams with long slot holes, short slot holes, and standard circular holes during the three-point bending test. The strain measurements were performed under both buckling and peak loads. Tensile strain was defined as positive strain, whereas compressive strain was defined as negative strain. The figure shows that as the load increased from the buckling load to the peak load, the strain distribution shifted from symmetric to asymmetric owing to the varying hole shapes on the left side of the mid-span. For the specimens with long slot holes, the tensile strain at the yield load was similar to that at the peak load. In contrast, for specimens with short slot holes and standard circular holes, there was a significant difference between the tensile strains at the yield and peak loads. The analysis suggests that in the long slot hole specimens, the bolts lose their pre-tightening force owing to slippage during the failure stage, resulting in the tensile yield stress approaching the ultimate tensile stress of the lower flange and steel plate. Meanwhile, the bolts in the short slot holes and standard circular holes provided a shear force to the lower flange of the steel beam, leading to fracture at the lower flange and the formation of a peak curve.

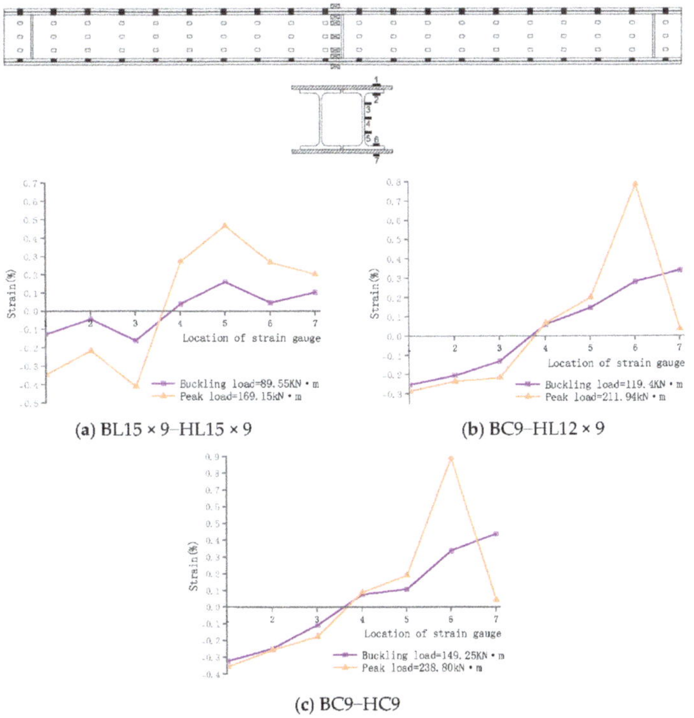

Figure 10. Strain distribution curves. Note: Positions 1–7 correspond to the strain gauge numbering, matching the strain curve numbers in the figure.

For a detailed analysis of this behaviour, Figure 11 illustrates the mechanism of the long slot hole specimens. As the mid-span load increases, the bolt pre-tightening force begins to release along a diagonal direction. When the vertical displacement caused by buckling at the mid-span becomes sufficiently large, the horizontal component of the bolt pre-tightening force cannot resist sliding along the beam length. Consequently, the bolts on both sides of the mid-span begin to slip toward the mid-span. This leads to a gradual expansion of the local plastic deformation zone around the bolt holes in the z-direction, with plastic deformation spreading from the flange surface toward the web.

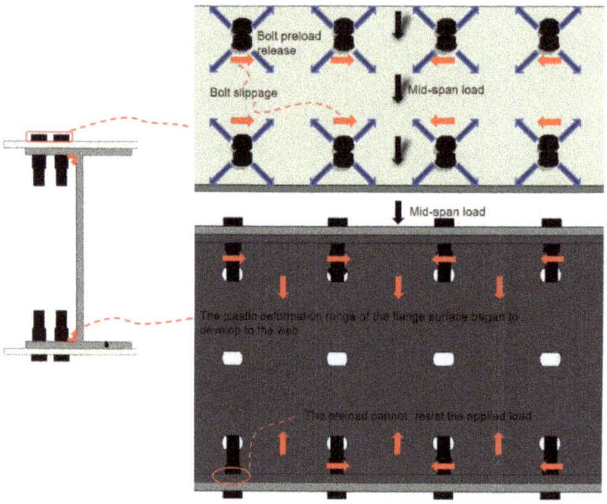

Figure 11. Transfer of forces from mid-span loads in prefabricated H-beams. Note: The red arrows represent the stress transfer path, while the black arrows indicate the application of external loads.

Additionally, the distribution of the compressive strain in the upper flange and tensile strain in the lower flange is largely consistent. Because of the introduction of holes in the web, the strain at the neutral axis of the H-shaped steel is not zero, which violates the assumption of a plane section. Moreover, as the size of the holes in the web increases, the nonlinearity of the strain distribution along the web becomes more pronounced.

4. Finite Element Model Validation and Parametric Study

4.1. Finite Element Model Development

To further investigate the impact of hole configurations on the mechanical performance of prefabricated H-shaped steel beam components, the ANSYS finite element analysis software was used to develop models for five specimens with different hole configurations. These models were used to simulate and analyse the buckling behaviour and bending capacity of the experimental specimens. The double-combined HM150 steel, connection plates, and bolts were modelled using Solid Brick 10-node 185 solid elements [33–35]. It is able to more accurately simulate material behaviour and structural responses, particularly capturing local effects such as stress concentrations and deformation patterns under complex geometries or loading conditions. Considering the balance between computational accuracy and efficiency, mapped meshing was applied uniformly across all the specimens. The grid size was set to 10 × 10 mm for the double-combined HM150 steel and connection plate elements, and the M8 bolt elements were divided into 5 mm mesh. To ensure accuracy, a mesh refinement was applied around the bolt holes and chamfered areas. The boundary conditions for the simply supported beam model constrained the translational displacement in the X, Y, and Z directions at one end of the model, and the translational displacement in the Y and Z directions at the other end; CONTA174 contact elements were used for the contact surfaces, and the target surfaces were modelled using TARGE170 target elements [36]. The contact stiffness, penetration tolerance, and friction coefficient of the contact surfaces were set as 0.1, 0.01, and 0.35, respectively.

The material behaviour was modelled using a multilinear kinematic hardening (MKIN) elastoplastic model and the von Mises yield criterion was used to predict the yielding of the components (unless otherwise specified, all subsequent references to stress refer to the von Mises stress). The size specifications and mechanical properties of the double-combined HM150 steel were selected according to the Design Code for Steel Structures [37]. The constitutive model for the Q235B steel was defined using a trilinear elastoplastic hardening model, as shown in Figure 12a. The 8.8-grade M8 high-strength bolt shanks and nuts were modelled as equivalent diameter cylinders, with the equivalent diameter selected according to High-Strength Large Hexagonal Head Bolts for Steel Structures [38]. The constitutive relationship of the bolts was defined using a bilinear elastoplastic hardening model, as shown in Figure 12b. The elastic modulus E for both the steel and bolts was set to 2.06 × 105 N/mm^2, with a Poisson's ratio of 0.3.

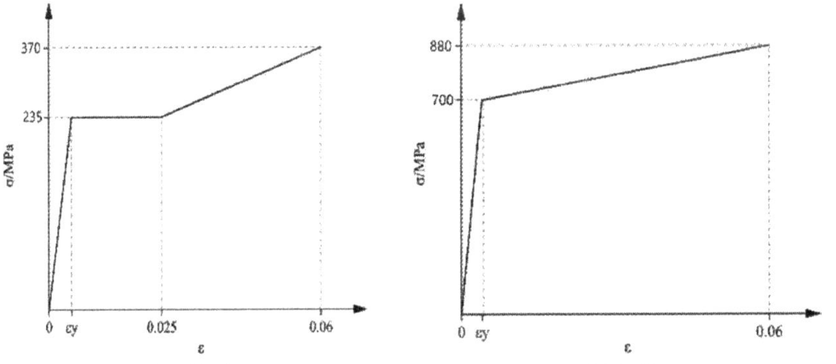

(**a**) Steel (**b**) bolt

Figure 12. Constitutive relationship diagram of the computational model.

4.2. Loading Conditions and Failure Modes

The loading conditions were consistent with those of earlier experiments. A pre-tightening force of 16 NM was applied to the 8.8-grade M8 high-strength bolts. To simulate the loading mode of the specimens, bending moments were applied in the XY plane to the nodes on the end faces of the coupled specimens and modelled using MASS21 mass elements. The numerical analysis was performed using step-by-step loading with automatic time increment settings for each subload step. In the first analysis step, the boundary constraints and initial bolt pre-tightening forces were applied. In the second step, a full pre-tightening force was applied to the bolts. In the third step, the pre-tightening force was locked in, and a bending moment was applied until significant deformation occurred in the model. A specimen was considered to have failed if the results no longer converged.

The results of the ultimate bending moment and failure modes obtained using ANSYS are listed in Tables 5 and 6, respectively. The ratio of the ultimate bending moment of the prefabricated H-shaped steel beam components to the mean value was 1.038, with a corresponding standard deviation (St.Dev) of 0.036. The maximum deviation between the ANSYS-predicted and experimental ultimate bending moments was 8.3%. The maximum y-direction displacement at the ultimate bending moment is 4.1 mm.

The failure modes obtained from ANSYS and the experimental results were compared using specimens BC9-HC11 and BL15 × 9-HL15 × 9 as examples. The results are shown in Figure 13a,c. The comparison indicates that the bending capacity and failure modes predicted by ANSYS closely matched the experimental results, validating the ANSYS model developed in this study. Therefore, the validated ANSYS model could be used for further parametric analyses of the specimens.

Table 5. Comparative analysis of ANSYS finite element simulation results.

Specimen	Test Results		ANSYS Results		Mp/Ma
	Mp/(kN·m)	Failure Mode	Ma/(kN·m)	Failure Mode	
BC9-HC9	238.80	FF-B	218.9	FF-B	1.091
BC9-HC11	199	FF-B	199	FF-B	1.000
BC9-HL12 × 9	211.94	FF-B	199	FF-B	1.065
BC9-HL15 × 9	175.12	FF-L	169.15	FF-L	1.035
BL15 × 9-HL15 × 9	169.15	FF-L	169.15	FF-L	1
Mean					1.038
St.Dev					0.036

Note: Mp represents the ultimate moment obtained from experiments, while Ma corresponds to the ultimate moment derived from ANSYS simulations. FF-L indicates failure due to flange buckling control, and FF-B represents the fracture of the lower flange plate governed by buckling control.

Table 6. Analysis of ANSYS finite element simulation results.

Specimen Name	Comparison of Deviations from the Respective Experimental Ultimate Moments /%	Reduction Factor of Ultimate Moment Capacity for the BC9-HC9 Component /%	Maximum y-Direction Displacement under Ultimate Moment Capacity /mm	Interpolation of the Maximum y-Direction Displacement for Each Respective Experiment /mm
BC9-HC9	8.3	--	79.13	+1.01
BC9-HC11	--	9.1	78.79	−0.04
BC9-HL12 × 9	6.1	9.1	79.01	+2.29
BC9-HL15 × 9	3.4	22.7	66.4	+0.97
BL15 × 9-HL15 × 9	--	22.7	46.31	+4.1

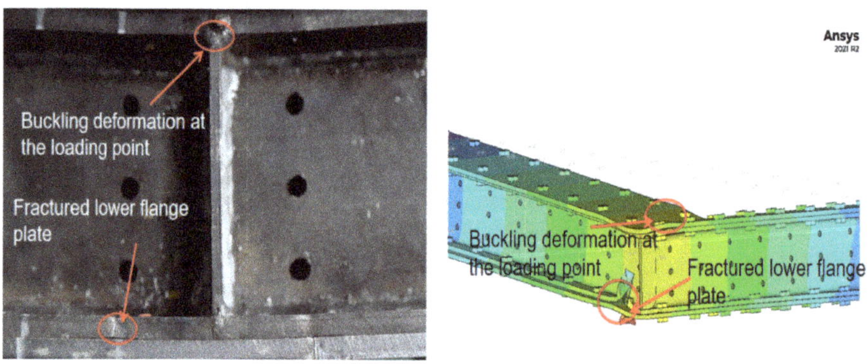

(**a**) BC9-HC11

Figure 13. *Cont.*

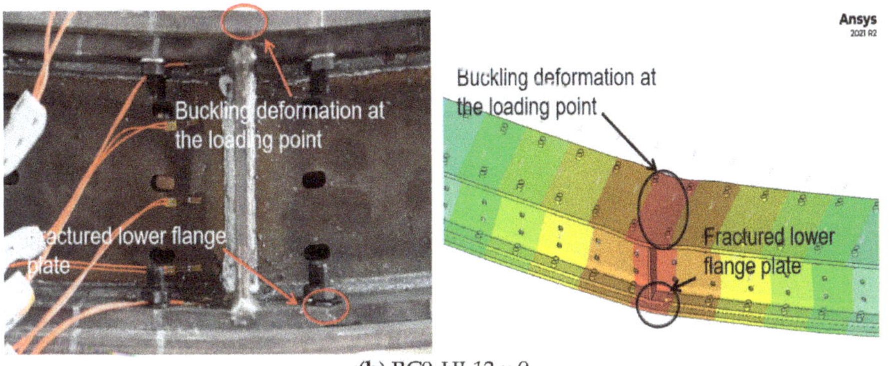

(**b**) BC9-HL12 × 9

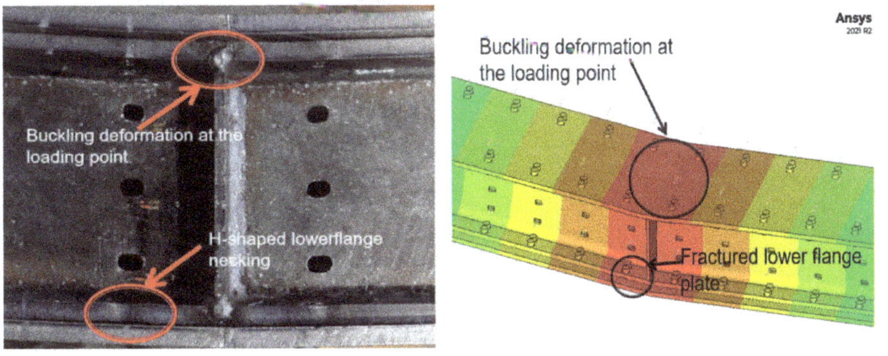

(**c**) BL15 × 9-HL15 × 9

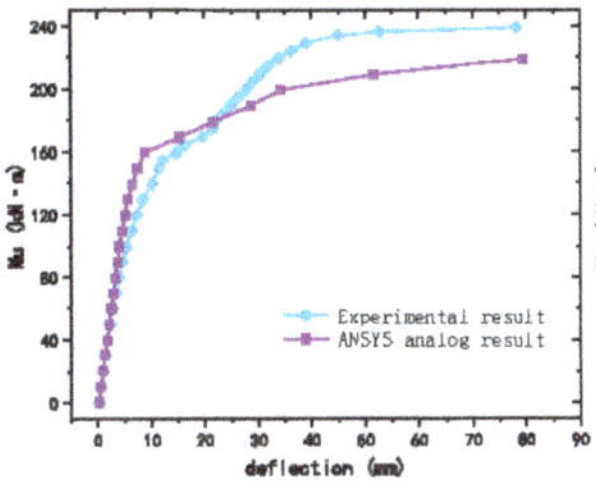

BC9-HC9

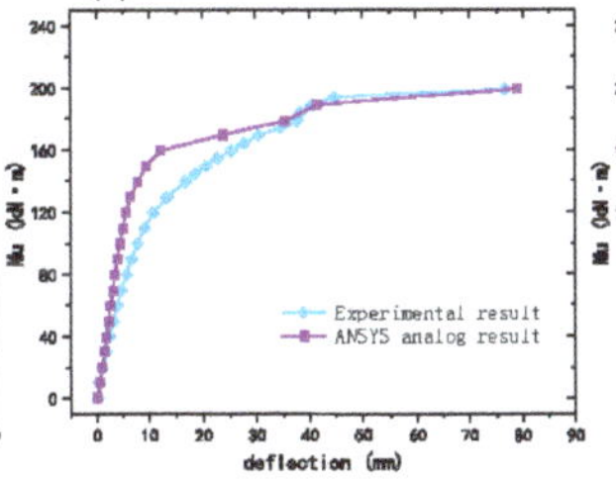

BC9-HC11

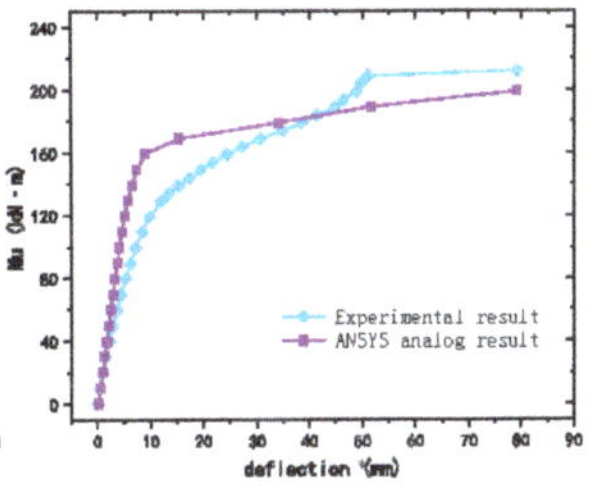

BC9-HC12× 9

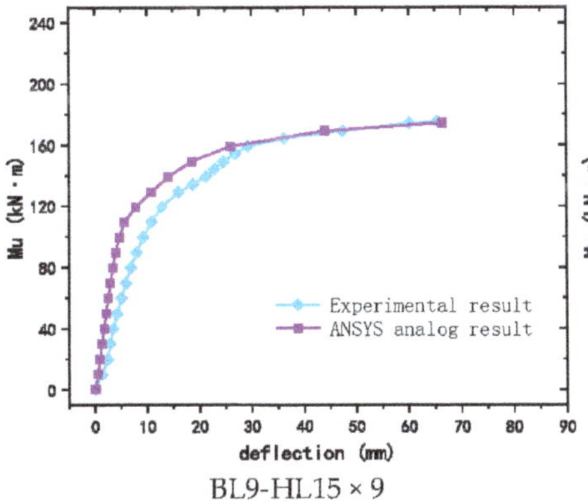

BL9-HL15 × 9

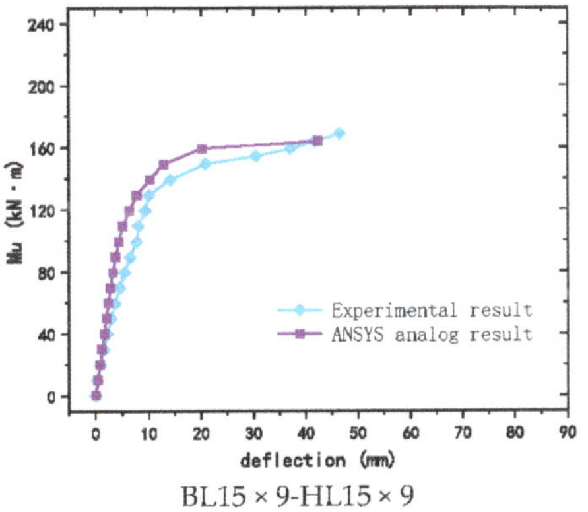

BL15 × 9-HL15 × 9

Figure 13. Comparison of failure modes between experimental results and ANSYS results.

4.3. Analysis of Bolt Pre-Tightening Force

The pre-tightening force of the bolts significantly affects the stability of the connections between components [39]. The preload of the bolts ensures that the structural components remain tightly clamped, providing sufficient frictional resistance to prevent relative movement under applied loads. After the pre-tightening force was applied, the compression of the components under bending moments gradually increased, leading to varying degrees of bolt loosening at different locations. At this point, the compressive force between the components was reduced from the full pre-tightening force to the residual pre-tightening force (F1/kN). The residual pre-tightening forces at the ultimate bending moment of the specimens are shown in Figure 11. UI, UO, LI, and LO represent the average residual pre-tightening forces for the bolts on the inner side of the upper flange, the outer side of the upper flange, the inner side of the lower flange, and the outer side of the lower flange, respectively.

From Figure 14, it is evident that the residual pre-tightening force of the upper flange bolts was significantly greater than that of the lower flange bolts. The loss in the pre-tightening force of the upper flange bolts ranged from approximately 5% to 33%. The compressive stress generated in the upper flange under the bending moment caused the components at the bolt connections to be in a bidirectional compression state. This resulted in only a slight elongation of the bolts owing to the Poisson effect, and ensured that the connection between the components remained highly reliable and tight under the residual pre-tightening force. In contrast, the lower flange bolts experienced a pre-tightening force loss of approximately 65% to 97%. The tensile stress in the lower flange under the bending moment placed the components at the bolt connections in a state of compression in one direction and tension in the other. The compression of the components under load was significant, and the reduction in the bolt length did not compensate for this compression, resulting in bolt loosening.

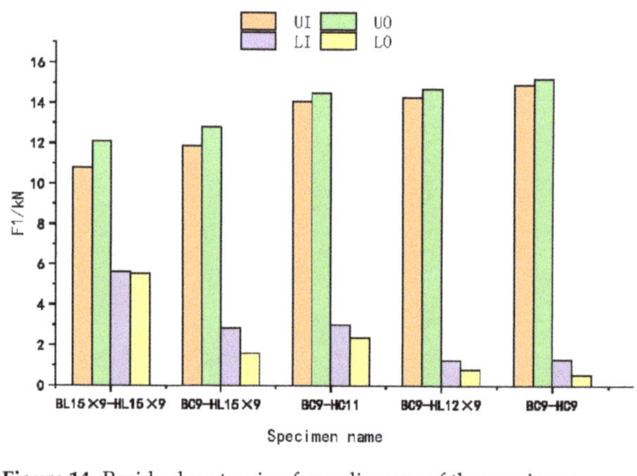

Figure 14. Residual pretension force diagram of the specimen.

The residual pre-tightening force of the bolts is linearly related to the ultimate bending moment of the specimens. From the comparison of single-variable specimens BC9-HC9 and BC9-HC11, it can be seen that when the bolt holes in the specimen are circular, the ultimate bending moment of the specimen increases as the hole radius decreases, while the residual pre-tightening force of the bolts decreases. However, for specimen BC9-HL12 × 9, the residual pre-tightening force loss under a higher bending moment is significantly lower than that for single-variable specimen BC9-HL15 × 9. This indicates that when the connection plate has standard circular holes, the bolt connections exhibit higher reliability than those with long slot holes. Therefore, reducing the hole size or increasing the contact area between the connection plate, H-shaped steel, and bolts can reduce the loss of the bolt pre-tightening force and enhance the reliability of the bolt connection.

4.4. Bolt Stress Analysis

For specimen BL15 × 9-HL15 × 9, where both the connection plate and H-shaped steel had long slot holes, the bolt head and nut were only in contact with the component at the long edges of the slot holes. During the application of the pre-tightening force, a very small area at the contact point between the bolt shank and the component reached the yield stress. At this stage, the stress distributions of the bolts in the upper and lower flanges are consistent, as shown in Figure 15a,g, respectively. After the load was applied and the specimen entered the elastic phase, the components relied on the static friction provided by the bolt pre-tightening force to coordinate the forces between them. During this phase, the deformation of the specimen was minimal and the loss of the bolt pre-tightening force was insignificant. The stress distributions in the bolts in the upper and lower flanges remained essentially the same, as shown in Figure 15b.

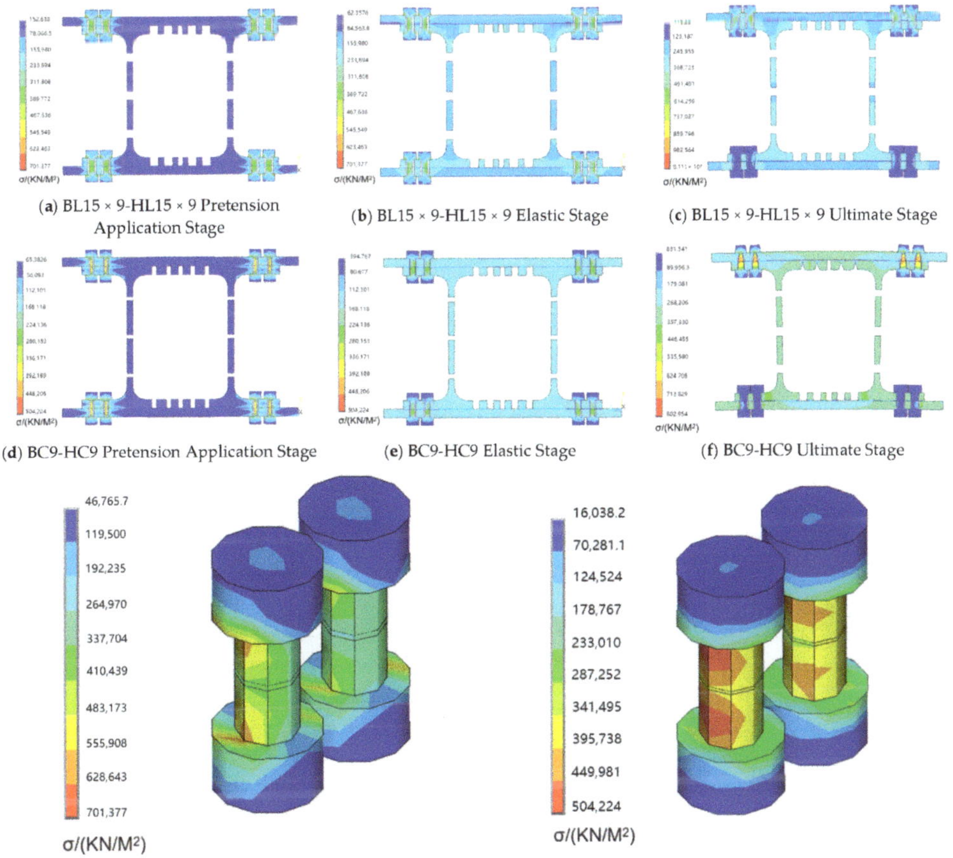

(**a**) BL15 × 9-HL15 × 9 Pretension Application Stage (**b**) BL15 × 9-HL15 × 9 Elastic Stage (**c**) BL15 × 9-HL15 × 9 Ultimate Stage

(**d**) BC9-HC9 Pretension Application Stage (**e**) BC9-HC9 Elastic Stage (**f**) BC9-HC9 Ultimate Stage

(**g**) Detailed Diagram of the Pretension Application Stage for BL15 × 9-HL15 × 9 (**h**) Detailed Diagram of the Pretension Application Stage for BC9-HC9

Figure 15. Cloud diagram of finite element analysis results for the bolt.

When the elastic limit bending moment was reached, the specimen entered the yield phase. The compression of the connection plate and the H-shaped steel flange in the compressed region under the bending moment gradually increased. The residual pre-tightening force of the bolts in the compressed region of the upper flange decreased slightly, and the corresponding bolt stress decreased. Simultaneously, the compression of the components in the tensile region began to increase rapidly. As the residual pre-tightening force of the lower flange bolts decreased sharply, the corresponding bolt stress also decreased. Upon reaching the ultimate bending moment, the stress in the lower flange bolts was significantly lower than that in the upper flange bolts, as shown in Figure 15c.

For specimen BC9-HC9, which featured standard circular bolt holes in the connection plates, the contact area between the bolt head and connection plate was larger than that in the long slot hole specimens. During the application of the pre-tightening force, the contact area between the bolt head and the connection plate did not reach the yield stress. However, in the H-shaped steel with long slot holes, a very small area at the contact point between the bolt shank end and component still reached the yield stress, as shown in Figure 15d,h.

After the load was applied, the stress distribution in the bolts followed a pattern similar to that of BL15 × 9-HL15 × 9. However, at the ultimate load stage, buckling deformation in the upper region of BC9-HC9 became more pronounced. The deformation of the standard circular holes caused additional compression on the upper flange bolts, increasing the stress near the bolts, as shown in Figure 15e,f. Because the loss in the pre-tightening force for BC9-HC9 was smaller than that for BL15 × 9-HL15 × 9, the bolt stress under the ultimate bending moment for BC9-HC9 was significantly higher than that for BL15 × 9-HL15 × 9.

4.5. Stress Analysis of the Connection Plates

Under the bending moment generated by the concentrated load, the bolts in the tension zone of the specimens loosened, leading to a more complex stress distribution. However, for the same hole type, the stress distributions in the connection plates of the lower flange were relatively consistent. Therefore, the stress diagrams of the BL15 × 9-HL15 × 9 and BC9-HC9 connection plates at different deformation stages are presented in Figure 11.

Figure 16a,b shows that during the pre-tightening force application stage, the range of influence of the pre-tightening force on the outer bolts of the flange was smaller than that on the inner bolts because of the width limitations of the H-shaped steel flange. Because the hole type BL15 × 9-HL15 × 9 is a long slot, the bolt heads and nuts only make contact with the components at the long edges of the slot, resulting in a slightly more radial distribution of the pre-tightening force influence compared to BC9-HC9.

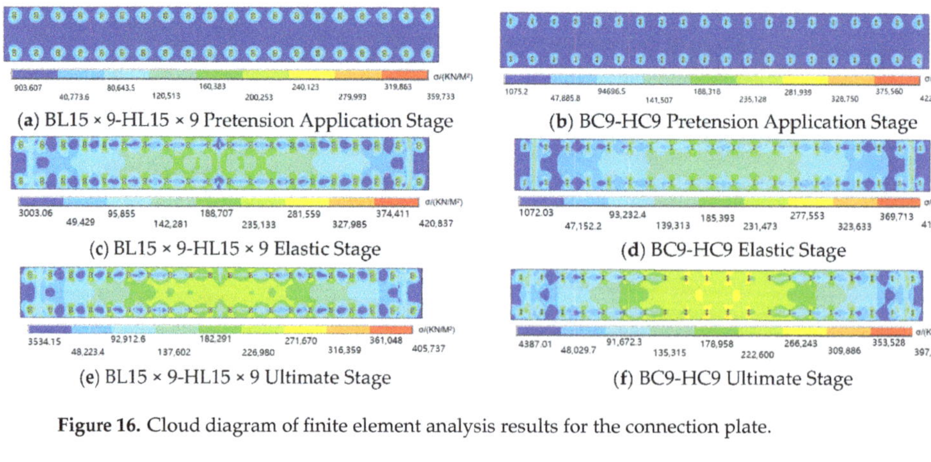

(a) BL15 × 9-HL15 × 9 Pretension Application Stage; (b) BC9-HC9 Pretension Application Stage; (c) BL15 × 9-HL15 × 9 Elastic Stage; (d) BC9-HC9 Elastic Stage; (e) BL15 × 9-HL15 × 9 Ultimate Stage; (f) BC9-HC9 Ultimate Stage

Figure 16. Cloud diagram of finite element analysis results for the connection plate.

Figure 16c,d shows that after the specimen entered the elastic phase, the connection plates began to exhibit diagonal high-stress bands along the bolt holes. Once these high-stress areas reach the yield stress, stress redistribution occurs, and the high-stress bands gradually expand outwards. Additionally, in BL15 × 9-HL15 × 9, compared to BC9-HC9, a distinct horizontal low-stress band appeared between the bolt holes along the x-axis direction. This can be attributed to the fact that in the elastic phase, the standard circular holes in the connection plates can maintain a higher level of residual bolt pre-tightening force compared to long slot holes, which helps coordinate the joint force between the connection plates and the H-shaped steel. However, as the mid-span deflection increased, the pre-tightening force loss in the tension zone bolts on both sides increased, preventing the connection plate and H-shaped steel from jointly resisting the tensile force generated by the bending moment. Consequently, the stress phenomenon in regions far from the mid-span became similar for both types of specimens.

Therefore, as shown in Figure 16e,f, after both specimens reached their ultimate stage, the diagonal high-stress bands formed during the elastic phase almost fully developed into full-section yielding near the mid-span. Because the bolts farther from the mid-span were still in the frictional force transfer phase in the ultimate state, horizontal low-stress bands between the bolt holes along the x-axis still existed. Moreover, it can be seen that in both hole types, ultra-high-stress bands are formed in the z-axis direction between the bolt holes at structurally weak points. Thus, under the ultimate bending moment, there was a risk of fracture at these locations.

5. Conclusions

A three-point bending test was conducted to study the effects of different bolt hole types on prefabricated H-shaped steel beams. The results indicate the following:

1. The variation in hole types between the H-shaped steel and connection plates leads to different failure modes, specifically mid-span flange buckling failure and lower flange fracture. The reduction in load-bearing capacity is closely related to the size and location of the holes.
2. The experimental results indicate that the ultimate bending moment capacity of specimens with long slotted holes and large circular holes decreased by 11–30% compared to specimens with standard circular holes. Specimens with standard circular holes demonstrated higher bending capacity and stiffness. Although the difference in bending capacity between double H-shaped steel specimens with long slotted holes and those with circular holes was less than 5%, the former exhibited greater mid-span deflection. Considering construction errors and convenience, the BC9-HL12×9 H-shaped steel beam is a reasonable choice.
3. Prefabricated H-shaped steel beams do not satisfy the plane section assumption. As the area of the web holes increases, the strain distribution in the web becomes more nonlinear. In the long slot hole specimens, the compressive strain distribution in the upper flange and the tensile strain distribution in the lower flange were largely consistent.
4. The ANSYS model was validated for failure modes and bending capacity, followed by a parametric analysis of the bolts and connection plates. The results indicated that the bolt pre-tightening force loss in the compression zone ranged from 5% to 33%, while in the tension zone, it ranged from 65% to 97%, with noticeable bolt loosening in the tension zone. The residual pre-tightening force in the specimens with standard circular holes was higher than that in those with long slotted holes under the same bending moment. Additionally, the connection plates between the z-direction bolt

holes in the lower flange of the H-shaped steel were structurally weak, posing a risk of tensile fracture.

Author Contributions: Conceptualisation, X.Z. and D.F.; methodology, X.Z., S.Y., and S.F.; software, D.F. and S.Y.; data curation, S.Y.; data recording, F.Z. and H.C.; preparation of the experiments, F.Z. and H.C.; writing—original draft preparation, S.Y.; writing—review and editing, X.Z. and S.F.; experiment, S.Y., D.F., F.Z., and H.C.; funding acquisition, X.Z. All authors have read and agreed to the published version of the manuscript.

Funding: This research is supported by the Major Scientific & Technological Innovation Projects of Shandong Province (No. 2021CXGC011204). The support is gratefully acknowledged.

Data Availability Statement: Data are not publicly available. The data may be made available upon request from the corresponding author.

Conflicts of Interest: Author Dawei Fan was employed by the company Qihang (Shandong) Investment and Construction Group Co., Ltd. The remaining authors declare that the research was conducted in the absence of any commercial or financial relationships that could be construed as a potential conflict of interest.

References

1. Huang, Z.; Zhou, H.; Tang, H.; Zhao, Y.; Lin, B. Carbon emissions of prefabricated steel structure components: A case study in China. *J. Clean. Prod.* **2023**, *406*, 137047. [CrossRef]
2. Tavares, V.; Soares, N.; Raposo, N.; Marques, P.; Freire, F. Prefabricated versus conventional construction: Comparing life-cycle impacts of alternative structural materials. *J. Build. Eng.* **2021**, *41*, 102705. [CrossRef]
3. Paudel, P.; Dulal, S.; Bhandari, M.; Tomar, A. Study on Pre-fabricated Modular and Steel Structures. *Int. J. Civ. Eng. (SSRG-IJCE)* **2016**, *3*, 7–14.
4. Chen, Z.; Khan, K.; Khan, A.; Javed, K.; Liu, J. Exploration of the multidirectional stability and response of prefabricated volumetric modular steel structures. *J. Constr. Steel Res.* **2021**, *184*, 106826. [CrossRef]
5. Zhou, J.; Li, Y.; Ren, D. Quantitative study on external benefits of prefabricated buildings: From perspectives of economy, environment, and society. *Sustain. Cities Soc.* **2022**, *86*, 104132. [CrossRef]
6. Shandong Provincial Highway and Bridge Construction Co., Ltd. Accessories for Enclosure System CN 202120376383.8, 2 November 2021.
7. American Society of Civil Engineers (ASCE), Specification for the Design of Coldformed Stainless Steel Structural Members, SEI/ASCE 8-02, Reston, Va. 2002. Available online: https://webstore.ansi.org/preview-pages/ASCE/preview_9780784405567.png?srsltid=AfmBOoqbuZnHS2jICrnKBIwRizA3uNkITxglkWlmyLd2veVLhyGw3U3S (accessed on 12 June 2024).
8. Swanson, J.A.; Leon, R.T. Bolted steel connections: Tests on T-stub components. *J. Struct. Eng.* **2000**, *126*, 50–56. [CrossRef]
9. Teh, L.H.; Clements, D.D.A. Block shear capacity of bolted connections in cold-reduced steel sheets. *J. Struct. Eng.* **2012**, *138*, 459–467. [CrossRef]
10. Feng, R.; Zhan, H.; Meng, S.; Zhu, J. Experiments on H-shaped high-strength steel beams with perforated web. *Eng. Struct.* **2018**, *177*, 374–394. [CrossRef]
11. Cavène, E.; Durif, S.; Bouchaïr, A.; Toussaint, E. Experimental study of slotted hole bolted cover-plate connection using full field measurement. In *Structures*; Elsevier: Amsterdam, The Netherlands, 2020; Volume 23, pp. 573–587.
12. *ANSI/AISC 360-22*; Specification for Structural Steel Buildings. American Institute of Steel Construction (AISC): Chicago, IL, USA, 2022.
13. Cristopher, D.M.; Anna, S.; Aaron, V. Experiments on Cold-Formed Steel C-Section Joists with Unstiffened Web Holes. *J. Struct. Eng.* **2013**, *139*, 659–704.
14. Pham, D.K.; Pham, C.H.; Pham, S.H.; Hancock, G.J. Experimental investigation of high strength cold-formed channel sections in shear with rectangular and slotted web openings. *J. Constr. Steel Res.* **2020**, *165*, 105889. [CrossRef]
15. Fang, Z.; Roy, K.; Chi, Y.; Chen, B.; Lim, J.B. Finite element analysis and proposed design rules for cold-formed stainless steel channels with web holes under end-one-flange loading. In *Structures*; Elsevier: Amsterdam, The Netherlands, 2021; Volume 34, pp. 2876–2899.
16. Nawar, M.T.; Arafa, I.T.; Elhosseiny, O. Numerical investigation on effective spans ranges of perforated steel beams. In *Structures*; Elsevier: Amsterdam, The Netherlands, 2020; Volume 25, pp. 398–410.

17. Sangeetha, P.; Revathi, S.M.; Sudhakar, V.; Swarnavarshini, D.; Sweatha, S. Behaviour of cold-formed steel hollow beam with perforation under flexural loading. *Mater. Today Proc.* **2021**, *38*, 3103–3109. [CrossRef]
18. Uzzaman, A.; Lim, J.B.; Nash, D.; Roy, K. Web crippling behaviour of cold-formed steel channel sections with edge-stiffened and unstiffened circular holes under interior-two-flange loading condition. *Thin-Walled Struct.* **2020**, *154*, 106813. [CrossRef]
19. Degtyareva, N.; Gatheeshgar, P.; Poologanathan, K.; Gunalan, S.; Shyha, I.; McIntosh, A. Local buckling strength and design of cold-formed steel beams with slotted perforations. *Thin-Walled Struct.* **2020**, *156*, 106951. [CrossRef]
20. Degtyareva, N.; Gatheeshgar, P.; Poologanathan, K.; Gunalan, S.; Tsavdaridis, K.D.; Napper, S. New distortional buckling design rules for slotted perforated cold-formed steel beams. *J. Constr. Steel Res.* **2020**, *168*, 106006. [CrossRef]
21. Wang, X.; Chen, W.; Li, J.; Yang, T. Fire Resistance Performance of Constrained H-Shaped Steel Columns with Uneven Vertical Temperature Distributions. *Buildings* **2024**, *14*, 2826. [CrossRef]
22. Weng, Y.; Li, S.; Yuan, X.; Xue, T. Research on Impact Resistance of H-shaped Steel Beam. In Proceedings of the International Conference on Green Building, Civil Engineering and Smart City, Singapore, 15–17 March 2022; Springer NatureSingapore: Singapore, 2022; pp. 1053–1061.
23. Karalar, M.; Dicleli, M. Effect of pile orientation on the fatigue performance of jointless bridge H-piles subjected to cyclic flexural strains. *Eng. Struct.* **2023**, *276*, 115385. [CrossRef]
24. Karalar, M.; Dicleli, M. Effect of thermal induced flexural strain cycles on the low cycle fatigue performance of integral bridge steel H-piles. *Eng. Struct.* **2016**, *124*, 388–404. [CrossRef]
25. Karalar, M.; Dicleli, M. Fatigue in jointless bridge H-piles under axial load and thermal movements. *J. Constr. Steel Res.* **2018**, *147*, 504–522. [CrossRef]
26. Karalar, M.; Dicleli, M. Low-cycle fatigue in steel H-piles of integral bridges; a comparative study of experimental testing and finite element simulation. *Steel Compos. Struct. Int. J.* **2020**, *34*, 35–51.
27. Xiao, F.; Zhu, W.; Meng, X.; Chen, G.S. Parameter identification of frame structures by considering shear deformation. *Int. J. Distrib. Sens. Netw.* **2023**, *2023*, 6631716. [CrossRef]
28. Xiao, F.; Meng, X.; Zhu, W.; Chen, G.S.; Yan, Y. Combined joint and member damage identification of semi-rigid frames with slender beams considering shear deformation. *Buildings* **2023**, *13*, 1631. [CrossRef]
29. Xiao, F.; Zhu, W.; Meng, X.; Chen, G.S. Parameter identification of structures with different connections using static responses. *Appl. Sci.* **2022**, *12*, 5896. [CrossRef]
30. *GB/T 228.1-2010*; Metallic Materials: Tensile Testing: Part 1: Method of Test at Room Temperature. General Administration of Quality Supervision, Inspection and Quarantine of the People's Republic of China: Beijing, China, 2012.
31. *GB/T 1041-2008*; Metallic Materials—Bend Test Method. China Standards Press: Beijing, China, 2008.
32. Karalar, M.; Dicleli, M. Effect of pile length on the low cycle fatigue performance of integral bridge steel H piles. In *Bridge Maintenance, Safety, Management, Life-Cycle Sustainability and Innovations*; CRC Press: Boca Raton, FL, USA, 2021; pp. 3888–3891.
33. ANSYS, Inc. *13.185. SOLID185—3D 8-Node Structural Solid*; ANSYS Help Documentation; ANSYS, Inc.: Canonsburg, PA, USA, 2024.
34. Sreejith, T.S.; Kaliveeran, V. Stress analysis of a member of jacket structure with different types of stiffeners. In *Dynamic Behavior of Soft and Hard Structures*; Velmurugan, R., Balaganesan, G., Kakur, N., Kanny, K., Eds.; Springer Nature: Berlin/Heidelberg, Germany, 2024.
35. Mendes, A.C.; Kolodziej, J.A.; Correia, H.J.D. Numerical modelling of wave-current loading on offshore jacket structures. *Ships Offshore Struct.* **2004**, *8*, 15–28.
36. Máleková, V.; Jendzelovsky, N. An analysis of contact elements of foundation structures. *Int. Virtual J. Sci. Tech. Innov. Ind.* **2012**, *7*, 61–65.
37. *GB 50017-2017*; Code for Design of Steel Structures. China Architecture & Building Press: Beijing, China, 2018; pp. 22–27.
38. *GB/T 1228-2006*; National Technical Committee of Standardization for Fasteners, High-strength Large Hexagon Head Bolts for Steel Structures. Standards Press of China: Beijing, China, 2006; pp. 2–4.
39. Kong, Q.; Li, Y.; Wang, S.; Yuan, C.; Sang, X. The influence of high-strength bolt preload loss on structural mechanical properties. *Eng. Struct.* **2022**, *271*, 114955. [CrossRef]

buildings

MDPI

Article

Study on the Shear Performance of the Interface between Post-Cast Epoxy Resin Concrete and Ordinary Concrete

Peiqi Chen [1,2], Hao Wang [1,2], Xiaojie Zhou [1,2,*] and Shilong Zhao [1,2]

[1] Tianjin Key Laboratory of Civil Buildings Protection and Reinforcement, Tianjin 300384, China; cpq@tcu.edu.cn (P.C.)

[2] School of Civil Engineering, Tianjin Chengjian University, Tianjin 300384, China

* Correspondence: zhouxj@tcu.edu.cn

Abstract: The interface of fresh-aged concrete represents a critical vulnerability within monolithic assembled monolithic concrete structures. In this paper, the shear performance of the interface between post-cast epoxy resin concrete and standard concrete is studied using experimental methods and finite element analysis. The objective is to furnish empirical data that support the broader adoption of epoxy resin concrete in assembled structures. A direct shear experiment of 19 Z-shaped samples and a computation of 20 finite element models were completed. The results from both experimental and computational analyses provided insights into several factors influencing the shear performance at the interface. These factors include the pre-cast part of concrete strength, the friction coefficient of the interface, the longitudinal reinforcement ratio at the interface, the compressive strength of concrete in the post-cast part, and confining stress. The findings indicate that utilizing epoxy resin concrete for post-cast material, roughing the interface, and setting keyways can enhance the shear performance of the interface so that it equals or even exceeds the cast-in situ sample. Optimal shear results are obtained when the compressive strength of the post-cast epoxy resin concrete closely matches that of the pre-cast conventional cement. Moreover, increasing the depth of the keyways rather than their width is more effective in improving the shear capacity of the sample. It is recommended that the depth of the keyway should be at least 30 mm, and its width should be no less than three times the depth. As the longitudinal reinforcement ratio at the interface increases, there is an enhancement in shear capacity coupled with a reduction in deformative performance. It is advisable to maintain this ratio below 1.0% to balance the strength and ductility effectively.

Keywords: epoxy resin concrete; interface; shear performance; experiment; finite element

1. Introduction

The interface regions within prefabricated concrete structures represent the most vulnerable sections in terms of shear resistance, but are also the critical sections of the fresh-aged concrete to transfer the load, significantly influencing the lifespan of the entire structure. Broad investigations have recently been carried out into the factors affecting the performance of these joints. These factors include the style of the joint, the processing approach of the interface, the setting of the keyway, the type and strength of the post-cast material, and the confining stress. A great deal of in-depth research has been carried out in recent years to address these issues.

Regarding the connection joint type and fresh-aged concrete interface treatment, Momayez et al. [1] showed that rough surface treatment can obtain greater bond shear strength. He et al. [2] observed a roughly linear correlation between bond shear strength and interface roughness, noting that the application of an interfacial agent markedly boosted the robustness of the junction. Julio et al. [3] explored various bond surface treatments, identifying sandblasting as the most effective method to increase bond shear strength, despite its lower construction efficiency; alternatively, keyways proved more practical for routine applications. Jang et al. [4] assessed the shear strength benefits of vertical joints; high-pressure, water-treated joints; and differently sized keyways, concluding that keyways substantially improved the joint shear strength. Reference [5] indicated optimal shear performance when a keyway with a depth of 30 mm was set. Yuan et al. [6] showed a decrease in shear capacity with an increase in the number of keyways, although larger keyways exhibited superior shear capacity compared to smaller ones. Yang et al. [7] studied the bond shear performance across different keyway designs on the UHPC-NC interface, finding that increased keyway width enhanced shear strength, whereas reduced spacing diminished it. Ye et al. [8,9] conducted shear experiments on key connections in pre-cast ultra-high-performance concrete segmental bridges, revealing that samples coated with epoxy resin binder at the keyway demonstrated relatively better ductility and higher stiffness than those with dry-key joints. Wang et al. [10] proposed a compact UHPC wet joint, and through both experimental and numerical simulation studies, demonstrated that this wet joint variant slightly outperformed the epoxy resin dry joint samples, affirming the wet joint's superior shear strength, robustness, and adhesion properties. Zhang et al. [11] found that augmenting the number or distribution density of dowel bars increased the stiffness and shear strength of the joint surface while decreasing the peak displacement. The reliability of their findings was confirmed through analysis using ABAQUS (2021) finite element software.

In terms of post-cast material types, Deng et al. [12] utilized high ductile concrete (HDC) to replace ordinary concrete in beam–column joints, discovering that HDC significantly enhanced both the deformation potential and energy absorption of the frame joints. Chen et al. [13] performed experiments and numerical simulations on monolithically cast frame beam–column joints and composite frame structures with post-cast ordinary and epoxy resin cement. Their findings indicate that the mechanical characteristics of beam–column joints with post-cast ordinary concrete were lesser than those of cast-in situ joints, while those reinforced with post-cast epoxy resin concrete matched or surpassed the in situ counterparts.

Regarding the effect of confining pressure on the shear strength, Shamass et al. [14] utilized the ABAQUS finite element software tool to model pre-cast concrete segmental box girder bridge nodes, identifying a linear correlation between the perimeter stress and the shear force of epoxy joints. Similarly, Gopal et al. [15] found that the shear capacity of UHPFRC grooved dry joints was enhanced with the application of greater confining stress.

Epoxy resin-based concrete is a compound material that is hardened and molded from a mixture of epoxy resin, a curing agent, and sand aggregate [16]. This substance demonstrates enhanced characteristics relative to standard silicate concrete in terms of tensile strength, crack resistance, material homogeneity, and integrity [17]. Given its robust interfacial bond strength with concrete, along with its thermal resistance, chemical durability, and mechanical characteristics [18], epoxy resin-based concrete is predominantly utilized in reinforcement projects, including concrete crack repair [18–22], as well as road construction. However, its application in building structures remains limited, presenting a broad spectrum of research and development opportunities for this engineering structural material. In 2019, Gil-Martin et al. [23] examined the mechanical properties of epoxy resin cement compared to tire rubber concrete in beam–column joints. Their findings indicated that, while the epoxy resin concrete beam–column joints showed commendable working efficiency, tire rubber cement was found to be inadequate for the central regions of connections. More recently, in 2021, El-Mandouh et al. [24] conducted an experimental investigation on the shear performance of 18 simply supported, super-reinforced epoxy resin concrete beams, demonstrating that these beams experienced less deformation and exhibited higher cracking and ultimate shear capacities, along with improved ductility, compared to beams composed of non-epoxy resin concrete. References [25,26] studied the mechanical properties of epoxy resin concrete members. These results provide valuable perspectives on the possible use of epoxy resin concrete in construction frameworks. However, the high viscosity and rapid hardening speed of epoxy resin concrete complicate the preparation process, which restricts its widespread application. Therefore, optimizing the preparation method of epoxy resin concrete remains a critical issue for future studies.

In summary, this paper advocates for the utilization of epoxy resin concrete as a post-cast material within assembled monolithic structures, focusing on the bond between aged and fresh concrete surfaces. Employing both experimental approaches and finite element analysis, this research aims to evaluate how different elements affect the shear performance of these concrete junctions. These elements encompass post-cast concrete strength class, friction coefficient of interface, confining stress, the longitudinal reinforcement ratio of interface, and the strength class of pre-cast concrete. The objective is to assess the shear behavior of the junction between post-cast epoxy resin concrete and standard concrete, thereby providing empirical data to support the use and advancement of epoxy resin concrete in assembled monolithic concrete frameworks. The research methodology and steps of this paper are depicted in Figure 1.

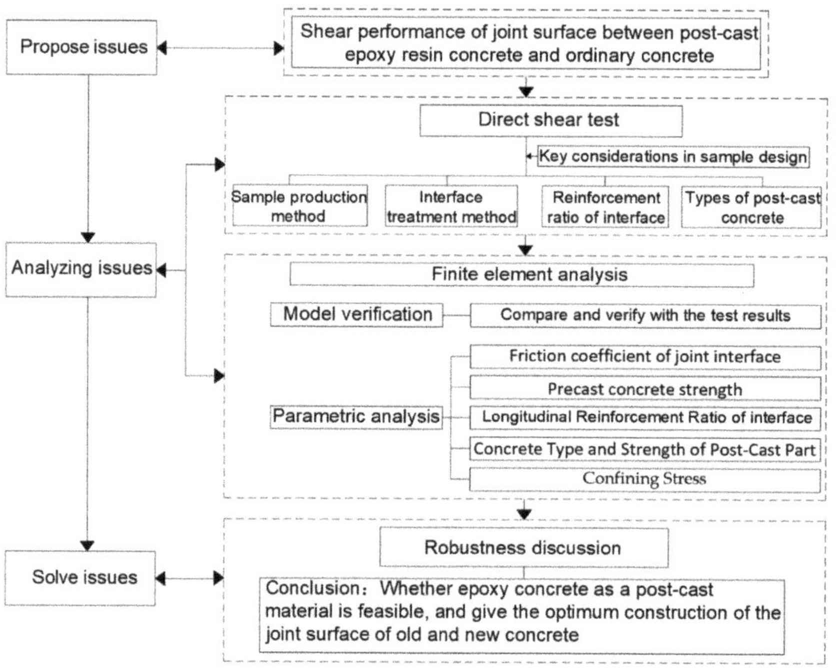

Figure 1. Flowchart of research work.

2. Experiment Profile

2.1. Experiment Piece Design

This study designed and fabricated a total of 19 Z-type straight shear samples. The set included two sets of cast-in situ contrasting samples (i.e., samples ZJ1 and ZJ2). Additionally, the experiment comprised four groups of samples featuring post-cast ordinary concrete and thirteen groups utilizing post-cast epoxy resin concrete. Details concerning the characteristics and configurations of these samples are provided in Table 1.

Table 1. Basic situation of the samples.

Serial Number	Sample Number	Interface Treatment Method	Type of Post-Cast Concrete	Longitudinal Reinforcement Ratio of Interface ρ/%
1	ZJ1	---	---	0
2	ZJ2	---	---	0.67(4Φ8)
3	EC01	No treatment	Epoxy resin concrete	0
4	EC02	No treatment	Epoxy resin concrete	0.67(4Φ8)
5	EC11	Roughening	Epoxy resin concrete	0
6	EC12	Roughening	Epoxy resin concrete	0.67(4Φ8)
7	EC13	Roughening	Epoxy resin concrete	1(4Φ10)
8	EC14	Roughening	Epoxy resin concrete	1.5(4Φ12)
9	CC11	Roughening	Ordinary concrete	0
10	CC12	Roughening	Ordinary concrete	0.67(4Φ8)
11	EC21–7025	Keyway (width: 70 mm, depth: 25 mm)	Epoxy resin concrete	0
12	EC22–7025	Keyway (width: 70 mm, depth: 25 mm)	Epoxy resin concrete	0.67(4Φ8)
13	EC22–7030	Keyway (width: 70 mm, depth: 30 mm)	Epoxy resin concrete	0.67(4Φ8)
14	EC22–8025	Keyway (width: 80 mm, depth: 25 mm)	Epoxy resin concrete	0.67(4Φ8)
15	EC22–8030	Keyway (width: 80 mm, depth: 30 mm)	Epoxy resin concrete	0.67(4Φ8)
16	EC22–9025	Keyway (width: 90 mm, depth: 25 mm)	Epoxy resin concrete	0.67(4Φ8)
17	EC22–9030	Keyway (width: 90 mm, depth: 30 mm)	Epoxy resin concrete	0.67(4Φ8)
18	CC21–7025	Keyway (width: 70 mm, depth: 25 mm)	Ordinary concrete	0
19	CC22–7025	Keyway (width: 70 mm, depth: 25 mm)	Ordinary concrete	0.67(4Φ8)

The basic composition of the assembled direct shear sample is depicted in Figure 2. Keyway extends horizontally across the entire interface, with the dimensions of the interface measuring $b \times h$ = 150 mm × 200 mm. Longitudinal reinforcement passes through the interface and is reliably anchored within both the pre-cast and post-cast parts, respectively. Detailed specifications concerning the dimensions and reinforcement layout of typical samples are illustrated in Figure 3.

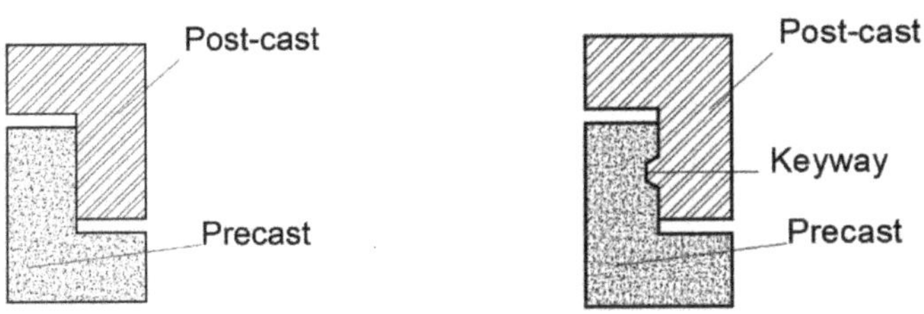

(**a**) Sample without keyway (**b**) Sample with keyway

Figure 2. Basic composition of assembled direct shear samples.

2.2. Basic Mechanical Properties of Material

During the production of the samples, C30-grade concrete was used for both pre-cast and post-cast plain concrete applications. The mixture proportions for the epoxy resin concrete are specified in Table 2. The compressive forces of both the ordinary and epoxy resin concretes were measured and are presented in Table 3. Here, the term "adjusted strength" refers to the strength values recalibrated for comparison using a standard cube sample measuring 150 mm × 150 mm × 15 mm. Additionally, the determined mechanical characteristics of the steel reinforcement are documented in Table 4.

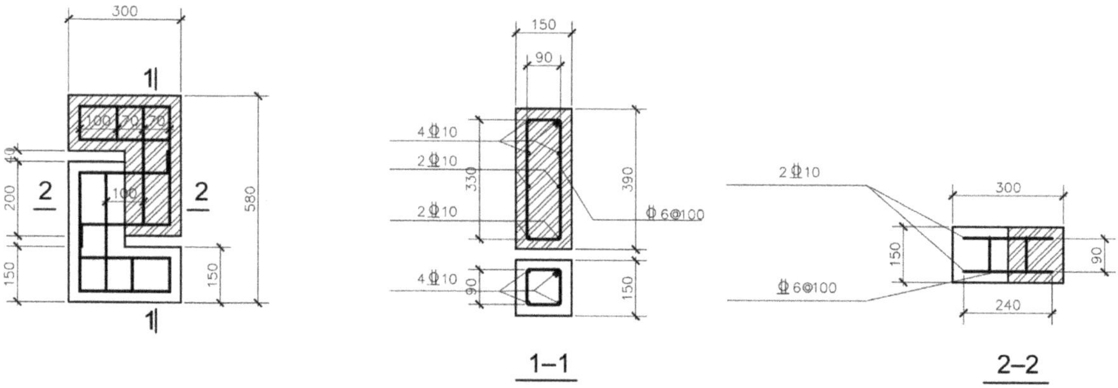

(**a**) Sample EC13

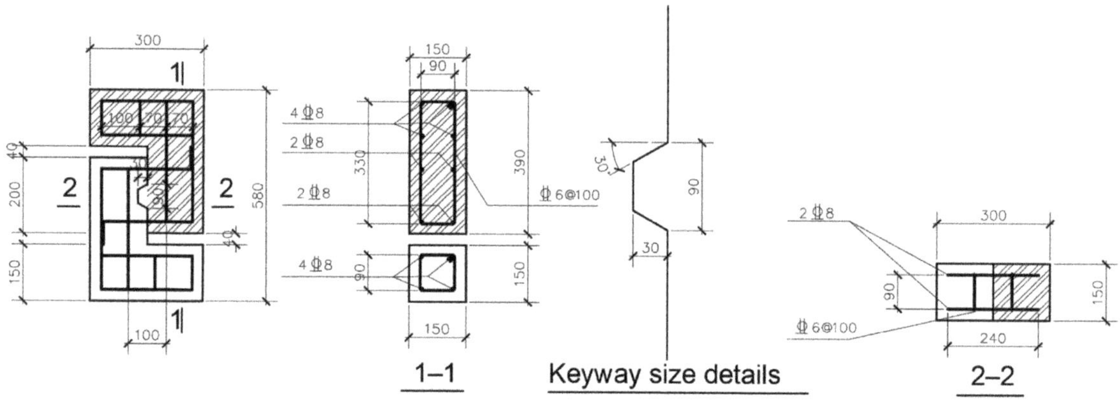

(**b**) Sample EC22–9030

Figure 3. Detailed dimensions and reinforcement of typical samples.

Table 2. Epoxy resin concrete mixing ratio.

Epoxy Resin (kg/m^3)	Curing Agent (kg/m^3)	Cement (kg/m^3)	Cobble (kg/m^3)	Sand (kg/m^3)	Diluent (kg/m^3)
800	320	1200	2800	3200	80

Table 3. Average cube compressive strength (MPa).

Materials	Sample Size (mm × mm)	Average Strength Value (MPa)	Adjustment of Strength (MPa)
Ordinary concrete (pre-cast section)	100 × 100 × 100	32.5	30.9
Ordinary concrete (post-cast portion)	100 × 100 × 100	35.4	33.6
Epoxy resin concrete (post-cast portion)	100 × 100 × 100	58.3	55.4

Table 4. Measured tensile strength of steel reinforcement (MPa).

Steel Reinforcement Type	Reinforcement Diameter (mm)	Average Yield Strength (MPa)	Elastic Modulus (MPa)
Longitudinal reinforcement	8	361.1	2.0×10^5
	10	367.8	2.0×10^5
	12	360.0	2.0×10^5
Stirrup	6	268.3	2.1×10^5

2.3. Loading Device and Measurement Solution

The configuration of the direct shear test apparatus is depicted in Figure 4. The loading equipment was 50 t hydraulic jack, while the steel beam of a 2000 t electro-hydraulic servo universal testing machine served as the vertical load counter frame. A transducer was placed above the jack to monitor loading metrics. To distribute the vertical load effectively, a 150 mm × 200 mm steel plate was located at the middle of both the top and bottom surfaces of the sample. This setup ensured that the central force applied during testing aligned with the shear surface (the interface of the aged and the fresh concrete), facilitating a pure shear condition. Initially, the sample underwent preloading to 30% of the anticipated cracking load, and the test was repeated several times to ensure uniform force application and the proper functioning of the loading apparatus. In the main loading phase, incremental loads of 10 kN were applied in a force-controlled manner until sample failure occurred.

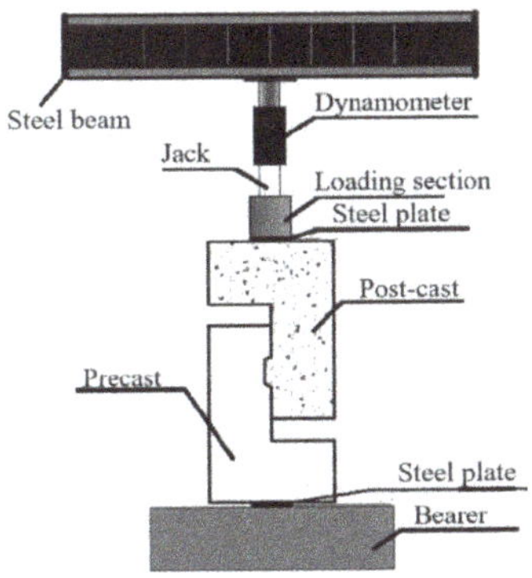

(**a**) Diagram of loading device

(**b**) Test loading device diagram

Figure 4. Direct shear test device.

The experimental protocol necessitated measuring the relative slip at the bonding surface between the two sides of the bonding surface throughout the loading phase. To eliminate the test errors, dual displacement gauges were strategically positioned on each side of the interface. The configuration of these displacement sensors is outlined in Figure 5. The relative slip across the interface was quantified by calculating the difference between the average readings from the displacement gauges on the left and right sides of the joint.

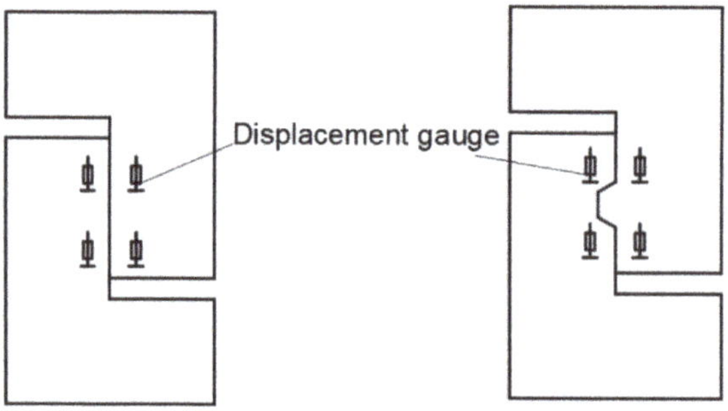

Figure 5. Displacement gauge layout.

3. Experimental Results and Analyses

3.1. Experimental Phenomena

(1) Cast-in situ samples ZJ1 and ZJ2

The cast-in situ unreinforced sample ZJ1 was damaged during the demolding process, indicating that the shear bearing capacity of the interface was extremely low.

For the cast-in situ sample ZJ2, with a longitudinal reinforcement ratio of 0.67%, the sample exhibited no signs of damage phenomena when subjected to vertical loads below 70 kN. The initial appearance of a vertical shear crack occurred at the middle and upper parts of the interface when the vertical load reached 70 kN. As the load increased, another vertical crack appeared in the middle and lower parts of the interface, eventually leading to full penetration of the cracks. Upon reaching the ultimate load of 94.3 kN, the relative displacement measured was 2.11 mm, followed by a decrease in the load. The final damage state of the sample, characterized by its distinct shear damage features, is illustrated in Figure 6.

Figure 6. The failure mode of sample ZJ2 (front).

(2) EC0 group assembled samples with untreated interface

For the unreinforced sample EC01, initially, no apparent damage was observed at the onset of loading. However, upon reaching a load of 53 kN, a sudden "bang" was heard and the sample split along the interface, separating the pre-cast from the post-cast parts. This led to a sharp decline in the bearing capacity, leading to the sample's failure. Apart from minor concrete spalling on one side of the prefabricated section, the rest of the sample exhibited significant damage, with a flat shear damage surface. This damage is depicted in Figure 7a as typical shear damage.

(**a**) Sample EC01 (front) (**b**) Sample EC02 (front)

Figure 7. Failure modes of EC0 group samples.

For sample EC02, with a longitudinal reinforcement ratio of 0.67%, there was no obvious change in the sample when the vertical load was 10–30 kN. However, when loaded to 40 kN, multiple fine vertical shear cracks began appearing on the interface. As loading continued, the number and width of these cracks gradually increased, accompanied by fine horizontal cracks at the positions of longitudinal reinforcements. The sample's bearing capacity peaked at 57 kN, after which the load began to decrease slowly. The interface cracks progressively widened and completely penetrated the sample, coupled with continuous spalling of the concrete. Following this, the load exhibited fluctuations and a gradual decline. The damage sustained by the sample is illustrated in Figure 7b, and is classified as typical shear damage.

(3) EC1 and CC1 assembled samples with roughened interfaces

The damage patterns of unreinforced samples EC11 and EC01 of post-cast epoxy resin concrete were similar, exhibiting typical shear damage (Figure 8a). Notably, the shear capacity of the former was slightly greater than that of the latter, indicating a favorable effect of the roughness of the interface on shear performance. Conversely, compared with EC11, the unreinforced samples of post-cast ordinary concrete CC11 sustained damage before formal loading, highlighting the extremely low shear capacity of the interface between the post-cast plain concrete and the pre-cast section.

The test phenomena of samples of the EC1 group (EC12~EC14) and EC02 samples showed similar damage patterns, predominantly occurring at the bond surface. With the increasing reinforcement ratio, both the cracking load and ultimate bearing capacity of the samples increased, while the corresponding displacement decreased. For samples EC13 and EC14, the concrete shielding layer at the upper part of the interface spalled off due to the action of longitudinal reinforcement pins. The damage patterns of these samples are depicted in Figure 8b–d.

The experimental phenomena observed for sample CC12 were similar to those of the group EC1. Comparatively, with the same reinforcement rate, the cracking loads for sample CC12 and EC12 were nearly identical, although the ultimate load for CC12 was slightly lower. This suggests greater shear friction between the epoxy resin concrete and the standard concrete. The failure mode of the sample is displayed in Figure 8e.

(**a**) Sample EC11 (back)

(**b**) Sample EC12 (back)

(**c**) Sample EC13 (back)

(**d**) Sample EC14 (front)

(**e**) Sample CC12 (front)

Figure 8. Failure modes of samples in EC1 group and CC1 group.

(4) EC2 and CC2 assembled samples with keyways on the interface

When loaded to 63 kN, the unreinforced sample EC21–7025 was split in two along the interface, where the keyway was sheared off and the damage surface appeared relatively flat, indicative of typical brittle shear damage, as shown in Figure 8a. Conversely, sample CC21–7025 sustained damage at the bond surface during handling, highlighting the minimal shear capacity of the interface between the post-cast ordinary concrete and the pre-cast concrete.

The damage process for the six reinforced samples in the EC22 group followed a similar pattern: initial cracking in tension was observed above the keyway, followed by local compressive vertical cracks forming below the keyway. As the load continued to rise, the number and extent of the cracks below the keyway expanded, and vertical shear cracks along the interface developed until the ultimate load was reached. After reaching this point, the load began to decrease, and the crushing of the concrete below the keyway became pronounced. The keyway appeared to be completely sheared when the width of the keyway root was small (e.g., 70 mm). Moreover, the maximum load of the sample increased with the depth of the keyway, but this also resulted in more severe damage to the prefabricated part. The damage modes for each sample are depicted in Figure 9b–g.

The damage process of sample CC22–7025 was similar to those of the six reinforced samples in the EC22 group, although it exhibited a lower ultimate load. During the damage phase, the keyway sheared, as illustrated in Figure 9h.

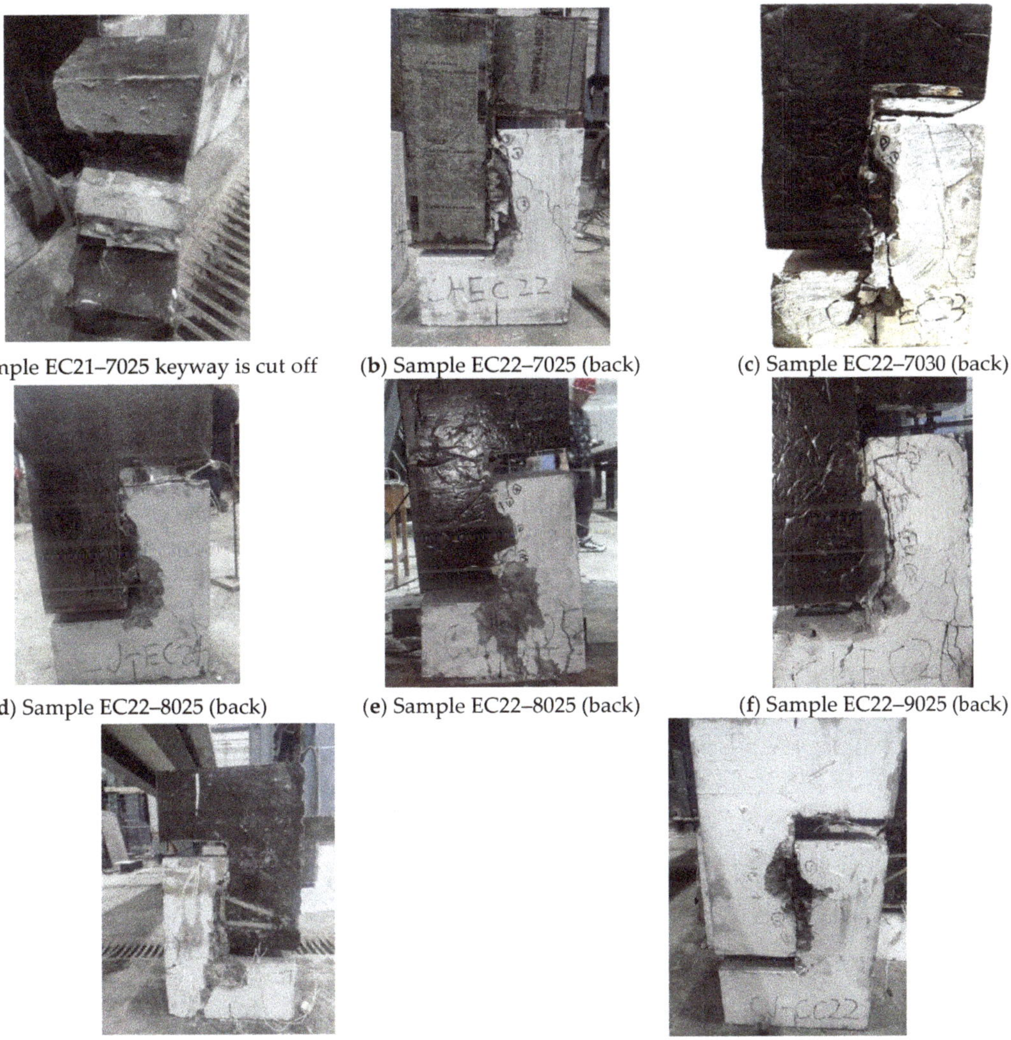

(**a**) Sample EC21–7025 keyway is cut off (**b**) Sample EC22–7025 (back) (**c**) Sample EC22–7030 (back)

(**d**) Sample EC22–8025 (back) (**e**) Sample EC22–8025 (back) (**f**) Sample EC22–9025 (back)

(**g**) Sample EC22–9030 (front) (**h**) Sample CC22–7025 (back)

Figure 9. Failure modes of samples in EC2 group and CC2 group.

From the examination of each sample's damage, several conclusions can be drawn: (1) The shear capacity of the interface was primarily derived from the bond and friction between the old and new concrete. Consequently, the shear capacity of the post-cast epoxy resin concrete samples was significantly greater than that of the post-cast ordinary concrete samples. At ultimate load, these samples typically exhibited a large shear slip and lost load-carrying capacity, characteristic of brittle shear damage. Appropriately configured longitudinal steel reinforcement substantially enhanced both the shear capacity and ductility of the interface. (2) Damage consistently occurred at the interface. Apart from the post-cast epoxy resin concrete and the samples with keyways on the interfaces, the pre-cast and post-cast portions of the rest of the samples remained relatively intact. However, in samples with keyways, damage to the pre-cast part was more severe, primarily due to the significant difference in durability between the post-cast epoxy resin concrete and the pre-cast concrete. When a strong keyway was embedded in a weaker pre-cast component, the interaction under load led to pronounced damage in the weaker pre-cast part. (3) In some cases, the sample loading may not have been fully geometrically centered, and torsional forces on the sample or loading-end stresses induced cracking. However, these effects were generally minor and did not significantly impact the experimental outcomes.

3.2. Load–Displacement Curve

The load–displacement curves of each sample are displayed in Figure 10. Among them, cast-in situ reinforcement sample ZJ2 serves as a comparison sample. The data of the experimental results are displayed in Table 5.

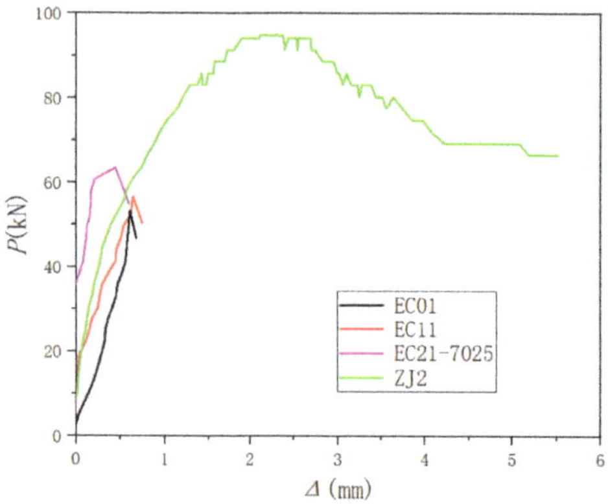

(**a**) Unreinforced group samples

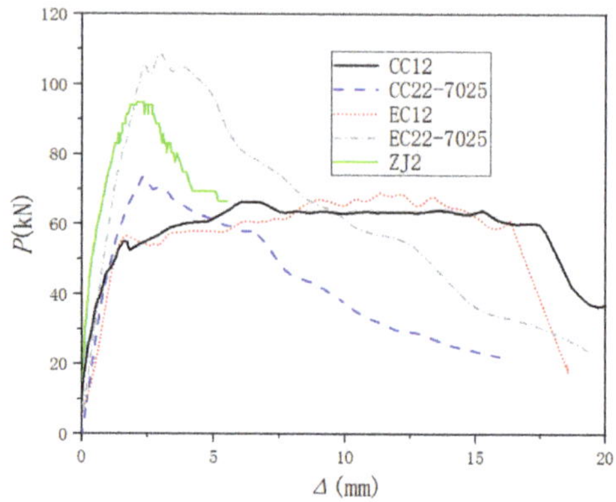

(**b**) EC group and CC group samples

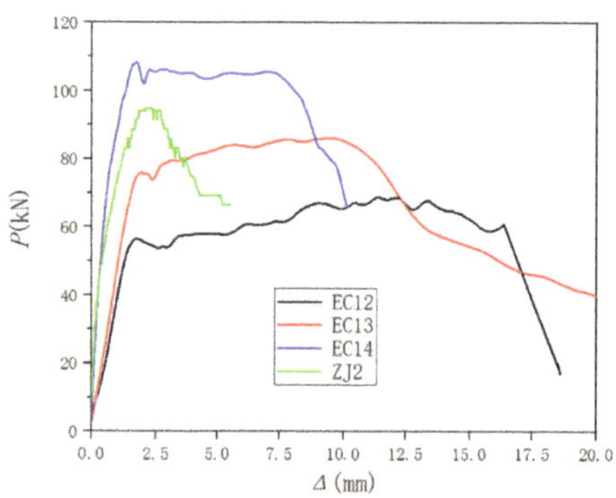

(**c**) EC1 group samples

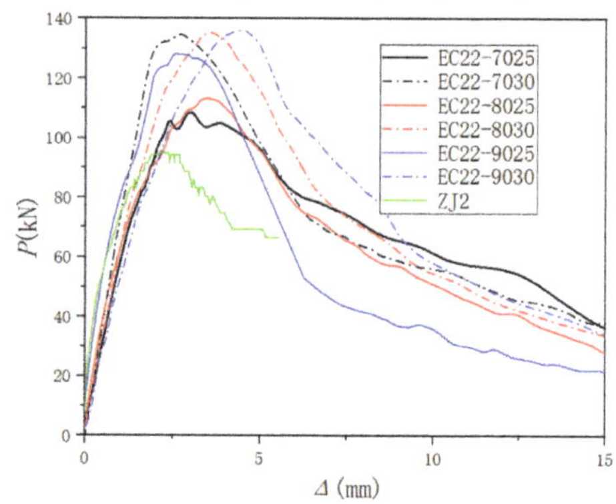

(**d**) EC2 group samples

Figure 10. Comparative analysis of load–displacement curves of each set of samples.

Table 5. Experimental data of each sample.

Sample Number	Interface Surface Treatment Method	Longitudinal Reinforcement Ratio of Joint Surface ρ/%	Cracking Load/kN	Peak Load/kN	Peak Load Displacement/mm
ZJ1	--	0	--	--	--
ZJ2	--	0.67(4Φ8)	70	94.3	2.11
EC01	No treatment	0	40	53.2	0.61
EC02	No treatment	0.67(4Φ8)	40	57.1	1.42
EC11	Roughening	0	40	55.2	0.65

Table 5. *Cont.*

Sample Number	Interface Surface Treatment Method	Longitudinal Reinforcement Ratio of Joint Surface ρ/%	Cracking Load/kN	Peak Load/kN	Peak Load Displacement/mm
EC12	Roughening	0.67(4Φ8)	40	56.6	1.75
EC13	Roughening	1.0(4Φ10)	40	76.0	1.95
EC14	Roughening	1.5(4Φ12)	40	108.4	1.75
CC11	Roughening	0	--	--	--
CC12	Roughening	0.67(4Φ8)	40	55.0	1.77
EC21–7025	Keyway (width: 70 mm, depth: 25 mm)	0	60	63.5	0.45
EC22–7025	Keyway (width: 70 mm, depth: 25 mm)	0.67(4Φ8)	50	108.3	2.88
EC22–7030	Keyway (width: 70 mm, depth: 30 mm)	0.67(4Φ8)	50	135.0	3.22
EC22–8025	Keyway (width: 80 mm, depth: 25 mm)	0.67(4Φ8)	50	113.8	3.21
EC22–8030	Keyway (width: 80 mm, depth: 30 mm)	0.67(4Φ8)	60	136.8	3.44
EC22–9025	Keyway (width: 90 mm, depth: 25 mm)	0.67(4Φ8)	60	126.2	2.80
EC22–9030	Keyway (width: 90 mm, depth: 30 mm)	0.67(4Φ8)	60	136.0	4.60
CC21–7025	Keyway (width: 70 mm, depth: 25 mm)	0	--	--	--
CC22–7025	Keyway (width: 70 mm, depth: 25 mm)	0.67(4Φ8)	20	74.6	2.18

(1) Analysis of unreinforced samples

The comparative analyses of the CC*1 and EC*1 groups of samples are depicted in Figure 10a. Among others the CC*1 group includes CC11 and CC21-7025, and the EC*1 group includes EC01, EC11 and EC21-7025.

Both the CC11 and CC21–7025 unreinforced samples made from post-cast ordinary concrete were damaged during the mold removal process, and consequently, no test data were available. In contrast, the three unreinforced samples from the EC*1 group, made from post-cast epoxy resin concrete, all exhibited brittle shear damage. These samples reached their shear capacity and then suddenly experienced shear failure, resulting in an instantaneous decrease in shear capacity. Among these, the sample with a roughened interface and a keyway exhibited the highest shear capacity, while the sample with an unroughened interface and no keyway displayed the lowest shear capacity, as illustrated in Figure 10a.

(2) Comparative analysis of types of post-cast concrete

The comparative analyses of the samples in the CC and EC groups are shown in Figure 10b.

The shear resistance of the CC12 and EC12 samples, which were only subjected to rough treatment on the interface, was derived from two main forces: the shear friction force at the interface and the dowel force exerted by the longitudinal reinforcement. Initially, the shear friction force dominated during the early loading stages. However, as this force waned, the dowel force from the longitudinal reinforcement became pivotal, providing consistent and stable shear resistance. This mechanical interplay is evident from the load–displacement curves for both samples, as shown in Figure 10b. A characteristic small peak occurred early in the loading process, followed by a slight decline in the curve, indicating the loss of shear friction at the interface. Subsequently, as the curve ascended once more, it coincided with significant bond slip, demonstrating that the dowel force of the longitudinal reinforcement assumed the primary role in shear resistance. This continued until the slip reached approximately 16 mm, at which point the curve abruptly fell, signaling the depletion of the interface's shear capacity. Therefore, although such samples displayed robust late-stage shear bearing and shear deformation capacity, excessive bond slip ultimately undermined the interface's structural utility. Consequently, the first peak point load observed should be taken as the shear bearing capacity for these samples. Additionally, Figure 9b reveals that, while the initial shear stiffness of the post-cast epoxy resin concrete sample EC12 was slightly smaller than that of the post-cast ordinary concrete sample CC12, its shear bearing capacity was slightly larger. This suggests that post-cast epoxy resin concrete offers a slight advantage in terms of shear friction at the interface over ordinary concrete. Nonetheless, this advantage is subtle, and the overall shear bearing capacity is primarily contingent upon the reinforcement rate of the interface.

The shear capacities of the samples CC22–7025 and EC22–7025, both featuring keyways on the interface, were derived from a tripartite system: the shear friction force at the interface, the shear bearing capacity offered by the keyway, and the dowel force from the longitudinal reinforcement. Within this system, the keyway played a crucial role in the stress mode of the cantilever deep beam. Empirical comparisons revealed a significant enhancement in mechanical performance due to the keyway integration. Specifically, the load-carrying capacity of the sample CC22–7025 showed a substantial increase of 35.3% compared to CC12, with the displacement at the peak load expanding by 23.2%.The sample EC22–7025 exhibited an even more pronounced increase of 91.3% in load-carrying capacity and a 64.6% increase in displacement at the peak load when compared to EC12. These observations underscore the superior initial shear stiffness and overall shear capacity of samples incorporating post-cast epoxy resin concrete, as well as their improved shear ductility. It is also illustrated in Figure 9b that the enhanced shear performance of post-cast epoxy resin concrete samples with keyways on the interface exceeded that of the monolithic cast samples. This highlights the beneficial impact of keyways on improving structural responses and durability, pointing towards the efficacy of keyway integration in complex stress environments.

(3) Comparative analysis of post-cast epoxy resin concrete samples with rough-surface-only and cast-in situ samples

The EC1 group was analyzed in comparison with the cast-in situ sample ZJ2; see Figure 10c.

As previously discussed, for samples without keyways on the interface, the load corresponding to the first peak point should be considered the definitive measure of the sample's shear bearing capacity. According to Figure 9c, as the longitudinal reinforcement rate on the interface increased, both the shear bearing capacity and the corresponding peak load displacement of the sample rose, indicating an enhancement in shear ductility. Specifically, the shear bearing capacity of sample EC12 (the longitudinal reinforcement rate of the interface was 1.5%) was the highest, 16.6% greater than that of the cast-in situ sample ZJ2. This demonstrates that the shear performance of the interface in post-cast epoxy resin concrete can match or surpass that of cast-in situ samples, provided that the reinforcement rate is sufficiently high. Although the shear performances of other samples in the EC1 group did not achieve the levels observed in the cast-in situ sample, their later-stage shear deformation capacity was superior, highlighting a distinct advantage of using post-cast epoxy resin cement from a constructional perspective.

(4) Comparative analysis of post-cast epoxy concrete samples with keyways on the interface and cast-in situ samples

The comparative analysis between the EC2 group and the cast-in situ sample ZJ2 is depicted in Figure 10d.

The load–displacement curves for each sample within the EC2 group exhibited similar patterns. As evidenced by the test curves of samples EC22–7025, EC22–8025, and EC22–9025, an increase in the width of the keyway root enhances the initial stiffness and shear load capacity of the samples. Nevertheless, this also results in a decrease in peak load–displacement, and the descending section of the curves becomes steeper, indicating a reduction in shear ductility. Further comparisons between pairs of samples, such as EC22–7025 and EC22–7030, EC22–8025 and EC22–8030, and EC22–9025 and EC22–9030, revealed that enhancing the keyway depth augments the shear load capacity. Samples with the narrowest keyway root widths showed the largest increases in load capacity. When examining the test curves of samples EC22–7030, EC22–7030, EC22–8030, and EC22–9030, all of which featured keyway depths of 30 mm, it is evident that there was minimal variation in the shear bearing capacity across these samples. However, samples with wider keyway roots generally displayed lower initial stiffness.

4. Finite Element Analysis

4.1. Constitutive Model of Materials

Concrete was modeled using a plastic damage model. For ordinary concrete, the constitutive equation aligned with the uniaxial stress–strain curve specified in the Code for the Design of Concrete Structures (GB 50010-2010) [27], as illustrated in Figure 11a. Epoxy resin concrete utilized the plastic damage model for concrete, with its constitutive relationship based on the comprehensive curve equation for compressive stress–strain fitted by the research group according to the test and specification [13]. This is de-tailed in Equation (1), The compressive stress–strain curve is displayed in Figure 10b. For the tensile stress–strain curve, epoxy resin concrete refers to that of ordinary concrete, adopting the same constitutive model. The stress–strain curves for the reinforcement were modeled using a bifold model, as depicted in Figure 10c. Table 6 presents the principal parameters of plastic damage for both plain and epoxy resin concrete.

$$y = \begin{cases} ax + (4.9 - 4.23a)x^2 + (-4.67 + 6.67a)x^3 + (-0.27 - 4.74a)x^4 + (1.07 + 1.27a)x^5, & 0 \le x \le 1 \\ \frac{x}{b(x-1)^2 + x}, & x > 1 \end{cases} \quad (1)$$

a and b are undetermined parameters with values in the following ranges:

$$\begin{cases} 0 < a < 1.0 \\ 1.0 < b < 10.0 \end{cases}$$

Table 6. Concrete plastic damage parameters.

Concrete Type	Dilation Angle	Eccentricity	f_{b0}/f_{c0}	K	Viscosity Parameter	Poisson Ratio	Elastic Modulus
Epoxy resin concrete	30°	0.1	1.16	0.6667	0.0002	0.45	12,000 Mpa
Ordinary concrete	30°	0.1	1.16	0.6667	0.0021	0.2	30,000 Mpa

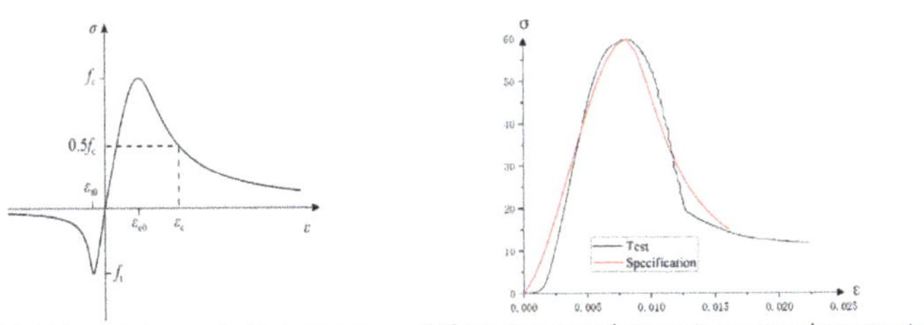

(**a**) Uniaxial stress–strain curve of ordinary concrete (**b**) Stress–strain curve of epoxy resin concrete under compression

Figure 11. *Cont.*

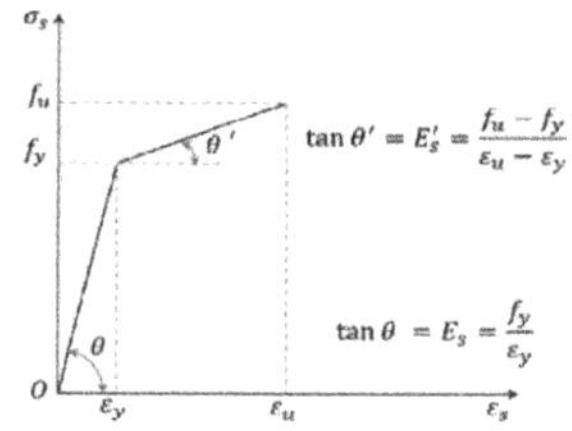

(c) Reinforcement stress–strain curve

Figure 11. Stress–strain curves of each material.

In this paper, the values for parameters a and b are set at 0.7 and 7, respectively.

4.2. Establishment of Model

In the models, both epoxy resin concrete and ordinary concrete were represented using C3D8R elements and T3D2 elements. The steel reinforcement was treated as fully embedded within the concrete, disregarding any bonding effects with the concrete. The meshing was conducted using a structured division method, with the cell meshing size set to 11 mm for both old and new concrete and to 15 mm for steel reinforcement cells, as depicted in Figure 12. The interface between fresh-aged concrete was modeled using a contact approach that directly considered the interaction between the two surfaces, and the Coulomb friction model was utilized to simulate the bonding effect at the interface. According to the provisions of both U.S. specifications [28] and the provisions of the EU specifications [29], the coefficient of friction μ should be taken as 1.0 for hardened concrete surfaces that are clean, free of slurry, and roughened to a concave and convex depth of at least 1/4 inches. For contact surfaces of rolled structural steel with shear pins or reinforcement, the coefficient of friction μ is set at 0.7. For contact surfaces that are roughened and have a roughness of $\geq$1.5 mm, the coefficient of friction μ ranges from 0.7 to 1.0. Following several trial computations, the friction coefficient of the joint surface between epoxy resin concrete and ordinary concrete was determined to be 0.8. For models incorporating keyways on the interfaces, the normal action was designated as "hard contact", which implies there is no limitation to the amount of pressure that can be transferred across the contact surfaces. When the contact pressure becomes negative, indicating the initiation of separation between the two surfaces, the contact constraints at the joints are released. In contrast, models without keyways in the joint feature "soft contact" for normal actions, set as "table", allowing for some degree of penetration between the two surfaces. This setting is particularly effective for simulating the interactions between old and new concrete, permitting penetration with a displacement of 0.01 mm when the pressure reaches 10 kN.

Boundary conditions of the model mirrored the boundary settings employed in the experimental setup, with the sample's bottom designated as a fixed constraint. A reference point was set at the midpoint of the top of the sample for coupling purposes, and displacement loading was carried out at this reference point to prevent the stress concentration in the loading beam during the loading process. The analysis was carried out in two stages. In the first stage, the translation in the Z direction and the rotational degrees of freedom in the X, Y, and Z directions at the reference point were restricted. In the second stage, a downward displacement of 19 mm was applied at the reference point. The loading system was a replica of that which was used in the test. Figure 13 illustrates the boundary conditions and loading methods under the finite element model.

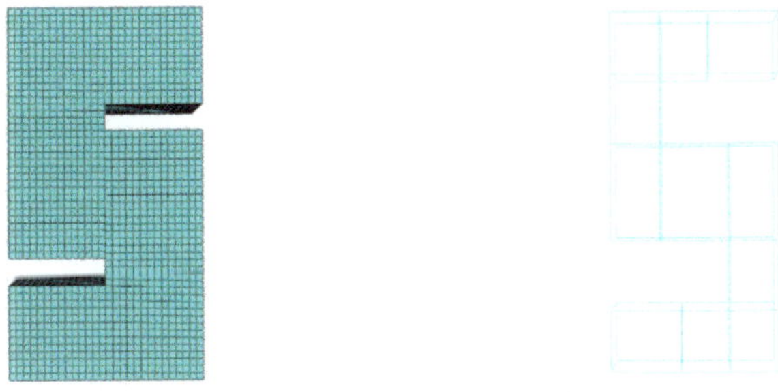

(a) Mesh division of concrete (b) Mesh division of reinforcement

Figure 12. Mesh division of the model (with EC12 as an example).

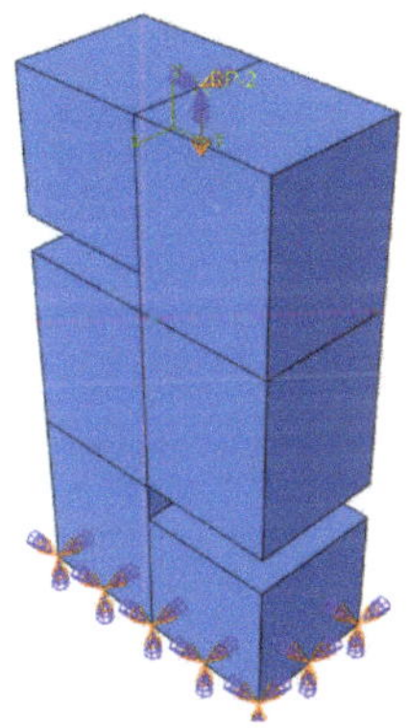

Figure 13. Boundary conditions and loading methods.

4.3. Model Verification

Samples EC12 and EC22–9030 were selected to confirm the precision of the finite element model. The distribution of the damage factor for concrete under compression served as a proxy for actual sample damage, which is depicted in Figure 14. Figure 14a indicates that the damage to sample EC12 was mainly concentrated in the ordinary concrete on the right side of the lower part of the concrete, which is basically in line with experimental observations. Similarly, as shown in Figure 14b, damage to sample EC22–9030 was primarily concentrated at the upper and lower sides of the keyway, with notable crushing of the concrete below the keyway, closely matching the test results.

Figure 15 and Table 7 present comparisons between the experimental results and the simulation results for the load–displacement curves of the samples. The simulation curves generally reflected the basic trends observed in the test curves. For sample EC12, the discrepancy between the simulated and actual peak loads was approximately 9.0%, and notably, the error in peak load displacement reached about 214%. In contrast, for sample EC22–9030, the error in peak load was around 1.1%, and the error in peak load displacement was 38.6%. These variations can largely be ascribed to the finite element simulation's omission of certain detrimental factors, including residual stresses and the adjustment of gaps in the test loading apparatus. On the other hand, the test results of the load–displacement curve of sample EC12 showed two peak points. In this paper, the load corresponding to the first peak point is taken as the peak load, which is different from the computation and value method in finite element simulation. If we consider comparing the average value of the load and displacement at the two peak points with the simulation results, the difference between the two was not significant. Despite these discrepancies, the finite element model successfully approximated the mechanical behavior of the direct shear samples on the interface between aged and fresh concrete.

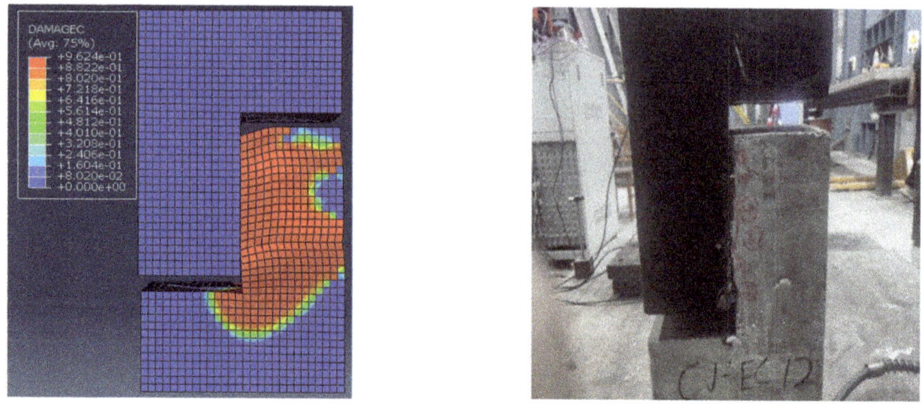

(**a**) EC12

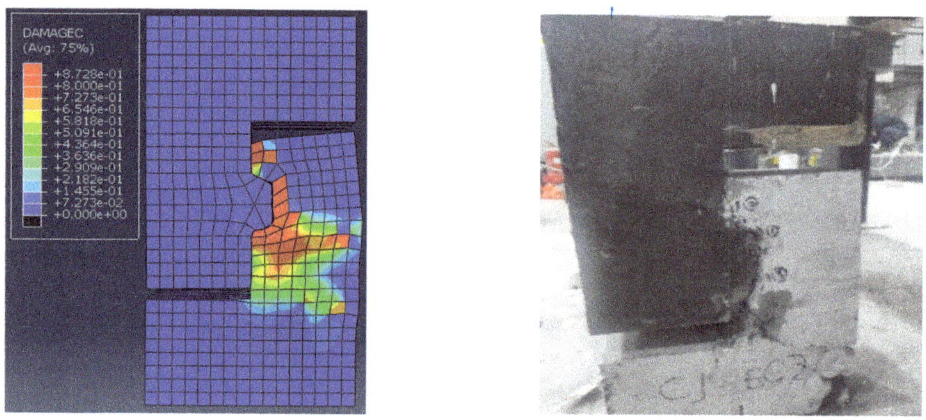

(**b**) EC22–9030

Figure 14. Comparison of failure modes of samples.

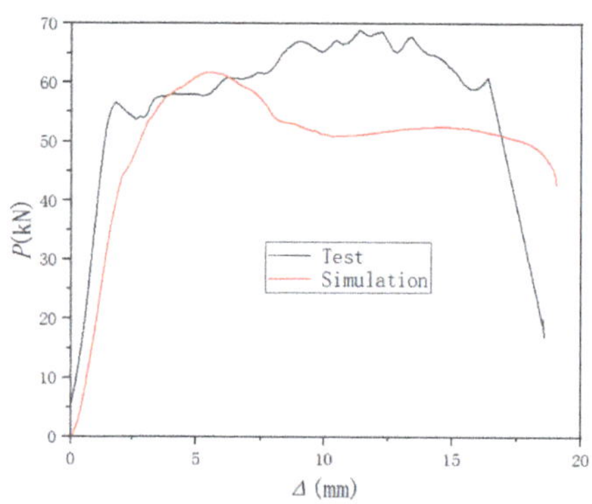

(**a**) EC12

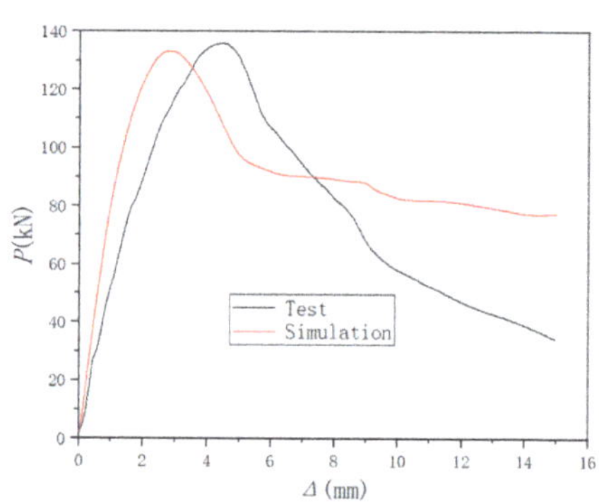

(**b**) EC22–9030

Figure 15. Comparative analysis of test outcomes and simulation outcomes of load–displacement curve.

Table 7. Comparative analysis of experimental and simulated values of shear capacity.

Sample Number	P_t/kN	Δ_t/mm	P_{nu}/kN	Δ_{nu}/mm	P_{nu}/P_t	Δ_{nu}/Δ_t
EC12	56.6	1.75	61.7	5.5	1.09	3.14
EC22-9030	136	4.6	134.5	2.9	0.99	0.63

Note: *Pt* and Δt represent the peak load and peak displacement from the test data, respectively. *P*nu and Δnu denote the peak load and peak displacement obtained from the numerical simulation.

4.4. *The Influence of Component Parameters on the Shear Performance of the Interface*

Recognizing the consensus on the beneficial effect of the keyway on the shear resistance of the interface, this paper mainly focuses on scenarios where no keyway is present on the interface. Based on the EC12, a variable parameter analysis using finite element modeling was conducted. This analysis ensured that certain parameters, such as the strength of the post-cast epoxy resin concrete and the stirrup ratio of the beam–column, remained constant. Conversely, five variables were systematically altered: the interface friction coefficient, the strength of the pre-cast concrete, the longitudinal reinforcement ratio, the confining stress, and both the type and strength of the post-cast material. A total of five groups of 19 finite element models were established, and the specific parameters for each sample are shown in Table 8.

Table 8. Parameter setting of direct shear model.

Sample Number	Coefficient of Friction	Pre-Cast Concrete Strength Grade	Reinforcement Ratio/%	Confining Stress/kN	Post-Cast Material Type
EC12	0.8	C30	0.67(4Φ8)	---	Epoxy resin concrete
EC1–0.7	0.7	C30	0.67(4Φ8)	---	Epoxy resin concrete
EC1–0.9	0.9	C30	0.67(4Φ8)	---	Epoxy resin concrete
EC1–1.0	1.0	C30	0.67(4Φ8)	---	Epoxy resin concrete
EC2–C20	0.8	C20	0.67(4Φ8)	---	Epoxy resin concrete
EC2–C40	0.8	C40	0.67(4Φ8)	---	Epoxy resin concrete
EC2–C50	0.8	C50	0.67(4Φ8)	---	Epoxy resin concrete
EC2–C60	0.8	C60	0.67(4Φ8)	---	Epoxy resin concrete
EC2–C70	0.8	C70	0.67(4Φ8)	---	Epoxy resin concrete
EC2–C80	0.8	C80	0.67(4Φ8)	---	Epoxy resin concrete
EC3–1.0	0.8	C30	1.0(6Φ8)	---	Epoxy resin concrete
EC3–1.34	0.8	C30	1.34(8Φ8)	---	Epoxy resin concrete
EC3–1.67	0.8	C30	1.67(10Φ8)	---	Epoxy resin concrete
CC4–C30	0.8	C30	0.67(4Φ8)	---	Ordinary concrete (C30)
CC4–C40	0.8	C30	0.67(4Φ8)	---	Ordinary concrete (C40)
CC4–C50	0.8	C30	0.67(4Φ8)	---	Ordinary concrete (C50)
EC5–5	0.8	C30	0.67(4Φ8)	5	Epoxy resin concrete
EC5–10	0.8	C30	0.67(4Φ8)	10	Epoxy resin concrete
EC5–15	0.8	C30	0.67(4Φ8)	15	Epoxy resin concrete
EC5–20	0.8	C30	0.67(4Φ8)	20	Epoxy resin concrete

4.4.1. Influence of Friction Coefficient of Joint Interface

The coefficients of the friction coefficients μ were derived from the guidelines specified by pertinent American and European standards, as illustrated in Tables 9 and 10.

Table 9. Code for design of concrete structures ACI318 [18].

Contact Surface Type	Coefficient of Friction μ
The cast-in situ concrete	1.4
The hardened concrete surface is clean, free of slurry, and the surface is roughened so that the concave and convex depth is not less than 1/4 inch	1.0
The surface of hardened concrete is clean, with no floating slurry but no roughening	0.6
Rolled structural steel with shear pins or rebars on the contact surface	0.7

Table 10. European standard model code 2010 [19].

Contact Surface Type	Roughness Rt	Coefficient of Friction μ
Smoothness	<1.5 mm	0.5~0.7
Rough	≥1.5 mm	0.7~1.0
Very rough	≥3 mm	1.0~1.4

Note: Rt denotes the total vertical discrepancy between the utmost peak and the deepest valley within the designated measurement zone.

Figure 16 shows the stress cloud diagram for the samples with varying friction coefficients at the peak load, as well as the damage factor distribution diagram at the end of the test. It is evident that the friction coefficient had a slight influence on the failure mode of the samples. The peak stress of the samples appeared at the interface of the fresh-aged concrete. The post-cast part made a major contribution to the stress of the samples, and the damage was primarily localized in the upper part of the pre-cast concrete. Figure 17 presents the *P*-Δ curves of each sample. It can be observed that, with the rise in the friction coefficient, the initial stiffness of the interface increased slightly. After entering the plastic stress stage, the friction coefficient had a clearer effect on the sample's load-bearing capacity as the friction coefficient rose from 0.7 to 0.8, and the sample's load-bearing capacity rose from 59.7 to 61.7 kN, with a growth rate of 3.3%. The friction coefficient rose from 0.8 to 0.9, and the load-bearing capacity rose from 61.7 to 63.0 kN, with an increase rate of 2.0%. The load-bearing capacity rose from 63.0 to 65.3 kN, with a growth rate of 3%. The growth rate was 2.0%. From 0.9 to 1.0, the load carrying capacity increased from 63.0 kN to 65.3 kN, with a growth rate of 3.5%. In conclusion, it can be seen that a larger friction coefficient can maintain the shear load capacity and stiffness of the sample to a certain extent, as well as slow down the stiffness degradation.

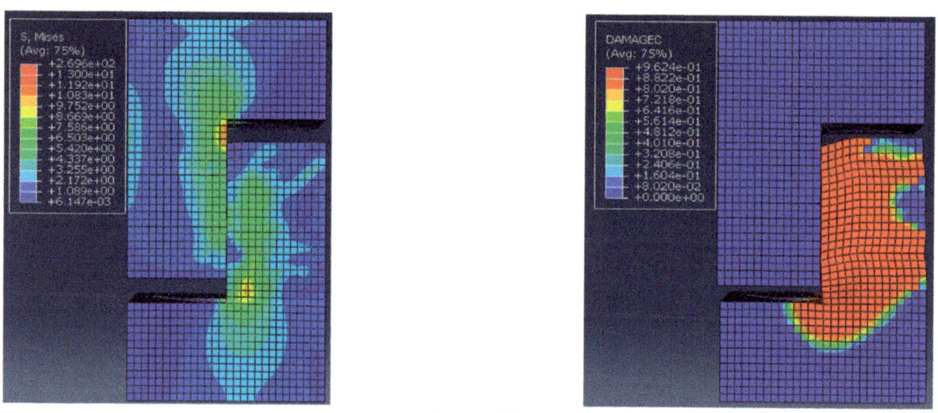

(**a**) EC12

Figure 16. *Cont.*

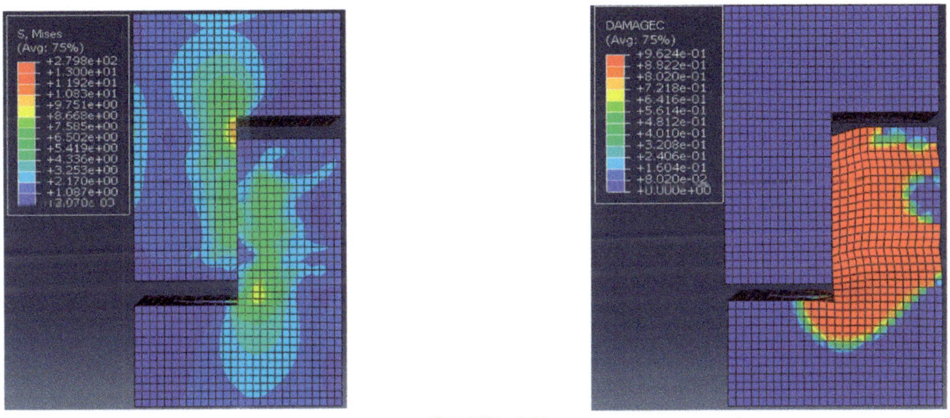

(**b**) EC1–0.7

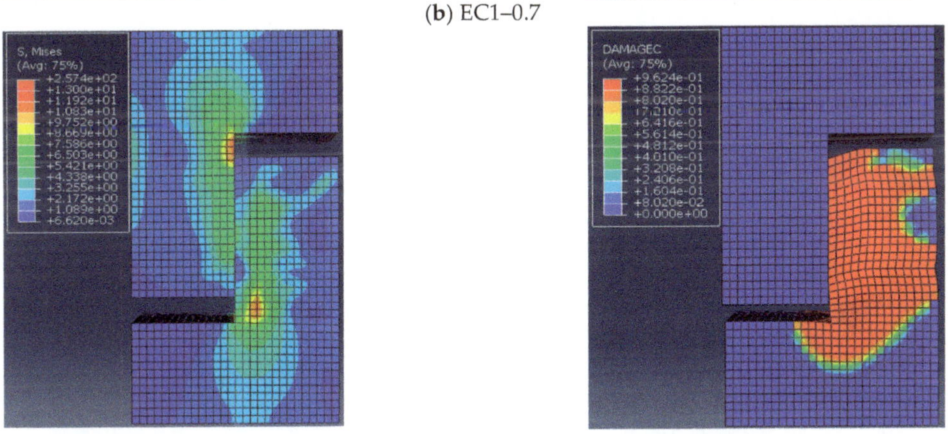

(**c**) EC1–0.9

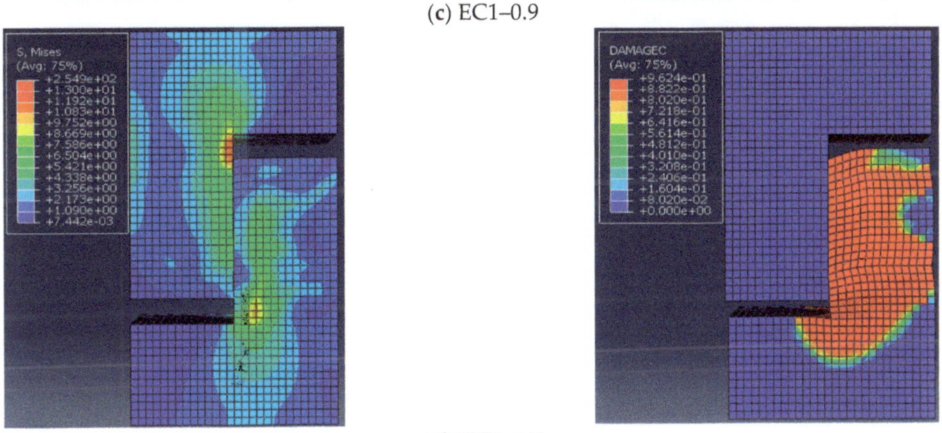

(**d**) EC1–1.0

Figure 16. Stress cloud diagram and damage factor distribution diagram of samples with various friction coefficients.

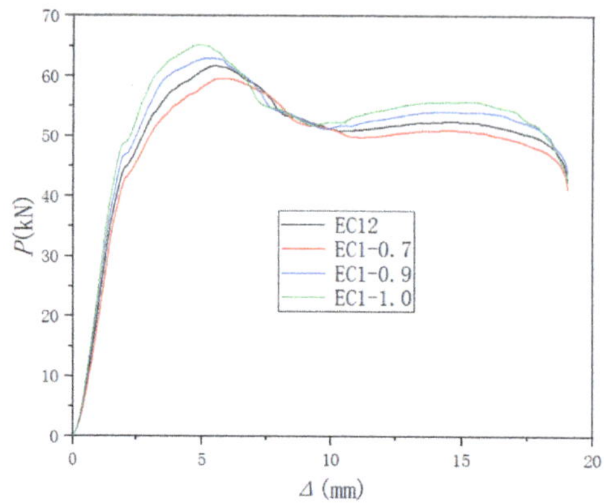

Figure 17. *P*-Δ curves of samples with different friction coefficients.

4.4.2. The Influence of Pre-Cast Concrete Strength

Figure 18 shows the stress cloud diagram and the damage factor distribution diagram at the end of the test for a typical model at peak load. These diagrams illustrate that the strength of the pre-cast concrete does not alter the failure mode of the interface. The peak stress in the sample occurred near the joint between the pre-cast and post-cast parts, with the post-cast section significantly adding to the overall stress. Figure 19 depicts the *P*-Δ curves of each sample, showing that while the concrete strength had minimal impact on the initial stiffness of the sample, it significantly affected the bearing capacity. As the concrete strength grade increased from C20 to C30, the bearing capacity rose from 56.6 kN to 61.7 kN, an increase of 8.3%. From C30 to C40, the capacity increased from 61.7 kN to 65.7 kN, an increase of 6.1%. From C40 to C50, it escalated from 65.7 kN to 68.4 kN, up by 3.9%. From C50 to C60, it rose from 68.4 kN to 70.7 kN, a rise of 3.3%. From C60 to C70, it increased from 70.7 kN to 72.1 kN, an increase of 1.9%. From C70 to C80, it climbed from 72.1 kN to 73.7 kN, an increase of 2.1%. Subsequently, as the concrete strength continued to increase, the rate of improvement in the sample's bearing capacity was diminished. Ultimately, the optimal performance of the sample is achieved when the strength of the post-cast epoxy resin concrete was close to or slightly exceeded that of the pre-cast concrete. This is attributed to the optimal interaction capabilities between the fresh and aged concrete at the interface, allowing for the most effective utilization of their combined mechanical properties.

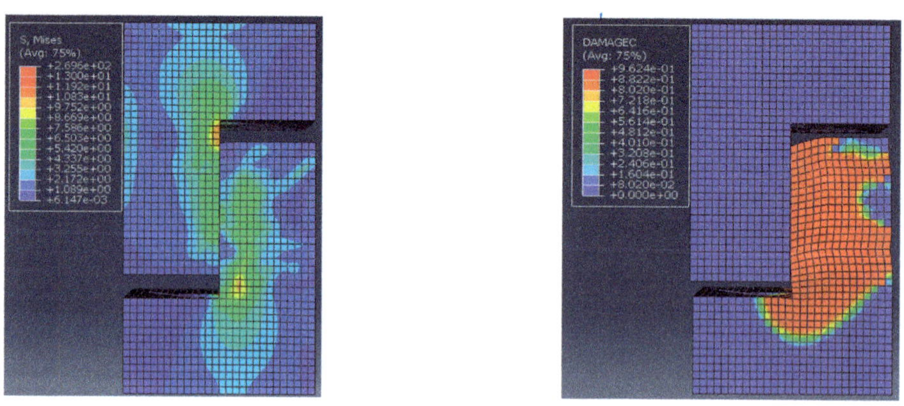

(**a**) EC12

Figure 18. *Cont.*

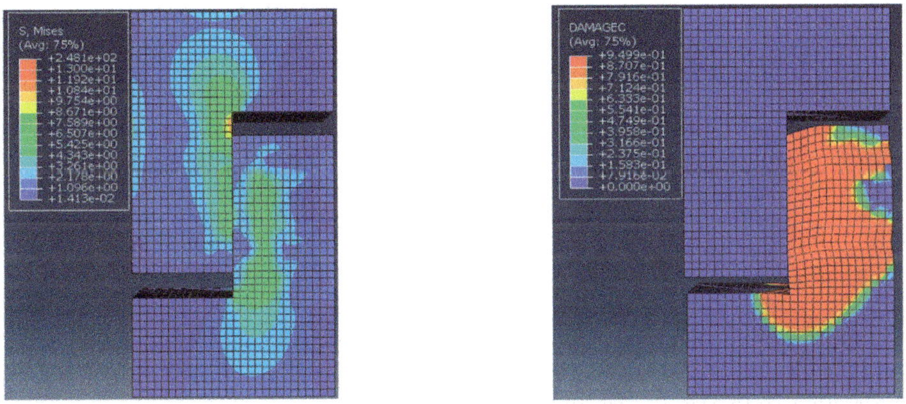

(**b**) EC2–C20

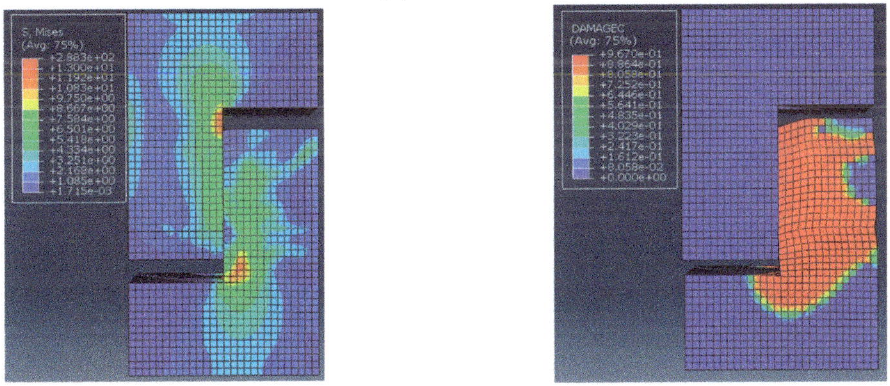

(**c**) EC2–C40

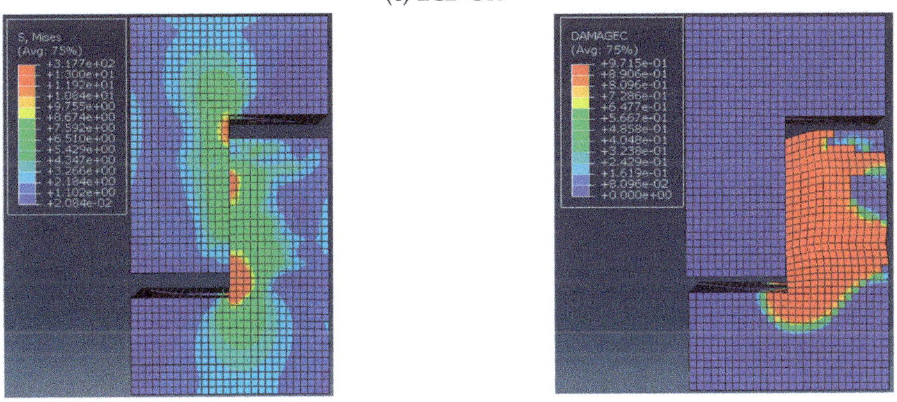

(**d**) EC2–C60

Figure 18. *Cont.*

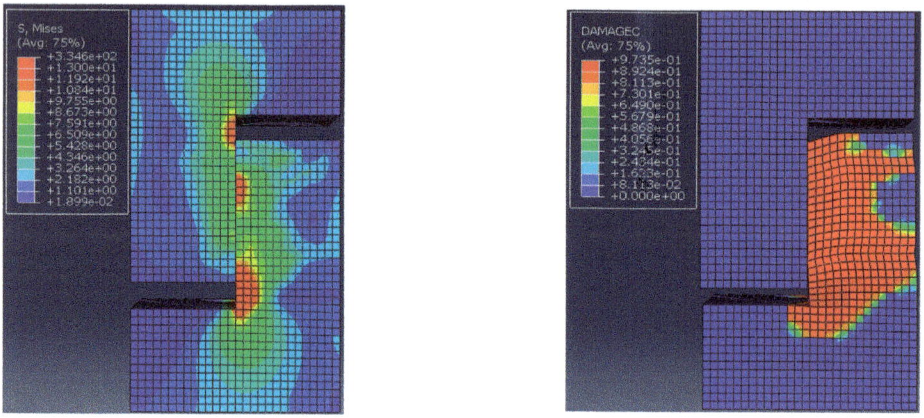

(**e**) EC2–C80

Figure 18. Stress cloud diagram and damage factor distribution diagram of various pre-cast concrete strength grade samples.

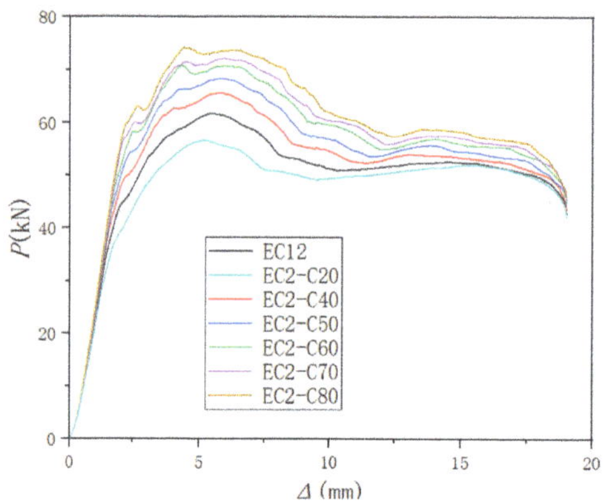

Figure 19. *P*-Δ curves of various pre-cast concrete strength grade samples.

4.4.3. The Impact of Longitudinal Reinforcement Ratio of Interface

Figure 20 displays the stress contour for each sample under peak load and the damage factor distribution map at the end of the test, varying according to the number of longitudinal reinforcements at the interspersed interface. Figure 21 presents the *P*-Δ curves for each sample, demonstrating that changes in reinforcement ratios did not alter the failure mode of the joint area, yet significantly impacted the bearing capacity. With the increase in the reinforcement ratio, the peak load increased, the peak displacement decreased, and the descending section of the curve became steeper, signifying accelerated degradation of the bearing capacity over the peak. For example, with a rise in the reinforcement ratio from 0.67% to 1.0%, the bearing capacity rose from 61.7 kN to 82.9 kN, achieving a growth rate of 25.6%. Increasing the reinforcement ratio further from 1.0% to 1.34%, the bearing capacity rose to 109.8 kN, and the growth rate was 24.5%. From 1.34% to 1.67%, it slightly decreased to 107.4 kN. This trend demonstrates that, while lower reinforcement ratios result in significant increases in bearing capacity, higher ratios do not continue to enhance the capacity, and instead lead to steeper post-peak curve descents and reduced ductility. This is primarily due to the interface's shear bearing capacity increasingly depending on the concrete's strength as the reinforcement ratio increases, thereby enhancing the brittle failure characteristic.

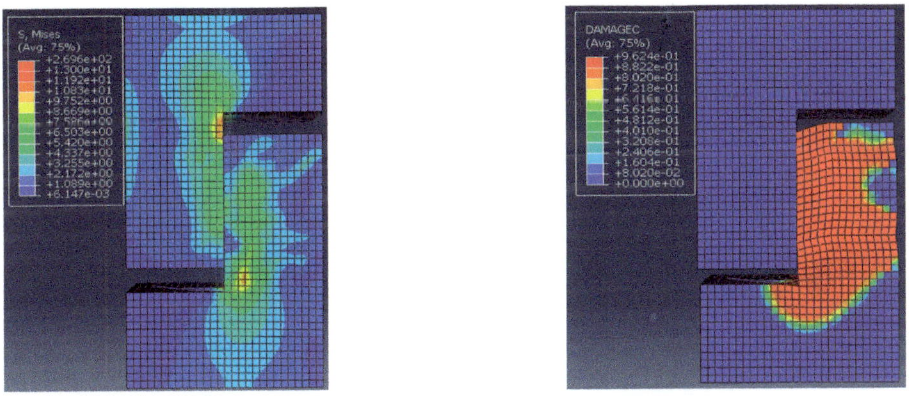

(**a**) EC12

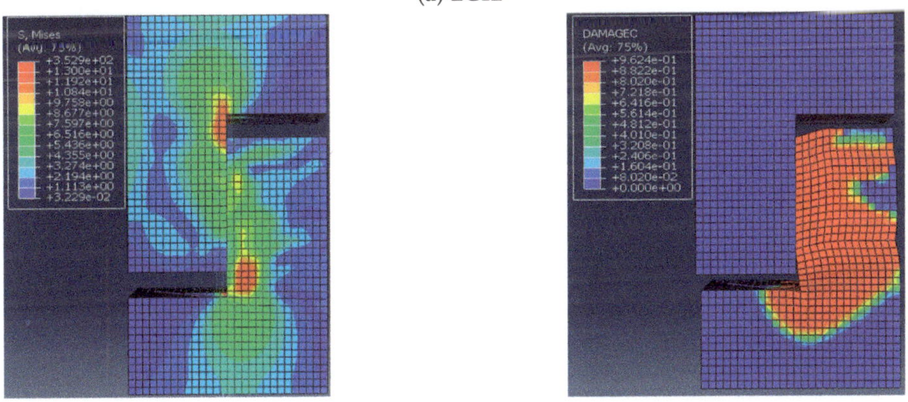

(**b**) EC3–1.0

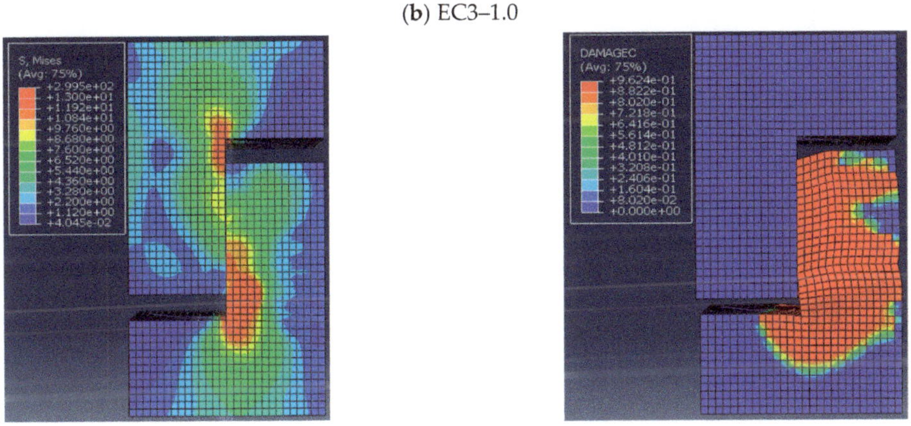

(**c**) EC3–1.34

Figure 20. *Cont.*

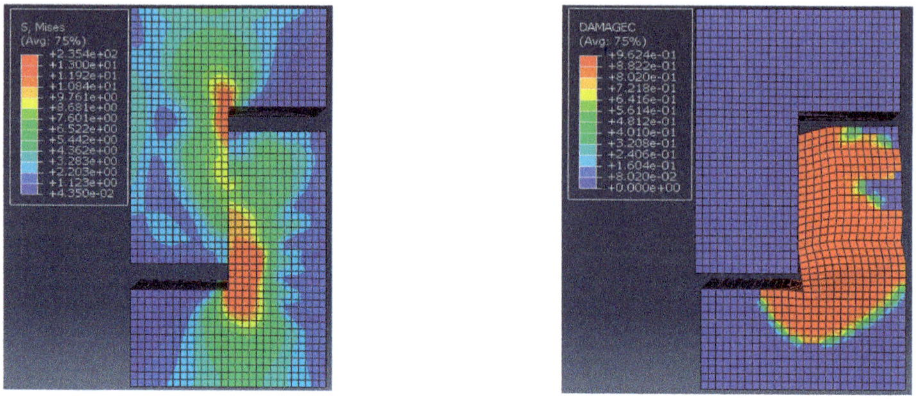

(**d**) EC3–1.67

Figure 20. Stress cloud diagram and damage factor distribution diagram of each sample, with various reinforcement ratios.

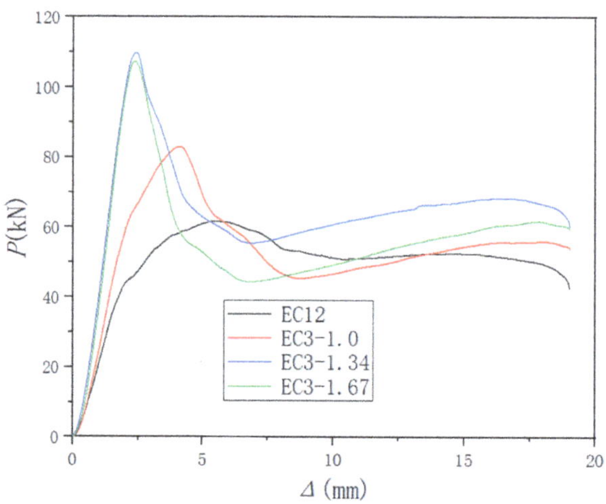

Figure 21. P-Δ curves of samples with various reinforcement ratios.

4.4.4. The Impact of Concrete Type and Strength of Post-Cast Part

Figure 22 presents the stress cloud diagrams and damage factor distribution diagrams for samples with varying strengths of post-cast ordinary concrete. Meanwhile, Figure 23 showcases the P-Δ curve of each sample. Notably, the damage observed in post-cast ordinary concrete samples is more obvious than in those using post-pouring epoxy resin concrete samples, affecting both old and new concrete. Analysis of Figure 23 reveals that the bearing capacities for samples EC12, CC4-C30, CC4-C40, and CC4-C50 were 61.7 kN, 48.1 kN, 53.0 kN, and 53.7 kN, respectively. Thus, as the strength of post-cast ordinary concrete progressively increased, so too did the bearing capacity of the samples, albeit without reaching the performance levels of samples that utilized epoxy resin concrete as the post-cast material. This outcome underscores the superior performance of epoxy resin materials in the post-cast phase of assembled structures, confirming their effectiveness in enhancing structural integrity and durability.

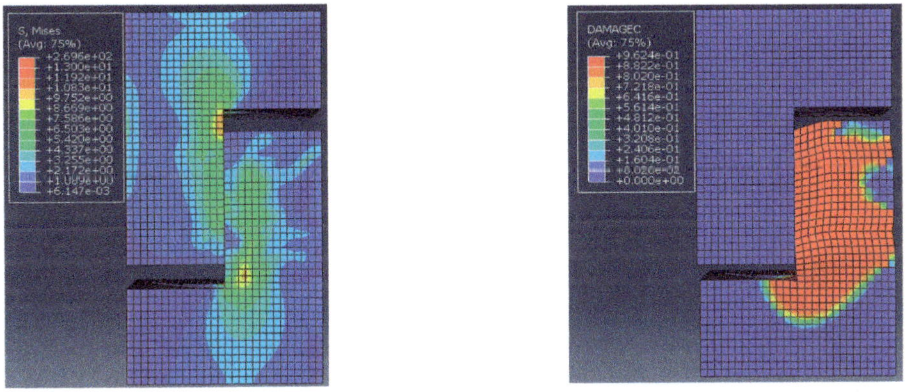

(**a**) EC12

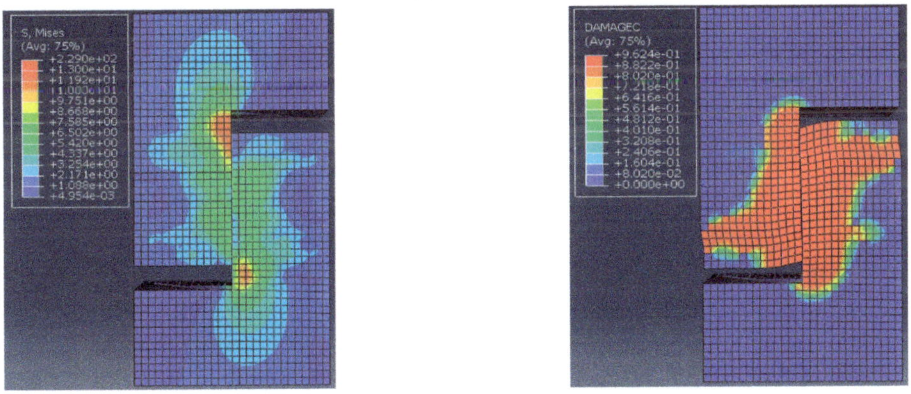

(**b**) CC4–C30

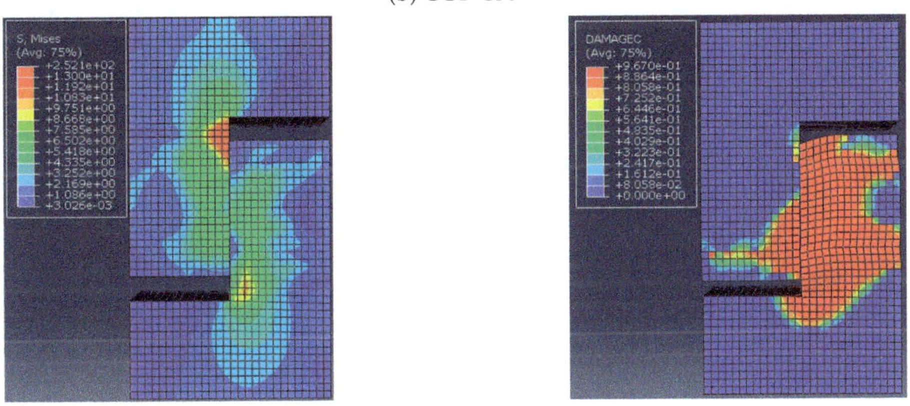

(**c**) CC4–C40

Figure 22. *Cont.*

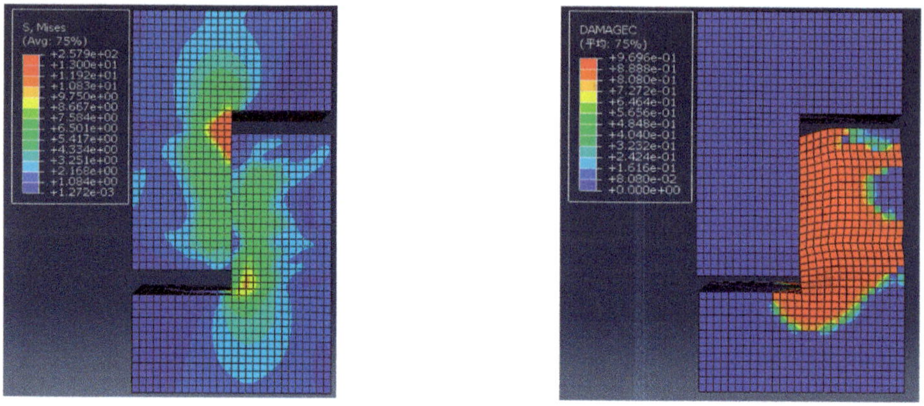

(**d**) CC4–C50

Figure 22. Stress cloud diagram and damage factor distribution diagram of various post-cast ordinary concrete strength samples.

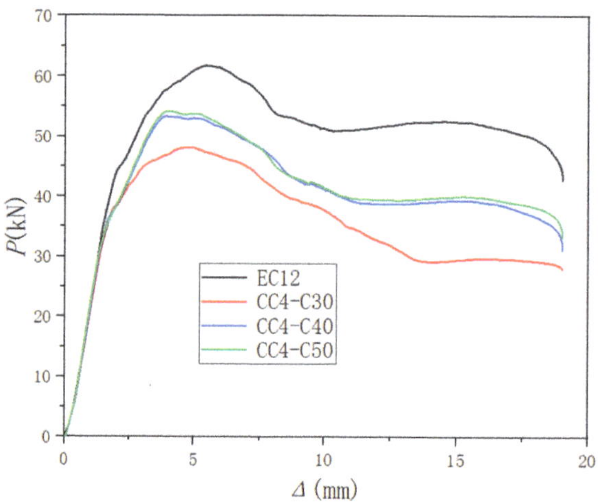

Figure 23. P-Δ curves of different post-cast ordinary concrete strength samples.

4.4.5. Effect of Confining Stress

Figure 24 displays the stress cloud diagrams and damage factor distribution diagrams for each sample at peak load under varying levels of confining stress. Figure 25 illustrates the P-Δ curves of these samples. At the first stage of loading, confining stress constrained the shear deformation of the samples, resulting in a steeper slope of the load–displacement curve and increased initial stiffness. As loading progressed, the ultimate load capacity of each sample rose, along with an increase in confining stress. Specifically, when the horizontal restraint force increased from 0 to 5 kN, the shear capacity increased from 61.7 kN to 71.2 kN, marking a growth rate of 13.3%. Further increases in the confining force from 5 kN to 10 kN enhanced the bearing capacity to 75.2 kN, with a growth rate of 5.3%. An increment from 10 kN to 15 kN increased the capacity to 78.9 kN, and the growth rate was 4.7%. When the horizontal restraint force increased from 15 kN to 20 kN, the capacity reached 78.9 kN to 83.1 kN, reflecting a growth rate of 5.1%.

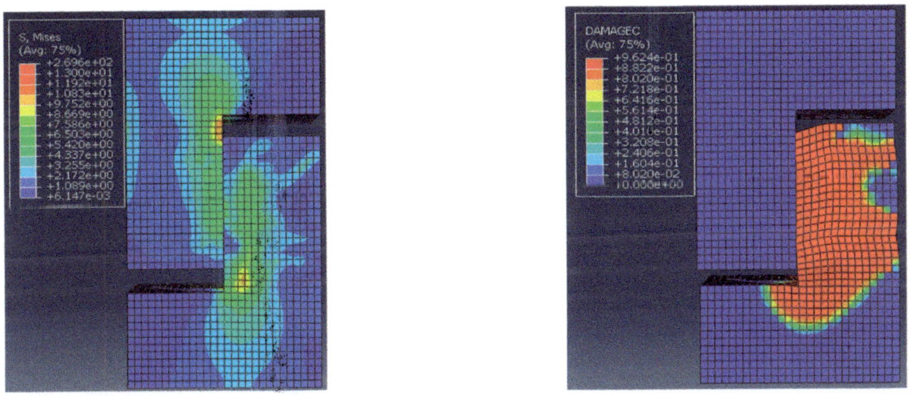

(**a**) EC12

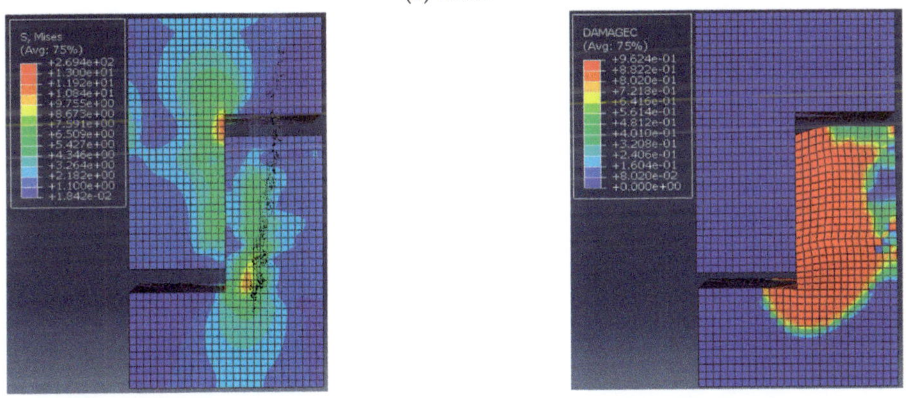

(**b**) EC5–5

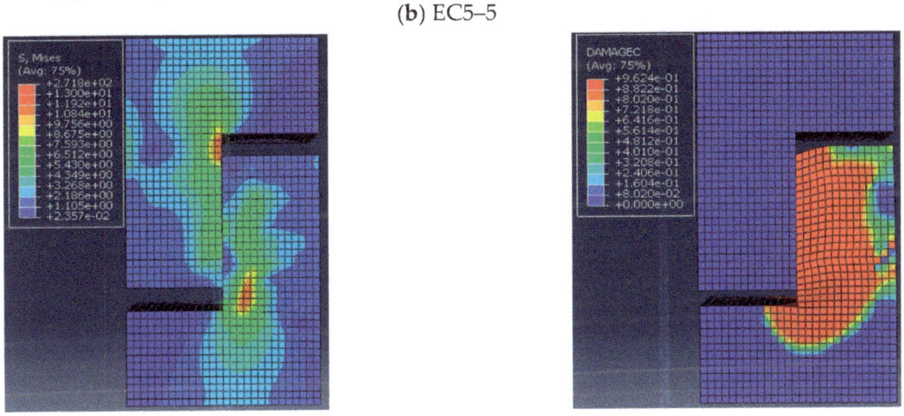

(**c**) EC5–10

Figure 24. *Cont.*

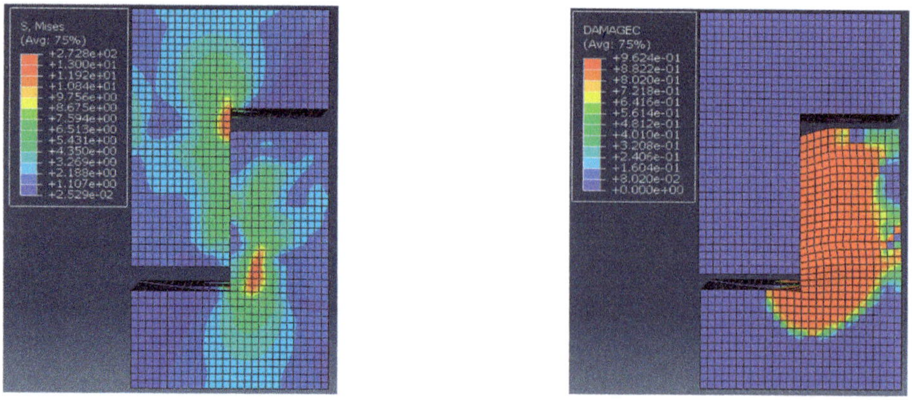

(**d**) EC5–15

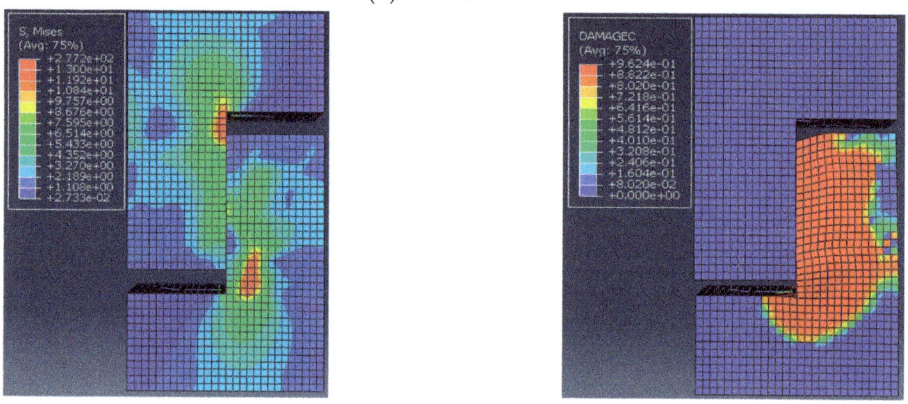

(**e**) EC5–20

Figure 24. Stress cloud diagram and damage factor distribution diagram of each sample with different confining stress levels.

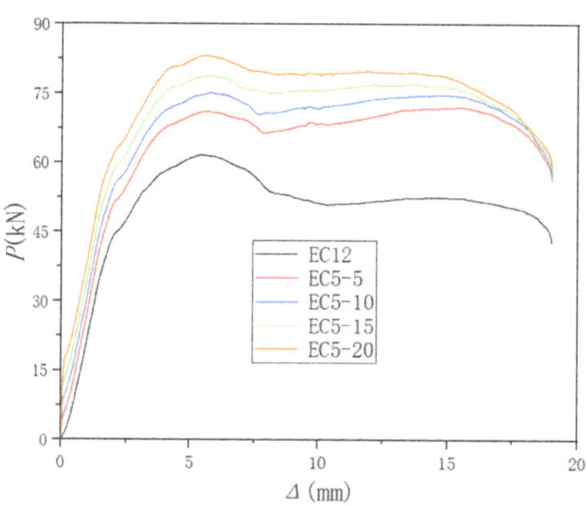

Figure 25. P-Δ curves of samples with different confining stress levels.

5. Conclusions

In this paper, epoxy resin concrete is used as the post-cast material for the direct shear performance testing, finite element simulation, and parameter analysis of fresh-aged concrete joints. The findings offer valuable insights into the broader application of epoxy resin concrete in urban construction:

(1) Epoxy resin concrete has excellent mechanical properties and bonds well with ordinary concrete. Utilizing it as a post-cast material and combining roughening of the interface with the inclusion of a keyway enhances the shear performance to levels comparable to or even surpassing those of cast-in situ samples. Therefore, it is suggested that epoxy resin concrete can be used as a post-cast material for assembled structures, and it is favored over ordinary concrete.

(2) When ordinary concrete is used as the post-cast material, increasing its strength enhances the shear resistance of the interface. Nevertheless, this resistance remains lower than that of both cast-in situ and post-cast epoxy resin concrete samples. Conversely, when the strength of the epoxy resin concrete closely matches that of the pre-cast concrete, the shear resistance of the interface is maximized. Regardless of whether ordinary concrete or epoxy resin concrete is used as the post-cast material, the absence of a keyway on the interface results in suboptimal shear resistance compared to cast-in-place constructions. Therefore, incorporating a keyway is crucial for achieving the desired structural performance. This implies that post-cast epoxy resin concrete, with strength comparable to pre-cast concrete and featuring a keyway joint surface, offers a superior solution for treating the interface between old and new concrete.

(3) The keyway is directly involved in the shear resistance of the interface. Enhancing the depth of the keyway more effectively enhances the shear bearing capacity than expanding its width, as it enlarges the shear working face. In accordance with applicable regulations, it is recommended that the depth of the keyway should be at least 30 mm, and the width of the keyway should not be less than three times the depth. Moreover, raising the friction coefficient of the interface not only boosts its shear bearing capacity, but also stabilizes its deformation performance, underscoring the importance of roughening the interface. The interface treatment methods that can be adopted include wetting treatment, application of interfacial agents, keyway setting, etc. Conversely, while raising the longitudinal reinforcement ratio enhances the shear capacity, it adversely affects the deformation performance, suggesting that this ratio should not exceed 1.0%. Furthermore, enhancing the confining stress improves both the shear bearing capacity and the stability of the deformation performance of the interface.

Author Contributions: Conceptualization, P.C.; Software, H.W. and S.Z.; Writing—original draft, H.W.; Writing—review & editing, P.C. and X.Z.; Project administration, P.C. and X.Z. All authors have read and agreed to the published version of the manuscript.

Funding: This research was funded by the National Natural Science Foundation of China, grant number 51678389.

Data Availability Statement: Data are contained within the article.

Conflicts of Interest: The authors declare no conflict of interest.

References

1. Momayez, A.; Ehsani, M.; Ramezanianpour, A.; Rajaie, H. Comparison of methods for evaluating bond strength between concrete substrate and repair materials. *Cem. Concr. Res.* **2005**, *35*, 748–757. [CrossRef]
2. He, Y.; Zhang, X.; Hooton, R.D.; Zhang, X. Effects of interface roughness and interface adhesion on new-to-old concrete bonding. *Constr. Build. Mater.* **2017**, *151*, 582–590. [CrossRef]
3. Julio, E.N.; Branco, F.A.; Silva, V.D. Concrete-to-concrete bond strength. Influence of the roughness of the substrate surface. *Constr. Build. Mater.* **2004**, *18*, 675–681. [CrossRef]

4. Jang, H.O.; Lee, H.S.; Cho, K.; Kim, J. Experimental study on shear performance of plain construction joints integrated with ultra-high performance concrete (UHPC). *Constr. Build. Mater.* **2017**, *152*, 16–23. [CrossRef]
5. Jiang, H.; Hu, Z.; Cao, Z.; Gao, X.; Tian, Y.; Sun, X. Experimental and numerical study on shear perfor-mance of externally prestressed precast UHPC segmental beams without stirrups. In *Structures*; Elsevier: Amsterdam, The Netherlands, 2022; Volume 46, pp. 1134–1153.
6. Yuan, A.; Zhao, X.; Lu, R. Experimental investigation on shear performance of fiber-reinforced high-strength concrete dry joints. *Eng. Struct.* **2020**, *223*, 111159. [CrossRef]
7. Yang, J.; Xia, J.; Cheng, C.; Wang, J.; Zhang, J.; Wang, G. Research on the bonding performance of UHPC–NC inter-faces with different sizes of grooves. *Front. Mater.* **2022**, *9*, 859717. [CrossRef]
8. Ye, M.; Li, L.; Li, H.; Zhou, C. Shear behavior of joints in precast UHPC segmental bridges under direct shear load-ing. *Constr. Build. Mater.* **2022**, *357*, 129212. [CrossRef]
9. Cao, J.; Wu, F.; Fan, X.; Lei, S.; Liu, L.; Wang, X.; Peng, K. Experimental and analytical studies on direct shear performance of UHPC keyed joints. *Eng. Struct.* **2024**, *311*, 118209. [CrossRef]
10. Wang, H.; Zhou, Z.; Zhang, Z.; Zou, Y.; Jiang, J.; Zeng, X. Experimental and numerical studies on shear behavior of prefabricated bridge deck slabs with compact UHPC wet joint. *Case Stud. Constr. Mater.* **2023**, *19*, e02362. [CrossRef]
11. Zhang, W.; Yan, B.; Ye, Y.; Yi, W. Direct shear test study on old and new concrete. *J. Build. Eng.* **2024**, *82*, 108391. [CrossRef]
12. Deng, M.K.; Ma, F.D.; Ye, W.; Yin, P.F. Experimental study on aseismic behavior of locally used HDC assembled frame beam-column joints. *Eng. Mech.* **2019**, *36*, 68–78.
13. Chen, P.; Xu, S.; Zhou, X.; Xu, D. An Experimental Study on Flexural-Shear Behavior of Composite Beams in Precast Frame Structures with Post-Cast Epoxy Resin Concrete. *Buildings* **2023**, *13*, 3137. [CrossRef]
14. Shamass, R.; Zhou, X.; Wu, Z. Numerical analysis of shear-off failure of keyed epoxied joints in precast concrete segmental bridges. *J. Bridge Eng.* **2017**, *22*, 04016108. [CrossRef]
15. Gopal, B.A.; Hejazi, F.; Hafezolghorani, M.; Voo, Y.L. Shear Strength of Dry and Epoxy Joints for Ultra-High-Performance Fiber-Reinforced Concrete. *ACI Struct. J.* **2020**, *117*, 279–288. [CrossRef]
16. Beeldens, A.N.N.E.; Monteny, J.; Vincke, E.; De Belie, N.; Van Gemert, D.; Taerwe, L.; Verstraete, W. Resistance to biogenic sulphuric acid corrosion of polymer-modified mortars. *Cem. Concr. Compos.* **2001**, *23*, 47–56. [CrossRef]
17. Qian, Y.; Li, Z.; Jin, Y.; Wang, R. Experimental Study on Axial Tension Members of a New Epoxy Resin Concrete. *Sci. Adv. Mater.* **2021**, *13*, 2005–2015. [CrossRef]
18. Xiang, Q.; Xiao, F. Applications of epoxy materials in pavement engineering. *Constr. Build. Mater.* **2020**, *235*, 117529. [CrossRef]
19. Rahman, M.M.; Akhtarul Islam, M. Application of epoxy resins in building materials: Progress and prospects. *Polym. Bull.* **2022**, *79*, 1949–1975. [CrossRef]
20. El-Hawary, M.; Al-Khaiat, H.; Fereig, S. Fereing Effect of Sea Water on Epoxy-repaired Concrete. *Cem. Concr. Compos.* **1998**, *20*, 41–52. [CrossRef]
21. Yu, T. *The Experimental Study of Repairing Concrete Cracks With Epoxy Resin Grouting*; Beijing University of Technology: Beijing, China, 2016. (In Chinese)
22. Zhang, W.; Guan, X.; Ren, J.; Gu, X. Experimental study on chloride permeability of concrete surface-treated with epoxy resin. *J. Build. Mater.* **2008**, *11*, 339–344.
23. Gil-Martín, L.M.; Rodríguez-Suesca, A.E.; Fernández-Ruiz, M.A.; Hernández-Montes, E. Cyclic behavior of RC beam-column joints with epoxy resin and ground tire rubber as partial cement replacement. *Constr. Build. Mater.* **2019**, *211*, 659–674. [CrossRef]
24. El-Mandouh, M.A.; Abd El-Maula, A.S. Shear strength of epoxy-modified reinforced concrete beams. *Innov. Infrastruct. Solut.* **2021**, *6*, 1–14. [CrossRef]
25. Jin, Y. *Research of Epoxy Resin Concrete (Mortar) Modified and Application*; Jilin Jianzhu University: Changchun, China, 2010.
26. Li, X. *Study on the Behavior of the Axial Compression Members of the New Epoxy Resin Concrete*; Jilin Jianzhu University: Changchun, China, 2019.
27. *GB 50010-2010*; Code for Design of Concrete Structures. China Architecture & Building Press: Beijing, China, 2015. (In Chinese)
28. Joint ACI-ASCE Committee 550. Emulating Castce Detailing in Precast Concrete Structures. 2001. Available online: http://civilwares.free.fr/ACI/MCP04/5501r_01.pdf (accessed on 1 December 2022).
29. *EN1992-1-2*; European Standard. Euro-Code2: Design of Concrete Structures-Part1: General Rules and Rules for Buildings. European Committee for Standardization: Bruxelles, Belgium, 2021.

buildings

MDPI

Article

Experimental Study on Seismic Performance of Prefabricated Monolithic Concrete–Polystyrene Panel Composite Wall Panels

Kaozhong Zhao [1,2], Zijia Fan [1,3,*], Yuming Zhang [1,*], Yufeng Xu [1,3] and Sihong Liu [1]

[1] School of Civil Engineering, Shandong Jianzhu University, Jinan 250101, China; zkz@sdjzu.edu.cn (K.Z.); 17865310248@163.com (Y.X.); 17852815378@163.com (S.L.)

[2] Key Laboratory of Building Structural Retrofitting and Underground Space Engineering, Shandong Jianzhu University, Ministry of Education, Jinan 250101, China

[3] Engineering Research Institute of Appraisal and Strengthening of Shandong Jianzhu University Co., Ltd., Jinan 250101, China

* Correspondence: fzj62127480@163.com (Z.F.); zym@sdjzu.edu.cn (Y.Z.)

Abstract: A normal composite wall panel is a structural component composed of polystyrene insulation boards and concrete surface layers reinforced with steel wire mesh. It can be entirely prefabricated in a factory or constructed with the concrete surface layers cast on-site. A novel prefabricated monolithic concrete–polystyrene panel composite wall panel (CPC wall panel) is proposed in this study. The CPC panel features a middle part that is prefabricated in the factory while the reinforced concrete regions at its two side ends are cast on-site. To evaluate the seismic performance of the wall panel, 18 CPC specimens were designed, manufactured, and quasi-statically tested, through which the structural behaviors, failure mode, and load-bearing capacity were studied. In addition, the influences of the height-to-width ratio and the vertical compressive stress level on the seismic performance of the CPC panels were also investigated. The test results showed that the connectors spaced at 400 mm × 500 mm could ensure the concrete layers on both sides of the polystyrene board worked collectively under seismic conditions. When subjected to lateral loads, the interface between the newly poured concrete and the existing concrete exhibited good bonding. Moreover, the failure mode of the CPC wall panel was largely correlated to the height-to-width ratio that, for specimens having four steel bars of 12 mm diameter and a height-to-width ratio greater than 1, the flexural failure was initially developed, followed by diagonal shear failure. In specimens with a height-to-width ratio of 1, flexural and diagonal shear failures occurred almost simultaneously. For specimens with a height-to-width ratio of less than 1, the final diagonal shear failure was predominant. The longitudinal reinforcing bars at the two ends of the CPC panels could effectively improve their lateral load-bearing capacity, with the enhancement influenced by the height-to-width ratio, the vertical load applied to the wall panel, and the cross-sectional area of the steel bars. In practice, the lateral load-bearing capacity of the CPC panel can be conservatively evaluated using the calculation method of the reinforced concrete shear walls. Finally, the ductility of the CPC specimens was affected by the height-to-width ratio and the axial compressive stress level, such that the specimens with a larger height-to-width ratio and lower axial compressive stress exhibited better ductility.

Keywords: prefabricated monolithic; reinforced concrete; composite wall panels; seismic performance; quasi-static test

1. Introduction

A composite wall panel is a component that allows the insertion of thermal insulation materials into the interior of a wall. The middle of the wall panel is composed of a thermal insulation material (such as polystyrene board), and the two sides of the polystyrene board are made of a steel mesh concrete surface layer or a mortar surface layer. Through the reasonable setting of connectors, the steel wire mesh concrete surface layers on both sides of the thermal insulation material can be connected as a whole to bear the entire force. This kind of wall panel not only improves the thermal insulation performance of the wall, but prevents the thermal insulation material in the external wall from falling off. Composite wall panels are divided into non-load-bearing wall panels for external maintenance structures only, and load-bearing wall panels used as vertical load-bearing components. There are more types of non-load bearing wall panels, such as Autoclaved Lightweight Aerated Concrete (ALC) wall panels in [1], Glass Fiber-Reinforced Cement (GRC) panels in [2], lightweight wall panels with a combination of foam concrete and cement fiber board developed by Fernando et al. [3], and lightweight sandwich wall panels consisting of Lightweight Foam Concrete (LFC) panels and Mineral Hydrated Foam Material (MHFM) insulation developed by Dong-Hyeon Shin et al. in [4]. Domestic and foreign scholars have carried out some research on the use of a composite wallboard as a vertical load-bearing component. In reference [5], a composite wallboard in which one side was a concrete load-bearing structure and the other side was a layer protecting the insulation board was studied. Under the action of a horizontal wind load, the protective layer, the insulation board, and the structure remained sound and did not fall off. Amran et al. [6] carried out a study on the vertical load-bearing capacity of sandwich composite wall panels faced with foam concrete, wherein both sides of the face contained a reinforcement mesh comprising 6 mm diameter steel trusses for tying. The study determined that the connection ensured the face sides of the wall panels were jointly loaded, and it was found that the wall panels' load-bearing capacity was related to the height-to-width ratio. Pavese et al. [7] conducted seismic performance tests on a structural model of a two-story house with concrete sandwich composite wall panels, with field-sprayed concrete containing a galvanized steel wire mesh on both sides of the insulated panels, and tensile reinforcing bars with a diameter of 8 mm and a spacing of 300 mm between the upper and lower wall panels. The tests showed that openings in the wall panels reduced their bearing capacity, and the structure showed good integrality.

References [8–13] studied various types of wall panels with different connectors, such as GFRP restraint connectors and grid-type shear connectors, and obtained the mechanical properties of the wall panels with various types of connectors. In addition, there have also been some studies on composite sandwich panels with different fibers added to concrete panels. In reference [14], the seismic performance of a thermal insulation composite wall panel with steel mesh ceramsite concrete containing alkali-resistant glass fiber was studied. The wall panel was cast-in-place, and the vertical reinforcement of the steel mesh was used for reliable anchoring. The results showed that the alkali-resistant fiber improved the crack resistance of the wall panel, and the wall panel showed good ductility. The study in [15] conducted seismic performance tests on a structural model of a two-story building constructed with ultra-high-performance concrete sandwich composite wall panels on the face, and the upper and lower wall panels were reliably connected using pre-embedded bolts. The tests showed that the structure achieved good seismic performance, but the connecting structure was complicated. References [16–18] studied a new type of precast concrete sandwich wall panel composed of two basalt fiber-reinforced polymer (FRP)-reinforced geopolymer concrete wythes and an insulation layer. The results showed that the larger the slenderness ratio was, the lower the axial bearing capacity of the wall panel was. Additionally, as the ratio of load eccentricity-to-sectional thickness increased, the axial load capacity decreased.

References [19–21] studied a load-bearing composite wall panel with a thickness of 120 mm and a concrete surface layer of 25 mm; the wall panels were all prefabricated, and it was found that when a vertical load was applied, the thinner surface layer was damaged as a result of local instability, and the concrete surface layer could not fully play its role. For example, the steel mesh in the surface layer was either discontinuous or the upper and lower ends were not reliably anchored, and no other structural measures were taken. Under lateral loading, when the height-to-width ratio of the wall panel was small, the horizontal cracks running along the discontinuous section of the steel mesh were opened, and the steel mesh could not fully play its role, as a result of which the bearing capacity of the component was low. In summary, the research on load-bearing composite wall panels can be divided into two types: one is the research on cast-in-place composite wall panels, whereby the surface concrete on both sides of the wall insulation panel is cast-in-place on-site, and it is reliably anchored via the vertical reinforcement of the surface steel mesh, or the structural steel reinforcement is set up to ensure that the wall panels are fixed at the upper and lower ends and can thus transfer the horizontal force. In the other case, the wall panels are all prefabricated in the factory and assembled on site, and the connections between the wall panels, as well as the fixings of the upper and lower ends, require special connectors. If the composite wall panel is cast-in-place, the construction process will be more costly and the cycle will be longer. If all the panels are prefabricated, the structure becomes complicated, limiting the transfer of the lateral load and resulting in poor seismic performance.

To overcome the limitations associated with fully prefabricated or fully cast-in-place composite wall panels, we have developed a new solution—the prefabricated monolithic composite wall panel. The thickness of the wall panels is 200 mm, the middle is composed of 100 mm thick polystyrene panels, and the two sides are composed of 50 mm thick steel wire mesh coated in concrete. A special new type of connector is used to connect the two sides of the steel wire mesh into a whole, and this can be fixed accurately to the position of the steel reinforcement mesh. Each wall panel's middle section is prefabricated in the factory and installed on-site. The steel wire mesh of the wall panel is discontinuous on the upper and lower layers without anchoring. The wall panels are mounted with a backing belt at the end and cast-in-place at the construction site, while the vertical reinforcement bars are set up on a post-pouring belt running through the upper and lower layers, and strong longitudinal reinforcement bars at the bottom of the structure are anchored into the foundation, forming a monolithic concrete–polystyrene composite load-bearing wall panel. No other connecting structures are inserted between and upon the upper and lower ends of the wall panels. In this paper, we designed and fabricated 18 specimens of prefabricated monolithic concrete–polystyrene panel composite wall panels. The seismic performance of the prefabricated monolithic composite wall panels under horizontal earthquake actions was investigated through quasi-static testing. The effects of the height-to-width ratio and vertical load on the seismic performance of the wall panels were analyzed, and the reliability of the connector arrangement was validated. The general flow chart for this is shown in Figure 1.

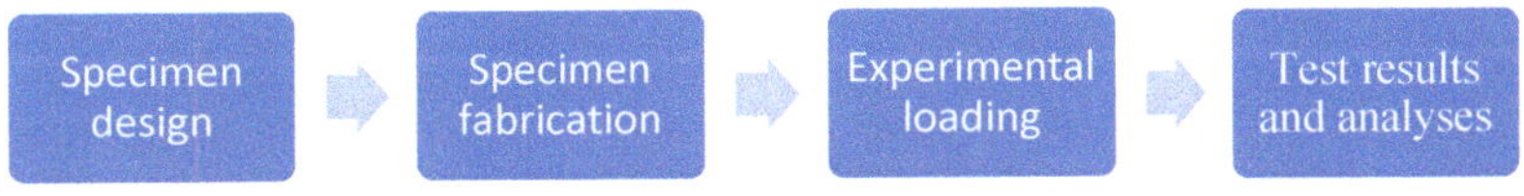

Figure 1. Flow chart.

2. Test Overview

2.1. Composition of Concrete–Polystyrene Composite Wall Panels

The CPC panel used in this study consists of a polystyrene board, steel wire mesh, and concrete facing applied on both sides, as shown in Figure 2. The wall panel has a thickness of 200 mm, with a 100 mm thick polystyrene board in the middle and a 50 mm thick concrete surface layer containing steel wire mesh on both sides. The two layers of steel wire mesh were fixed using special connectors, which not only conveniently connects the polystyrene boards with the steel mesh, but also accurately fixes the position of the steel mesh. This ensures that the concrete surface layer and the internal insulation sandwich layer can work effectively as a cohesive unit. The diameter of the steel mesh reinforcement is 3 mm, and the mesh size is 75 mm. The wall panel is prefabricated in specific factories and cast-in-place at both ends. Vertical reinforcement bars are incorporated through the upper and lower layers of the cast-in-place parts, forming an integrated composite wall panel structure.

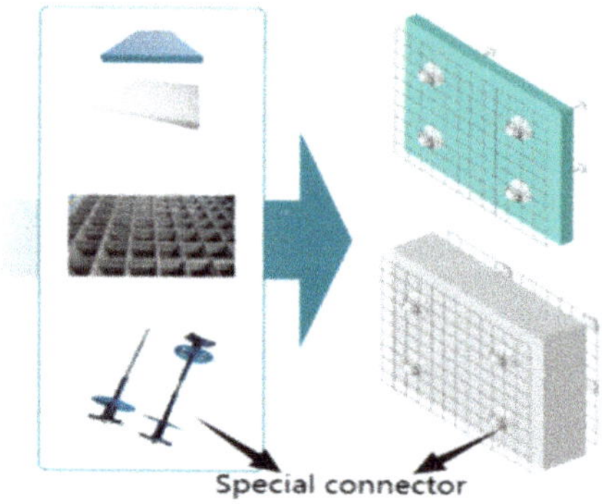

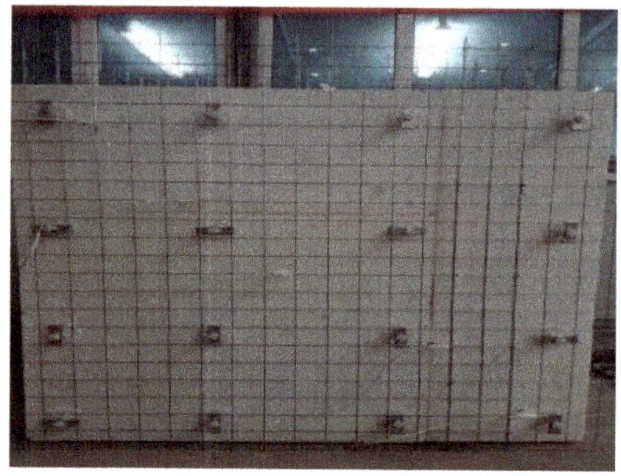

Figure 2. The structure of CPC plate.

Three different height-to-width ratios were tested in designing the specimens, with specific dimensions of 1500 × 1000 mm, 1500 × 1500 mm, and 1000 × 2000 mm, respectively. Vertical compressive stresses of 1.20 MPa, 0.80 MPa, and 0.40 MPa were applied to investigate the effects of vertical pressure on the seismic performance of the wall panels. A total of nine groups of 18 specimens was tested. The numbers and requirements of the specimens are shown in Table 1.

Table 1. Specimen number and requirements.

Groups	Specimen Number	Size of Specimen (H × W) mm	Vertical Compressive Stress (MPa)	Vertical Load (kN)
I	A1-1 A1-2	1500 × 1000	0.40	80
II	A2-1 A2-2	1500 × 1000	0.80	160
III	A3-1 A3-2	1500 × 1000	1.20	240
IV	B1-1 B1-2	1500 × 1500	0.40	120
V	B2-1 B2-2	1500 × 1500	0.80	240
VI	B3-1 B3-2	1500 × 1500	1.20	360
VII	C1-1 C1-2	1000 × 2000	0.40	160
VIII	C2-1 C2-2	1000 × 2000	0.80	320
IX	C3-1 C3-2	1000 × 2000	1.20	480

The specimens comprised a top beam and a floor beam to facilitate the application of vertical and lateral loads, with the anchoring of the specimens to the test stand. The top beam had a cross-section of 200 × 200 mm with 4C16 longitudinal reinforcement and A8@100 hoop reinforcement. The floor beam had a cross-section of 500 × 500 mm, with 8C25 longitudinal bars and A10@100 stirrups. The middle of the specimens was prefabricated, and the two ends comprised 200 mm of cast-in-place concrete. Each end of the specimens was equipped with 4C12 vertical reinforcement, with the reinforcement anchored in the top and bottom beams. The connectors were spaced vertically at 400 mm and horizontally at 500 mm.

The specific section size and reinforcement of the specimen are shown in Figure 3.

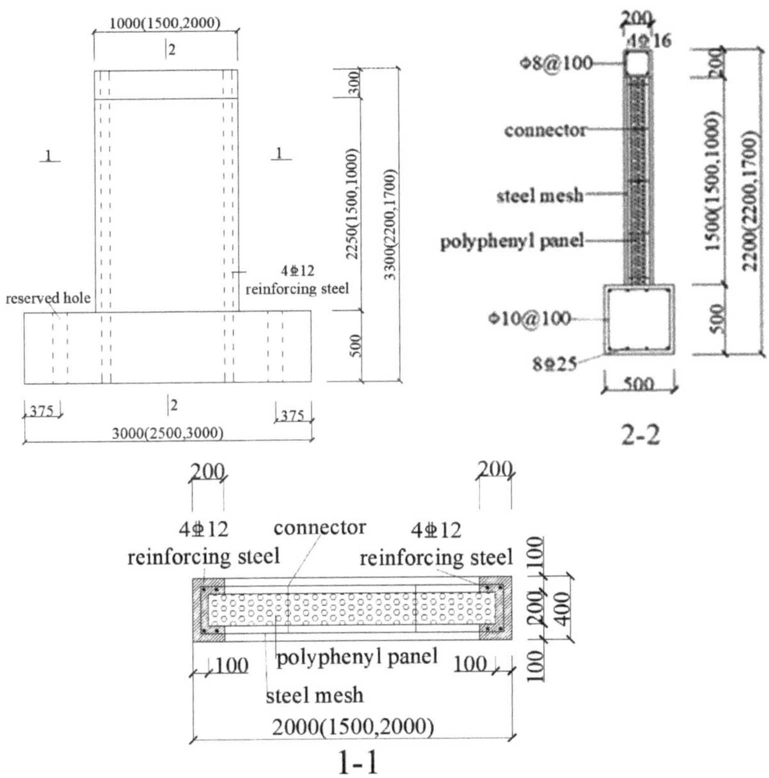

Note : The shadow part in the picture is cast-in-place, and the reinforcement is anchored into the bottom beam.

Figure 3. Construction drawing of A-type (B-type, C-type) specimens.

The concrete strength grade of the composite wall panel, top beam, and ground beam was C30, and the measured cube compressive strength of the composite wall panel concrete was 30.9 MPa. The yield strength of the C12 vertical reinforcement was measured to be 428 MPa, and the ultimate strength was 563 MPa. The ultimate strength of the steel wire mesh reinforcement was 650 MPa, and the conditional yield strength was 553 MPa (0.85 times the ultimate strength).

2.2. Loading Test

A vertical load was applied using a hydraulic pressure stabilizing loading system. During the test, a vertical load was first applied up to the required load value, which was kept constant during the application of the lateral load. An MTS electro-hydraulic servo loading system was used to apply the reciprocating lateral load, and the test loading device is shown in Figure 4.

Figure 4. Test loading device.

This study employed a displacement-controlled low-cycle reversed loading method, which involved the repetitive loading and unloading of the specimens in both positive and negative directions. The three sizes of specimens were loaded with a rate of increase of 0.2 mm until cracking occurred. After cracking, the specimen with a height-to-width ratio of 1.5:1 was loaded with a displacement increment of 1.0 mm, the specimen with a height-to-width ratio of 1:1 was loaded with a displacement increment of 0.8 mm, and the specimen with a height-to-width ratio of 1:2 was loaded with a displacement increment of 0.4 mm. Each displacement level was cycled once, as depicted in Figure 5. The test was stopped when the specimen's horizontal bearing capacity had reduced to 85% of its ultimate load-bearing capacity after cycling.

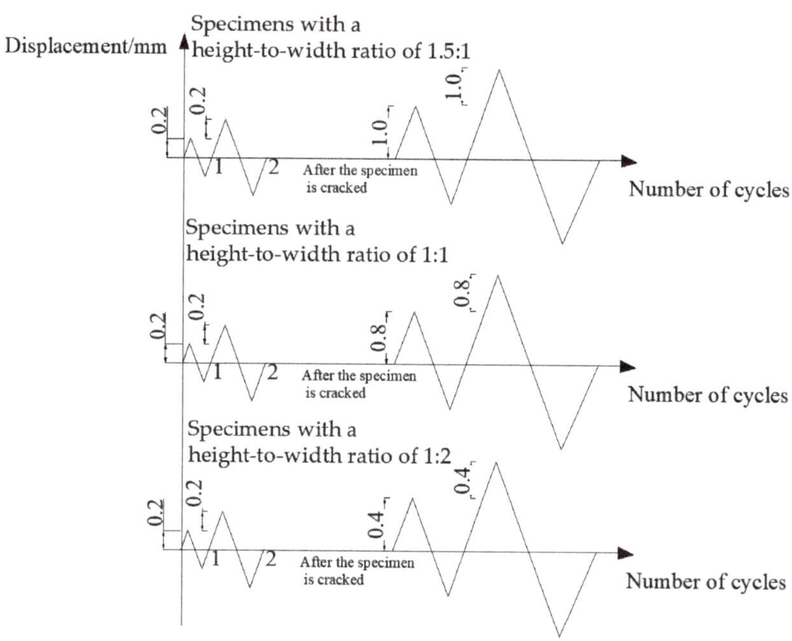

Figure 5. The relationship between the displacement and the number of cycles.

2.3. Layout of Measurement Points and Data Collection

The lateral load magnitude was automatically measured using the MTS loading system. External displacement gauges were placed in the centers of the top and bottom beams of the specimens, and data were collected using a high-performance data collector (TDS-540) to determine the real displacement of the specimen under lateral loading. Steel reinforcement strain gauges were arranged on the horizontal steel wire mesh and the reinforcing steel bar; the former were placed at five measurement points spaced evenly in the diagonal direction across the specimen, and the strain gauges on the reinforcing steel were placed at the bottom of the specimen. The arrangements of the displacement and reinforcement strain measurement points are shown in Figure 6.

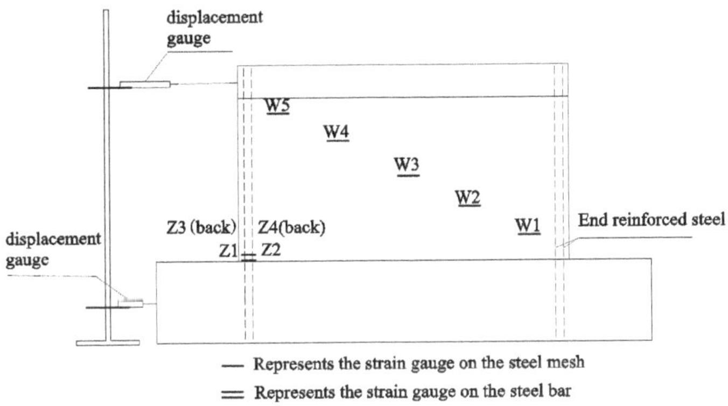

Figure 6. Layout of measuring points.

3. Test Results and Analyses

3.1. Failure Mode of Specimens

The modes of initial crack development in each specimen throughout the entire loading process were generally similar, with horizontal cracks emerging first under lateral loading. As the load increased, new horizontal cracks formed and gradually developed into oblique cracks. The reason for this was that the main tensile stress was generated in the specimen under both vertical and lateral loads. The main tensile stress at the ends was vertical, while the middle main tensile stress was at an angle with the cross-section of the specimen. When the main tensile stress exceeds the tensile strength of the concrete, cracks occur. Therefore, horizontal cracks were generated at the ends of the specimen, while oblique cracks developed in the middle. By observing the entire test phenomenon, it can be seen that specimens with different height-to-width ratios exhibited different failure modes. In the specimen with a height-to-width ratio of 1.5:1, the longitudinal reinforcements reached their yield point first, and noticeable horizontal cracks were observed at the bottom of the specimen. Subsequently, the concrete in the compression zone was crushed, marking the specimen's maximum bearing capacity. With the application of further load, a prominent main oblique crack formed diagonally along the specimen, significantly reducing its bearing capacity. Consequently, the specimen exhibited flexural failure initially, followed by diagonal shear failure.

When the specimen with a height-to-width ratio of 1:1 approached its ultimate load-bearing capacity, the longitudinal reinforcements in the ends reached their yield strength, leading to the emergence of vertical cracks in the compression zone and signs of concrete crushing. Finally, the main oblique cracks that emerged in the diagonal direction along the specimen caused shear failure in the oblique section. Although the final damage mode was oblique cross-section shear failure, observations from the test indicated the longitudinal reinforcement in the specimen reached its maximum bearing capacity and also yielded, and the compression zone of the concrete showed signs of crushing. These observations suggest that flexural failure had occurred or was on the verge of occurring. Specimens with a height-to-width ratio of 1:2 eventually suffered shear failure due to the development of oblique cracks. When the specimens reached their ultimate horizontal bearing capacity, some of the end reinforcing bars yielded, but the concrete in the compression zone was not crushed. The final failure mode was diagonal shear failure, and no flexural failure occurred. After shear failure, further loading resulted in increased displacement, but the horizontal bearing capacity significantly decreased; no new cracks were generated. The original cracks gradually became wider or more extended, and the concrete in the corner of the wall panel became crushed and dislodged, exposing the steel mesh and longitudinal reinforcement.

The specimens exhibited two types of failure modes, namely, flexural failure and diagonal shear failure. The analysis of the reasons behind the different failure modes was as follows: For specimens with a height-to-width ratio of 1.5:1, under the applied vertical load condition, the shear resistance under lateral load exceeded the bending capacity of the normal section, resulting in the occurrence of flexural failure first. For specimens with a height-to-width ratio of 1:1, the shear resistance was approximately equal to the bending capacity of the normal section. Consequently, flexural failure and oblique section failure occurred almost simultaneously. In the case of specimens with a height-to-width ratio of 1:2, the bending capacity of the normal section significantly exceeded the shear bearing capacity. As a result, oblique section failure occurred while flexural failure did not occur.

Under the action of quasi-static cyclic loading, the concrete surface layers on both sides of the polystyrene board remained whole throughout the testing process. The prefabricated and cast-in-place concrete were well combined without separation, indicating that the concrete surface layer and the polystyrene board became a single unit via the fixation effect of the steel mesh and the pull bolt connector. The pull bolt connector balanced the vertical and lateral loads of the surface layers on the left and right sides, ensuring an overall force equilibrium and the cohesive behavior of the wall panel during seismic events.

The typical failure mode of the specimens is shown in Figure 7.

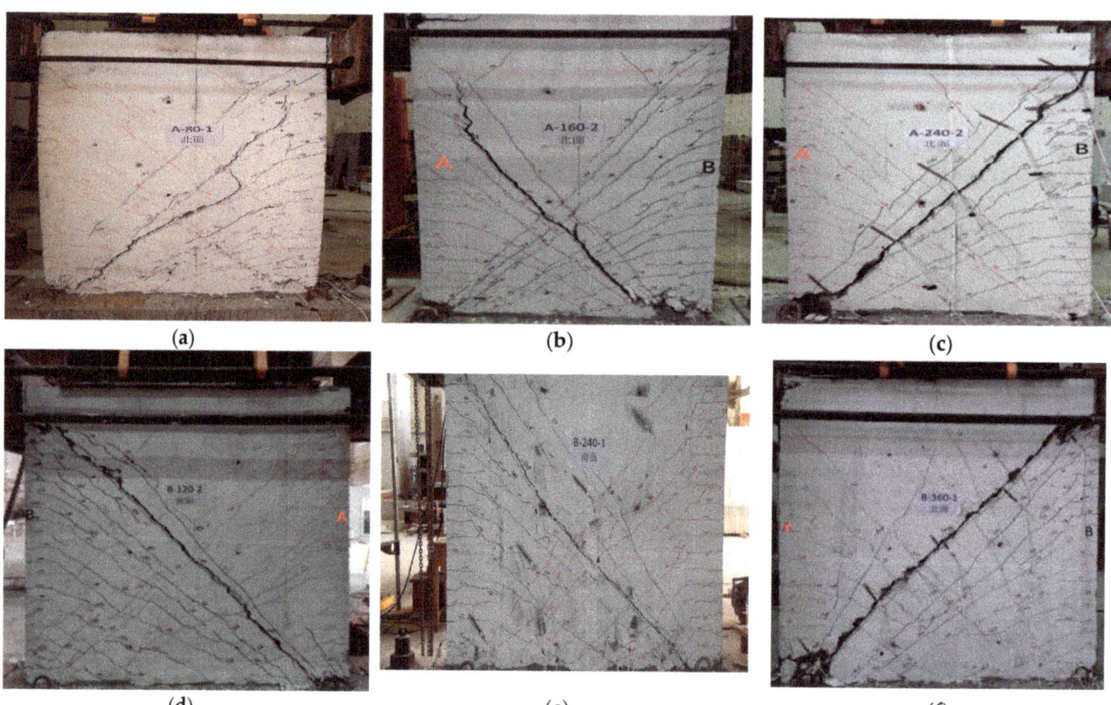

Figure 7. *Cont.*

(g) (h) (i)

Figure 7. Specimen failure mode: (**a**) specimen A1-1, (**b**) specimen A2-2, (**c**) specimen A3-2, (**d**) specimen B1-2, (**e**) specimen B2-1, (**f**) specimen B3-1, (**g**) specimen C1-1, (**h**) specimen C2-2 (**i**) specimen C3-1.

3.2. Hysteresis Curve and Energy Consumption

The hysteresis curves for each specimen are shown in Figure 8.

The hysteresis curves of the specimens generally adopted an inverted S shape, and there was a large shear slip. After reaching the ultimate load-bearing capacity, the stiffness and bearing capacity suddenly dropped. The reason was that the specimens all eventually underwent shear failure. Specimens with small height-to-width ratios and large axial pressure showed poor energy dissipation performances.

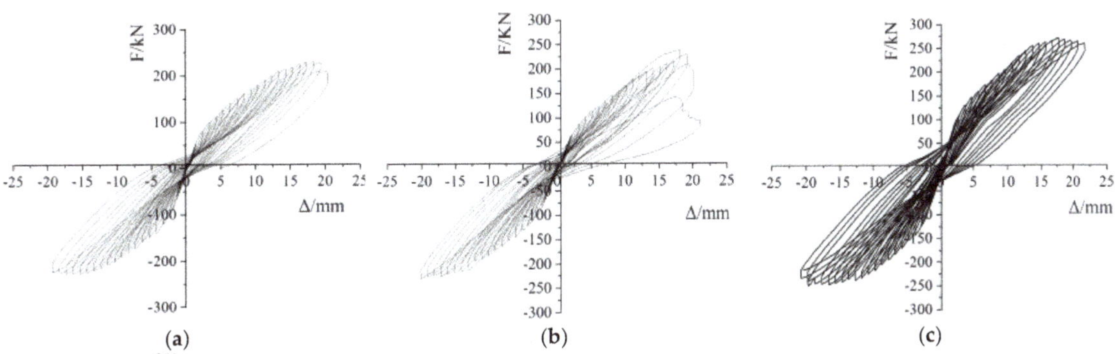

(a) (b) (c)

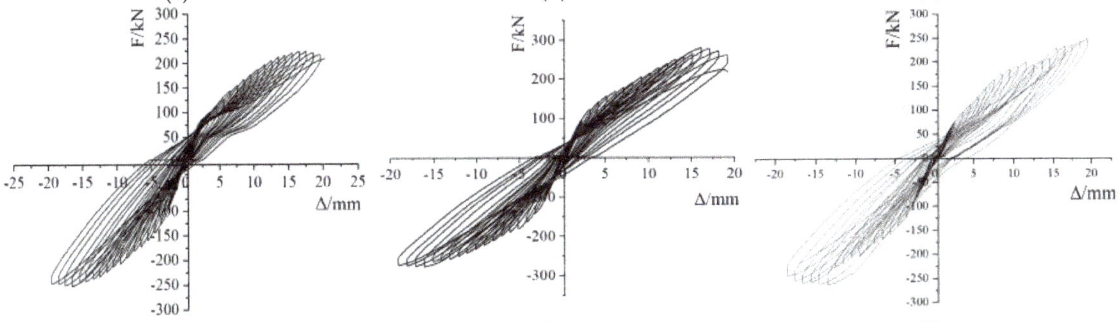

(d) (e) (f)

Figure 8. *Cont.*

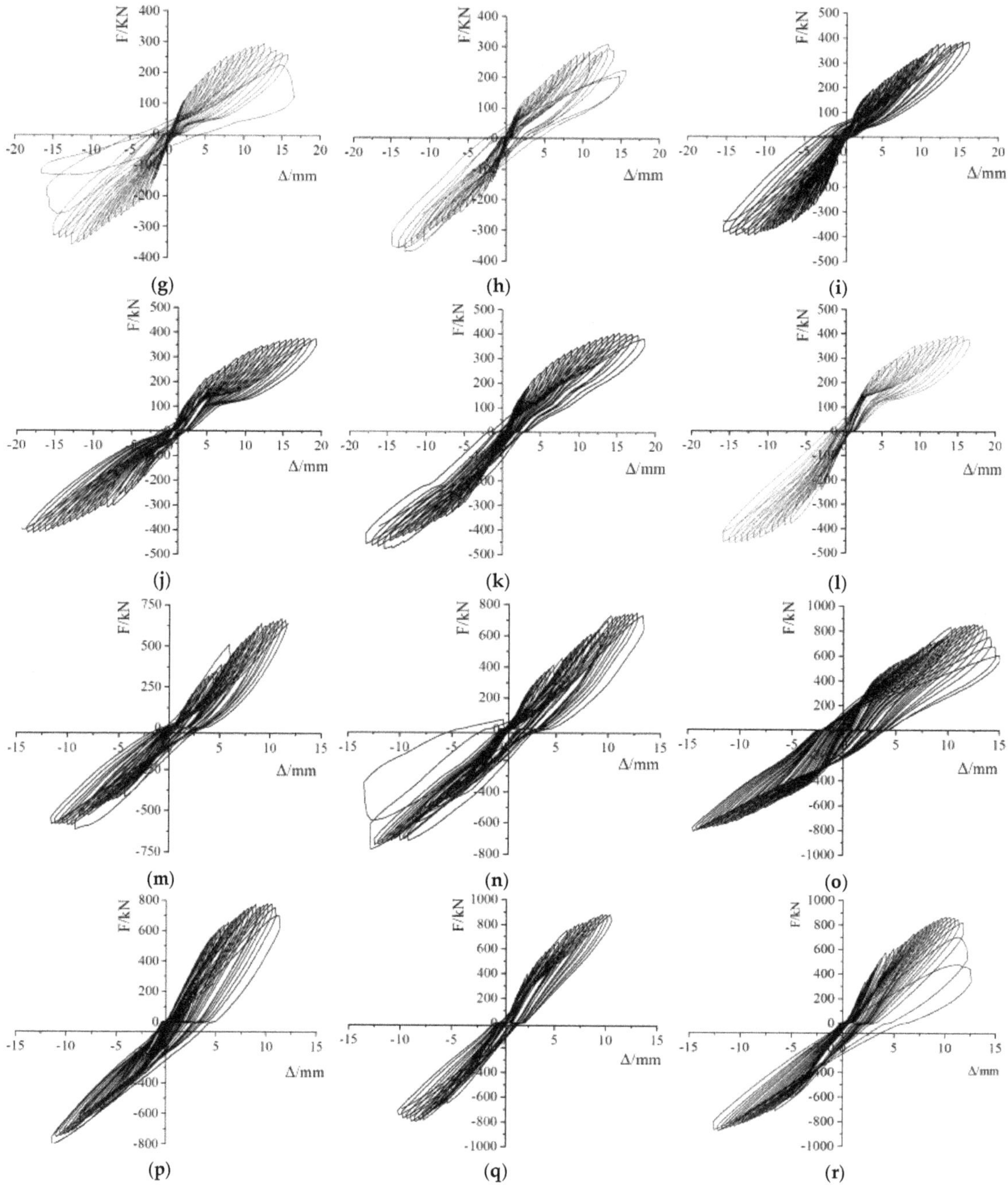

Figure 8. Hysteresis curves for (**a**) specimen A1-1, (**b**) specimen A1-2, (**c**) specimen A2-1, (**d**) specimen A2-2, (**e**) specimen A3-1, (**f**) specimen A3-2, (**g**) specimen B1-1, (**h**) specimen B1-2, (**i**) specimen B2-1, (**j**) specimen B2-2, (**k**) specimen B3-1, (**l**) specimen B3-2, (**m**) specimenC1-1, (**n**) specimenC1-2, (**o**) specimen C2-1, (**p**) specimenC2-2, (**q**) specimen C3-1, (**r**) specimen C3-2.

3.3. Ductility Analysis of Specimen

Ductility refers to the ability of materials, components, or structures to deform without significant reductions in post-yield bearing capacity under loading or other indirect effects. In this paper, the displacement ductility coefficient was used to analyze the ductility of our composite panel, which is calculated according to Formula (1):

$$\mu = \Delta_u / \Delta_y \tag{1}$$

where Δ_u is the ultimate displacement and Δ_y is the yield displacement.

The test results regarding the ductility coefficient of each specimen are shown in Table 2.

Table 2. Ductility coefficient test results of specimens.

Groups	Specimen Number	Ductility Coefficient r	Average Ductility Coefficient
I	A1-1	6.3	6.6
	A1-2	6.9	
II	A2-1	5.2	5.0
	A2-2	4.8	
III	A3-1	5.1	4.9
	A3-2	4.7	
IV	B1-1	4.2	4.6
	B1-2	4.9	
V	B2-1	3.9	4.1
	B2-2	4.3	
VI	B3-1	4.0	3.7
	B3-2	3.4	
VII	C1-1	2.7	2.8
	C1-2	2.8	
VIII	C2-1	3.3	3.1
	C2-2	2.8	
IX	C3-1	1.7	2.2
	C3-2	2.6	

The ductility coefficient of specimens with the same height-to-width ratio was averaged to establish the relationship between the specimen ductility and the height-to-width ratio. Similarly, the ductility coefficient of specimens with the same vertical compressive stress was averaged to determine the relationship between the specimen ductility and the vertical compressive stress. This relationship is illustrated in Figure 9.

It could be seen from Table 2 and Figure 9 that the greater the height-to-width ratio of the specimen, the better its ductility, and the greater the vertical pressure, the worse the ductility. The tests showed that specimens with a large height-to-width ratio experienced flexural failure, while those with a small height-to-width ratio exhibited diagonal shear failure. When the normal section was destroyed, the reinforcing steel bar yielded first, inducing a large deformation. The deformation ability of the component was strong, indicating good ductility. The deformation capacity of shear failure is related to the performance of the horizontal steel wire mesh. As the mesh was composed of cold-drawn low-carbon steel wire, its ductility was poor, meaning specimens with small height-to-width ratios have poor ductility.

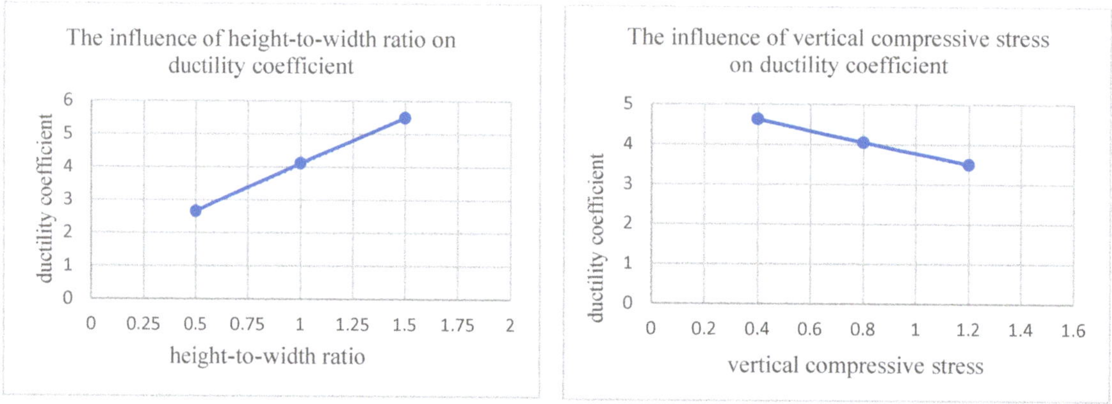

Figure 9. The influence of various factors on the ductility coefficient.

3.4. Analysis of Reinforcement Strain Test Results

Figure 10 shows the results of strain testing applied to the reinforcing bars positioned at the ends of some specimens.

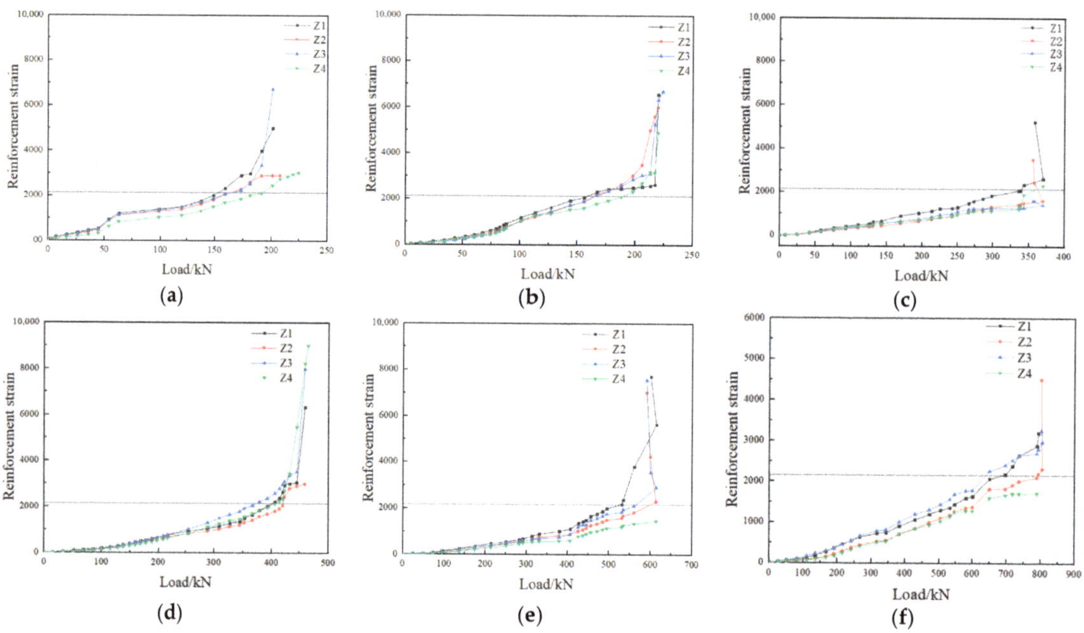

Figure 10. Reinforcement strain of specimens: (**a**) specimen A1-1, (**b**) specimen A2-2, (**c**) specimen B1-2, (**d**) specimen B3-1, (**e**) specimen C1-1, (**f**) specimen C3-1.

Figure 10 shows that for specimens with a height-to-width ratio of 1.5:1 the steel bar yielded before it reached its ultimate load-bearing capacity. For specimens with height-to-width ratios of 1:1 and 1:2, when the ultimate load-bearing capacity was reached, the longitudinal reinforcement had also reached its yield state. Therefore, the longitudinal reinforcing steel played a significant role.

The results of strain tests applied to the horizontal steel wire in a typical specimen are shown in Figure 11.

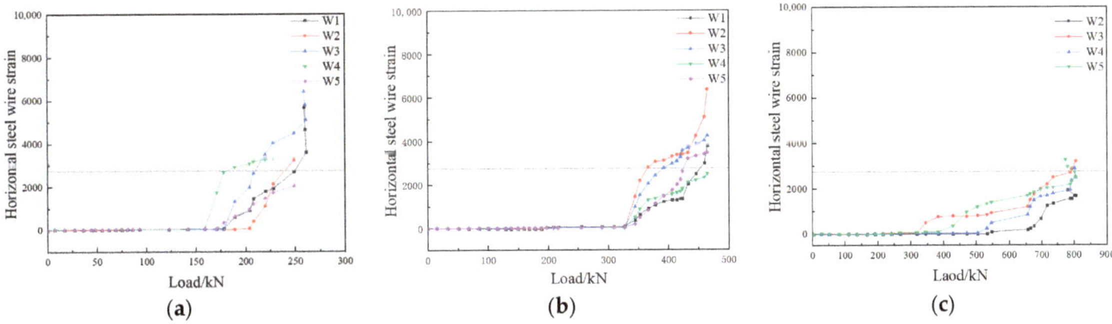

Figure 11. Horizontal steel wire strain of specimens: (**a**) specimen A3−2, (**b**) specimen B3−1, (**c**) specimen C3−1.

Figure 11 shows that the strain tests applied to the horizontal steel wire mesh exhibited relatively large discreteness, and uneven stress could be observed, which was related to the location of the measuring point. The final strain measured at the crack was larger there. After oblique cracking occurred in the specimen, the strain in the steel wire at the position of the crack underwent a sudden change and reached its conditional yield point. At this point, even individual steel wires were broken, indicating that the steel wire mesh was subjected to a relatively large horizontal seismic force.

3.5. Test Results and Analysis of Lateral Bearing Capacity

The average values of ultimate load in the push and pull directions for each specimen were taken as the lateral bearing capacity of each specimen, and the average value of the lateral bearing capacity of two specimens from each group was taken as the lateral bearing capacity of that group; the results of the lateral bearing capacity tests are shown in Table 3. The test results showed that the lateral bearing capacity of a specimen was not only related to its cross-sectional area, but also to its height-to-width ratio and the vertical compressive stress applied. For specimens with the same height-to-width ratio, the higher the vertical pressure, the greater the lateral bearing capacity. The reason for this phenomenon can be analyzed as follows: Under the same lateral load, a larger vertical pressure results in a larger compression zone, which in turn reduces the tensile stress in the longitudinal reinforcement. Consequently, the reinforcement is less likely to yield, and the bending capacity of the specimen's cross-section is improved.

Table 3. The ultimate load-bearing capacity test value and theoretical calculation value of the specimen.

Groups	I	II	III	IV	V	VI	VII	VIII	IX
Experimental value V_{su} (kN)	231.40	248.45	266.55	334.30	392.45	431.55	696.25	815.05	858.80
The theoretical value of lateral bearing capacity is calculated by Formulas (2)–(4) V_{u1} (kN)	141.47	165.41	187.54	245.99	299.86	349.65	545.08	688.74	821.51
The theoretical value of lateral bearing capacity is calculated by Formula (5) V_{u2} (kN)	226.14	236.54	246.94	351.19	366.79	382.39	476.24	497.04	517.84
The theoretical value of lateral bearing capacity of specimens V_u (kN)	141.47	165.41	187.54	245.99	299.86	349.65	476.24	497.04	517.84
V_{su}/V_u	1.64	1.50	1.42	1.36	1.31	1.23	1.46	1.64	1.66

Reference [20] carried out a seismic performance test on similar composite wall panels. Some specimens had a steel wire mesh anchorage, while others did not, and additional reinforcing bars were not provided in the wall panels. Through comparative test results, a prefabricated monolithic composite wall panel with reinforced steel bars, as presented in this paper, was compared to a wall panel without reinforced steel bars but with steel wire mesh anchorage. The lateral bearing capacity of the specimen with a height-to-width ratio of 1.5:1 increased by approximately 130%, while the lateral bearing capacity of the specimen with a height-to-width ratio of 1:1 increased by approximately 70%. Compared to specimens without anchorage and reinforcement, the lateral bearing capacity can be increased by approximately 160% when the height-to-width ratio is 1:2.

From the failure mode of the specimen, it was known that the precast and cast-in-place concrete were well combined without separation. The failure modes of the specimens resembled those of general reinforced concrete shear walls, meaning the bearing capacities of the former could be calculated with reference to the latter. In the calculation, only the effect of the concrete surface layer was considered, and the wall thickness was set as 100 mm. Due to the discontinuity between the upper and lower layers of vertical wires in the mesh and the lack of effective anchorage, the role of the wire mesh was not considered in the calculation of the bearing capacity of the normal section, and only the end reinforcing steel was considered. The calculation of shear in the oblique section should take into account the role of the horizontal steel wire in the steel wire mesh. Reference [22] provided the formula for calculating the bearing capacity of a reinforced concrete shear wall; the normal section bearing capacity of shear wall was calculated according to Equations (2)–(4), and the oblique section's bearing capacity was calculated according to Equation (5). The test results showed that these equations were applicable to this specific situation.

$$N = A_s' f_y' - A_s \sigma_s + \alpha f_c b_w x \tag{2}$$

$$N(e + h_{wo} - \frac{h_w}{2}) = A_s' f_y' (h_{wo} - \alpha_s') + \alpha f_c b_w x (h_{wo} - \frac{x}{2}) \tag{3}$$

$$e_0 = \frac{M}{N} = \frac{V_U H_W}{N} \tag{4}$$

$$V_U \leq \frac{1}{\lambda - 0.5}\left(0.5 f_t b_w h_{w0} + 0.13 N \frac{A_W}{A}\right) + f_{yh} \frac{A_{sh}}{S} h_{w0} \tag{5}$$

When calculating the bearing capacities of wall panels according to the above formula, λ is the shear–span ratio, and $\lambda = \frac{H}{h_{wo}}$; when $\lambda \leq 1.5$, then λ = 1.5. V_u is the ultimate load-bearing capacity of the specimen. f_c is the axial compression strength of concrete (MPa), and f_c = 23.48 MPa. f_t is the axial tensile strength of concrete (MPa), and f_t = 2.56 MPa. b_w is wall panel thickness, which we gave a value of 100 mm. h_{wo} is the effective width of the wall panels; when $h_{wo} \geq H$, we assumed that $h_{wo} = H$. H is the specimen height. f_{yh} is the conditional yield strength of the steel wire (MPa), and $f_{yh} = 533$ (MPa); A_{sh} is the cross-sectional area of the horizontal steel wire in the mesh with the same cross-section (mm^2); $A_{sh} = 14.14$ mm^2. S is the horizontal reinforcement spacing; S = 70 mm. f_y' is the yield strength of the reinforcement; f_y' = 428 MPa. σ_s is the tensile reinforcement stress; if the steel bar yields, $\sigma_s = f_y = 428$ MPa. N is the vertical load applied to the wall panel. x is the height of the compression zone.

The lateral load bearing capacity was calculated using Formulas (2)–(4) and the theoretical values of lateral load bearing capacity were calculated using Formula (5) for each group of specimens that is shown in Table 3. The theoretical values of lateral bearing capacity for specimens should take the minimum value of the two, and the results are shown in Table 3.

As Table 3 shows, for the specimen with a height-to-width ratio of 1.5:1, the theoretical value of the lateral load capacity calculated for the normal section was much smaller than the theoretical value calculated for shear resistance, which was consistent with the failure mode of the specimen. The specimen first showed flexural failure, followed by diagonal shear damage. For a specimen with a height-to-width ratio of 1:1, the theoretical values of lateral bearing capacity calculated for its normal section and shear resistance were not very different, so diagonal shear failure and flexural failure occurred at almost the same time. For a specimen with a height-to-width ratio of 1:2, the theoretical value of the lateral bearing capacity calculated for the normal section was much larger than the theoretical value of the shear bearing capacity calculated according to the shear resistance, resulting in the specimen exhibiting only diagonal shear failure. According to Table 3, when considering the composite wall panel as a reinforced concrete shear wall, the calculated lateral load-bearing capacity based on the literature [22] was significantly lower than the experimental values. Thus, in practice, the lateral load-bearing capacity of the CPC panel can be conservatively evaluated using the calculation method of the reinforced concrete shear walls. In addition, it was known from the test results that the theoretical values of a specimen with a height-to-width ratio of 1.5:1 and 1:2 differed from their experimental values. The reason for this was that the theoretical bearing capacity of a specimen with a ratio of 1.5:1 was calculated according to the normal section, without considering the compression effect of the steel wire mesh. The strain test results on the reinforcing steel bar showed that when the specimen reached its ultimate load-bearing capacity, the reinforcing steel bar entered the strengthening stage. In addition, a 50 mm area of non-insulation material was present at the end of the specimen, which contributed to its bearing capacity. The specimen with a height-to-width ratio of 1:2 underwent shear failure. It was found that the partial horizontal steel wires of the steel mesh broke during shear failure. The theoretical value of the shear-bearing capacity, calculated according to the conditional yield strength, was small. The shear span ratio of the specimen was 0.5; thus, we can infer that when the shear span ratio is small, the lateral bearing capacity of the wallboard, calculated according to [22], will be sufficient.

From Table 3, it can be seen that the experimental values of each specimen were much higher than the theoretical values calculated according to [22], with large errors. The reason for this is that the load-bearing behavior of composite wall panels is complex, and there are many factors that affect their lateral load-bearing, especially for components with small height-to-width ratios that may experience shear failure. The shear effect can affect the load-bearing capacity of specimens [23–26]. Reference [22] is a technical standard for practical engineering design, and the calculation formula is based on experimental results with a certain safety margin. In particular, Formula (5) takes the lower envelope of the experimental results. Additionally, when calculating theoretical values according to [22], the strength of the steel reinforcement was taken as the yield strength, and when the specimens reached their ultimate load-bearing capacity, it exceeded the yield strength, indicating that the material strength used in theoretical calculations was slightly lower than expected. Additionally, the role of the insulation board was not considered, resulting in a significant discrepancy between the experimental and theoretical values. For practical engineering applications, the composite wall panels studied in this topic can be calculated based on [22] to meet the requirements and be considered safe.

4. Conclusions

Compared with a cast-in-place composite wall panel, the prefabricated monolithic composite wall panel can shorten the construction period and reduce the project cost. Compared with a prefabricated wall panel, it has better seismic performance. Through the quasi-static testing and theoretical analysis of 18 prefabricated monolithic composite wall panels, the following conclusions have been obtained:

(1) The connectors spaced at 400 mm × 500 mm could ensure the concrete layers on both sides of the polystyrene board work collectively under seismic conditions In practical engineering, the spacing of connectors should not be greater than 400 mm.

(2) The failure mode of the specimens is related to their height-to-width ratio and the number of reinforced steel bars. When 4C12 reinforced steel bars were set at both ends of a specimen with a height-to-width ratio of 1.5:1 and subjected to lateral loading,

the longitudinal reinforcement at the end yielded first, followed by crushing of the concrete in the compression zone. Flexural failure occurred first, followed by diagonal shear failure in the diagonal direction along the main oblique cracks. For a specimen with an aspect ratio of 1:1, the shear failure caused by oblique cracks, the flexural failure caused by the yielding of tensile steel bars, and the crushing of concrete in the compression zone all occurred at almost the same time. A specimen with a height-to-width ratio of 1:2 ultimately underwent diagonal shear failure without flexural failure.

(3) The installation of strengthened longitudinal reinforcements at the ends of the wall panels improved their lateral bearing capacity and deformation resistance, preventing them from slipping along the bottom and losing their bearing capacity. The magnitude of the raise was related to the height-to-width ratio of the wall panels, the vertical load on the wall panels, and the area of the reinforced steel bar. When the end reinforcing steel bars with 4C12 were installed, compared with the composite wall panel without reinforcing steel bars but with steel wire mesh anchorage measures, the lateral bearing capacity of the specimen with a height-to-width ratio of 1.5:1 increased by approximately 130%, while the lateral bearing capacity of the specimen with a height-to-width ratio of 1:1 increased by approximately 70%. Furthermore, compared to the specimens without anchorage and reinforcement, the lateral bearing capacity could be increased by approximately 160% when the height-to-width ratio is 1:2.

(4) The lateral load-bearing capacity of the CPC panel can be conservatively evaluated using the calculation method of the reinforced concrete shear walls, with measured values exceeding theoretical values by 20–60%.

(5) Specimens with large height-to-width ratios underwent flexural failure and exhibited good ductility, while specimens with small height-to-width ratios underwent diagonal shear failure and demonstrated poor ductility.

5. Future Work

Based on the above conclusions, in the future, we can study the seismic performance of this type of wall panel by exploring conditions such as adding different fibers, varying the specifications of the steel wire mesh, and investigating different connection methods between concrete panels and polystyrene boards. Shaking table tests can also be performed on the overall structural model using this type of wall panel. Additionally, the theoretical models proposed in references [27–29] regarding the behavior of cracked plate-like structures under dynamic loading are helpful for studying the vibration characteristics of structures. We also consider conducting research in this area in the future.

Author Contributions: Conceptualization, K.Z.; validation, Y.Z. and S.L.; investigation, Z.F., Y.X. and S.L.; data curation, Z.F. and Y.X.; writing—original draft preparation, Z.F.; writing—review and editing, K.Z., Y.Z., Y.X. and S.L.; supervision, K.Z. and Y.Z.; project administration, K.Z. All authors have read and agreed to the published version of the manuscript.

Funding: This research received no external funding.

Data Availability Statement: Data are contained within the article.

Conflicts of Interest: Authors Zijia Fan and Yufeng Xu were employed by the company Engineering Research Institute of Appraisal and Strengthening of Shandong Jianzhu University Co., Ltd. The remaining authors declare that the research was conducted in the absence of any commercial or financial relationships that could be construed as a potential conflict of interest.

References

1. Li, H.; Wang, D.; Peng, L.; Zhou, J.; Zhang, C.; Wei, Q. Experimental study on wind resistance of autoclaved aerated concrete thermal insulation decorative composite exterior wallboard. *Build. Sci.* **2023**, *39*, 160–167. (In Chinese) [CrossRef]
2. Zang, K.; Wang, Y. The advantages of GRC board exterior insulation composite wall. *Brick* **2007**, *05*, 55–56. (In Chinese) [CrossRef]
3. Fernando, P.L.N.; Jayasinghe, M.T.R.; Jayasinghe, C. Structural feasibility of Expanded Polystyrene (EPS) based lightweight concrete sandwich wall panels. *Constr. Build. Mater* **2017**, *139*, 45–51. [CrossRef]

4. Shin, D.H.; Kim, H.J. Composite effects of shear connectors used for lightweight-foamed-concrete sandwich wall panels. *J. Build. Eng.* **2020**, *29*, 101108. [CrossRef]
5. Lei, S.; Wu, Z.; Zhang, C.; Huo, X.; Li, D. Experimental study on flexural performance of precast lightweight concrete thermal insulation exterior wallboard. *Build. Technol.* **2023**, *54*, 1611–1616. (In Chinese)
6. Amran, Y.M.; Ali, A.A.; Rashid, R.S.; Hejazi, F.; Safiee, N.A. Structural behavior of axially loaded precast foamed concrete sandwich panels. *Constr. Build. Mater.* **2016**, *107*, 307–320. [CrossRef]
7. Pavese, A., Bournas, D.A. Experimental assessment of the seismic performance of a prefabricated concrete structural wall system. *Eng. Struct.* **2011**, *33*, 2049–2062. [CrossRef]
8. He, Z.Z.; Pan, P.; Xiao, G.Q.; Shen, S.D.; Ren, J.Y. Test and analysis on axial performances of GFRP restraint connectors for sandwich insulation wall panels. *J. Build. Eng.* **2022**, *45*, 103457. [CrossRef]
9. Sylaj, V.; Fam, A. UHPC sandwich panels with GFRP shear connectors tested under combined bending and axial loads. *Eng. Struct.* **2021**, *248*, 113287. [CrossRef]
10. Pan, P.; He, Z.; Wang, H.; Kang, Y. Experimental investigation of C-shaped glass-fiber-reinforced polymer connectors for sandwich insulation wall panels. *Eng. Struct.* **2022**, *250*, 113462. [CrossRef]
11. Choi, I.; Kim, J.; Kim, D.; Park, J. Effects of grid-type shear connector arrangements used for insulated concrete sandwich wall panels with a low aspect ratio. *J. Build. Eng.* **2022**, *46*, 103754. [CrossRef]
12. Yan, M.; Wang, L.-G.; Chen, B.-L. Shear resistance and deflection prediction of steel–concrete–steel sandwich panel with headed stud connectors. *Structures* **2023**, *54*, 1690–1704.
13. Lou, X.; Xue, W.; Bai, H.; Li, Y.; Huang, Q. Shear behavior of stainless-steel plate connectors for insulated precast concrete sandwich panels. *Structures* **2022**, *44*, 1046–1056. [CrossRef]
14. Ma, S.; Hou, D.; Bao, P.; Wang, D. Influence of alkali-resistant glass fiber on seismic performance of precast ceramsite concrete sandwich wall panels. *Structures* **2022**, *38*, 94–107. [CrossRef]
15. Ding, R.; Sun, Y.T.; Nie, X.; Chen, D.Q. Experimental study on seismic behaviour of an unreinforced precast wall-slab structure based on UHPC sandwich panels. *J. Build. Eng.* **2023**, *68*, 106197. [CrossRef]
16. Kumar, S.; Chen, B.; Xu, Y.; Dai, J.G. Axial-flexural behavior of FRP grid-reinforced geopolymer concrete sandwich wall panels enabled with FRP connectors. *J. Build. Eng.* **2022**, *47*, 103907. [CrossRef]
17. Huang, J.Q.; Dai, J.G. Flexural performance of precast geopolymer concrete sandwich panel enabled by FRP connector. *Compos. Struct.* **2020**, *248*, 112563. [CrossRef]
18. Kumar, S.; Chen, B.; Xu, Y.; Dai, J.G. Structural behavior of FRP grid reinforced geopolymer concrete sandwich wall panels subjected to concentric axial loading. *Compos. Struct.* **2021**, *270*, 114117. [CrossRef]
19. Sun, Q. Study on the Compressive Performance of Concrete Sandwich Composite Wall Panels. Ph.D. Thesis, Shandong Jianzhu University, Jinan, China, 2019. (In Chinese).
20. Zhao, K.; Liu, M.; Huang, L. Experimental study on seismic performance of concrete sandwich composite slabs. *Build. Struct.* **2023**, *53*, 47–52. (In Chinese) [CrossRef]
21. Zhao, K.; Li, J.; Wei, X. Experimental Study on the Mechanical Properties of Concrete-Polystyrene Composite Wallboard. *J. Shandong Jianzhu Univ.* **2023**, *38*, 9–16. (In Chinese)
22. *JGJ3-2010;* Technical Specification for Concrete Structures of Tall Buildings. Construction Industry Press: Beijing, China, 2010. (In Chinese)
23. Chen, C.S. Nonlinear vibration of a shear deformable functionally graded plate. *Compos. Struct.* **2005**, *68*, 295–302. [CrossRef]
24. Kiani, K. Nonlocal and shear effects on column buckling of single-layered membranes from stocky single-walled carbon nanotubes. *Compos. Part B Eng.* **2015**, *79*, 535–552. [CrossRef]
25. Kiani, K. Free vibration of in-plane-aligned membranes of single-walled carbon nanotubes in the presence of in-plane-unidirectional magnetic fields. *J. Vib. Control.* **2016**, *22*, 3736–3766. [CrossRef]
26. Kiani, K.; Pakdaman, H. Nonlocal vibrations and potential instability of monolayers from double-walled carbon nanotubes subjected to temperature gradients. *Int. J. Mech. Sci.* **2018**, *144*, 576–599. [CrossRef]
27. Liew, K.M.; Hung, K.C.; Lim, M.K. A solution method for analysis of cracked plates under vibration. *Eng. Fract. Mech.* **1994**, *48*, 393–404. [CrossRef]
28. Nikkhoo, A.; Banihashemi, S.; Kiani, K. On non-stationary response of cracked thin rectangular plates acted upon by a moving random force. *Sci. Iran.* **2023**. [CrossRef]
29. Nikkhoo, A.; Banihashemi, S.; Kiani, K. Parametric investigations on dynamics of cracked thin rectangular plates, excited by a moving mass. *Sci. Iran.* **2023**, *30*, 860–876. [CrossRef]

buildings

MDPI

Article

Experimental Evaluation of Precast Concrete Walls with High-Tension Bolted Vertical Joints for Enhanced Ductile Behavior

Kyo Young Moon [1], Sung Jig Kim [1,*] and Kihak Lee [2]

[1] Department of Architectural Engineering, Keimyung University, Daegu 42601, Republic of Korea; mky9404@gmail.com

[2] Department of Architectural Engineering, Sejong University, Seoul 05006, Republic of Korea; kihaklee@sejong.ac.kr

* Correspondence: sjkim4@kmu.ac.kr; Tel.: +82-53-580-5273

Abstract: This paper presents an experimental program investigating the seismic performance of Precast Concrete (PC) walls incorporating a novel vertical joint featuring high-tension bolts. The proposed joint aims to enhance both the constructability and ductile behavior of PC walls while eliminating defects due to bolt exposure associated with conventional dry joints. To evaluate the shear performance of the joints, three specimens were constructed: a cast-in-place concrete wall, a PC wall with the conventional wet joint, and a PC wall with the proposed joint. Direct shear tests revealed a substantial improvement in the ductile behavior of the proposed vertical joint, demonstrating gradual and controlled failure, even at high strains. Static cyclic loading tests further assessed the seismic performance of two PC walls: one with the conventional wet joint and the other with the high-tension bolted vertical joint. The PC wall with the proposed joint exhibited a substantial 20% increase in ultimate displacement compared to the control specimen, along with demonstrably improved crack control and reduced damage. Therefore, it is concluded that although the strength and stiffness decreased due to its design characteristics, the proposed vertical joint could mitigate damage and improve the ductility capacity of the PC wall.

Keywords: precast concrete wall; vertical joint; dry joint; shear performance; seismic performance

1. Introduction

Precast Concrete (PC) elements are commonly prefabricated at manufacturing plants and transported to construction sites, where they are connected by grouting or welding to form an assembled monolithic structure [1]. Thus, the PC system is an efficient method that could allow for a shortened construction period and ensure uniformity and high quality of structural members compared to a conventional cast-in-place concrete system [2–5]. In the case of structural walls, which are primary lateral force-resisting elements, the size of each member is substantial, and thus, the construction of this member requires a significant time compared to other structural members. Utilizing the PC wall system makes it possible to shorten the construction period, thereby leading to a reduction in construction costs. However, due to the limitations of transportation and erection equipment, the width of PC walls is often smaller than that of a typical cast-in-place concrete wall. Thus, the assembly of PC wall elements is required, along with a substantial number of joint connections. The structural behavior of these connections could have a significant impact on the overall performance of PC wall elements [1]. Therefore, the design of these connections needs to consider various factors, including strength, stiffness, and energy dissipation capacity, to ensure practicality and reliability [6]. Inadequate design of PC wall joints, particularly if they have insufficient shear strength and ductility, can lead to the initiation of cracks and damages, resulting in a degradation in overall structural performance.

Generally, there are two types of vertical joints used for connecting PC wall elements: wet and dry joints. Existing research suggests that wet joints are more effective in ensuring the integrity and stiffness of the overall structure [7–9]. This enhanced performance stems from the installation of connecting materials and subsequent concrete or mortar casting, resulting in superior shear transfer capacity. However, compared to dry joints, wet joints are more complex to construct, pose challenges in quality control, and incur higher costs [10]. Traditional wet joints often involve processing dowel rebars into U-shapes. However, manual fabrication becomes increasingly cumbersome as the diameter of the rebars increases [11]. Moreover, the widely used vertical joint with wire loops may exhibit insufficient ductility, leading to a brittle failure mode [12]. Consequently, this deficiency may result in undesirable deformation and damage at the joint, compromising its continuity and potentially jeopardizing overall structural integrity. Therefore, to mitigate this risk under earthquake loads, efficiently controlling and dissipating energy within the joint is crucial [13].

Typical dry joints connect PC walls using welding bolts, plates, and other components. This approach offers improved structural performance compared to wet joints, particularly in terms of ductility and energy-dissipation capacity [14–17]. Dry joints generally facilitate construction due to their ease of assembly, but they require specific attention to maintain their performance. Exposed elements like bolts and plates are prone to corrosion and loss after assembly, necessitating finishing treatments or periodic maintenance. Welded plate connections, a popular dry joint method, provide excellent shear resistance but exhibit relatively lower ductility capacity [18]. Additionally, some dry joints utilize complex shapes with the aim of incorporating damping effects in the connection [2,5,19].

Therefore, to address the limitation of existing vertical joints, this study presents a novel vertical joint for enhanced constructability and ductile behavior in PC walls. The proposed joint utilizes high-tension bolts for efficient assembly of wall panels. To assess its shear performance, test specimens were constructed and rigorously evaluated under direct shear loading. Based on these results, large-scale wall specimens incorporating the developed vertical joint were fabricated. The seismic performance of these PC walls was subsequently assessed through static cyclic loading tests.

2. Proposed Vertical Joint

A novel vertical joint with high-tension bolts has been developed to enhance the ductile behavior of PC walls and address corrosion issues prevalent in existing dry joint methods. An overview of the vertical joint with high-tension bolts is illustrated in Figure 1. The proposed system involves pre-forming a joint space within the PC walls using formworks before concrete casting. The size of this space is determined by the length of the high-tension bolts, which are subsequently fastened within this prepared joint. Non-shrink grout is then injected to complete the assembly.

To address potential cracking and bearing failure during bolt installation and under load, a prefabricated plate is embedded within the wall on the bolt-receiving side. It also helps prevent bearing failure at the connection point where stress concentrates when loads are applied to the wall. The embedded plate is made of SS400 material and includes holes for fastening high-tension bolts. In addition, on the surface of the plate that contacts the concrete, there is a welded shear connector. This is designed to enhance the integrity of the concrete of the wall. Figure 2 provides a detailed illustration of the embedded plate. To ensure the high-tension bolts remain functional before the failure of PC walls, the specification and quantity of the high-tension bolts were selected based on the calculated shear strength. F10T-M20 standard bolts with a length of 180 mm were utilized for direct shear tests, while F10T-M22 standard bolts of the same length were employed for the static cyclic tests.

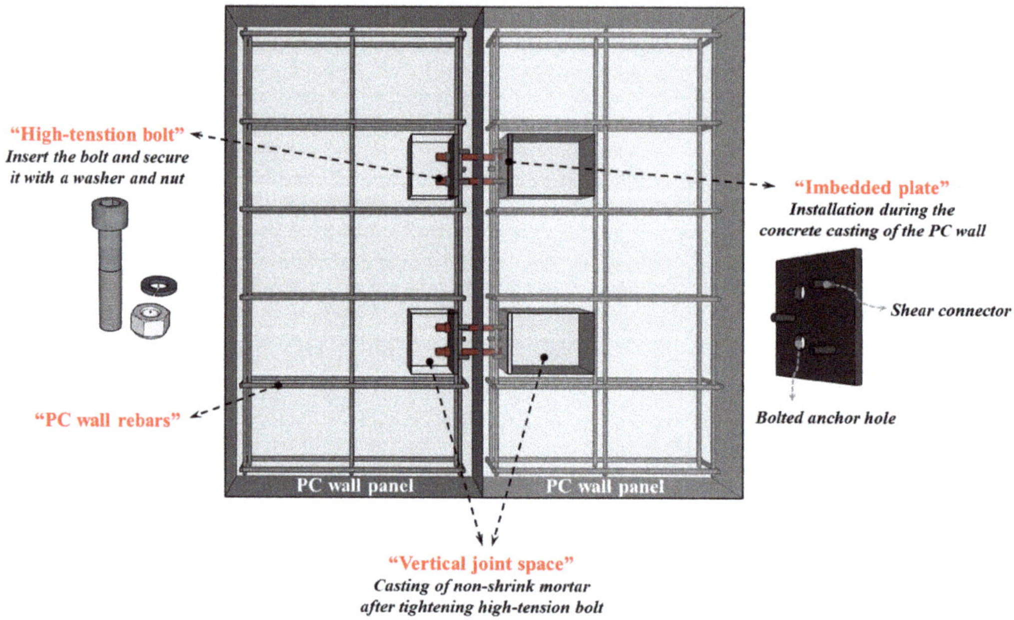

Figure 1. Connection with high-tension bolts.

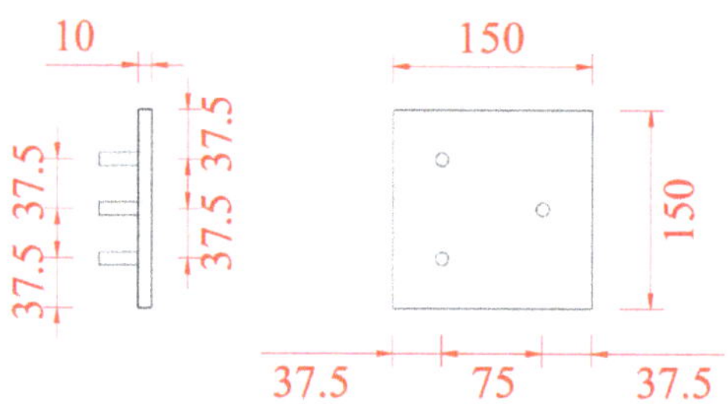

Figure 2. Details of embedded plate (unit: mm).

3. Shear Performance of the Proposed Vertical Joint

3.1. Details of Specimens

As detailed in Table 1 and Figure 3, three specimens were designed and constructed to evaluate the shear performance of the proposed vertical joint: a cast-in-place concrete wall (S-CIPW), a PC wall with the conventional wet joint (S-PWR), and a PC wall with the proposed joint (S-PWB). PC wall specimens feature two joints connecting three PC wall panels, with the load applied to the central panel's upper section. The width of the wall panels varies for each specimen due to the different applied slit sizes. However, for consistent experimental comparison, their overall assembled length is designed to be 2430 mm. The thickness and height of all wall panels are 200 mm and 800 mm, respectively. Additionally, all specimens have identical horizontal and vertical reinforcement ratios of 0.53% and 0.26%, respectively.

Table 1. Specimen details and shear tests.

Specimen	Type of Vertical Connection	Geometric Configuration of Wall Panel (mm)					Rebar Ratio (%)	
		Slit	Width		Thickness	Height	Vertical	Horizontal
			Center	Side				
S-CIPW	-	-	810	810	200	800	0.26	0.53
S-PWR	Transverse rebars	20	810	790	200	800	0.26	0.53
S-PWB	High-tension bolts	-	810	810	200	800	0.26	0.53

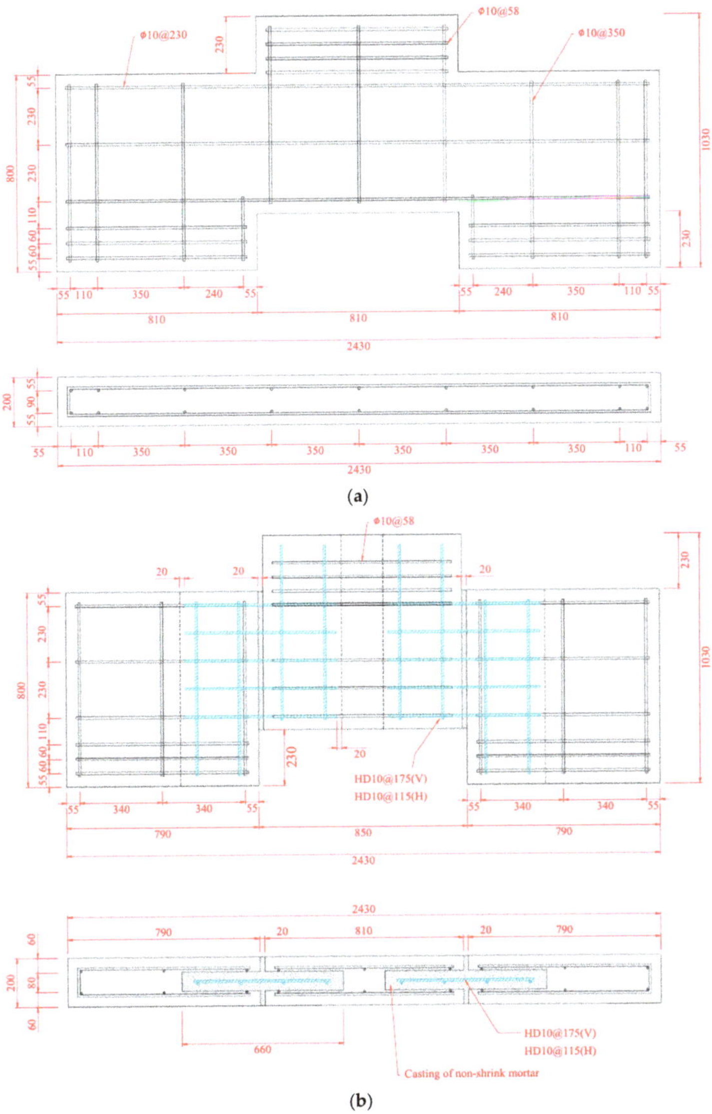

Figure 3. *Cont.*

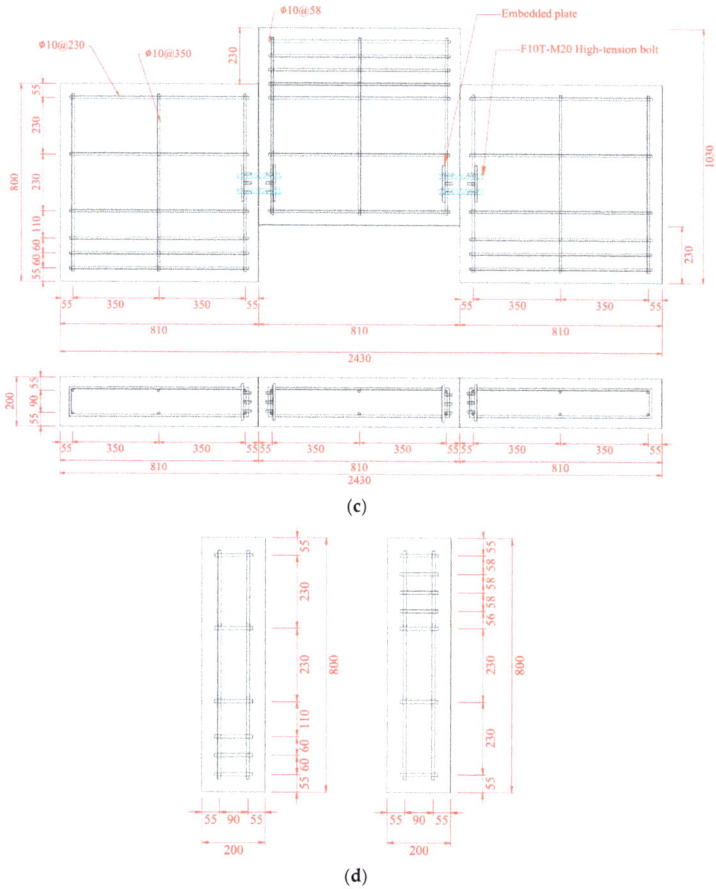

Figure 3. Elevation and section of specimens—shear tests (unit: mm): (**a**) S-CIPW; (**b**) S-PWR; (**c**) S-PWB; (**d**) Cross-section of all specimens.

S-CIPW serves as the reference specimen for evaluating the performance of the PC wall with the proposed vertical joint. It comprises three wall panels cast simultaneously as a single unit. The S-PWR specimen was designed to represent an improved version of the conventional wet joint for PC walls. Structural defects in the existing wet joint of the PC wall can frequently occur due to interference between the installed transverse reinforcement and insufficient transverse rebar ratio. In addition, the wet joint, as described previously, is often susceptible to significant shear cracking and abrupt force reduction at peak load. To address these vulnerabilities, the S-PWR employs a widened joint slit (20 mm) and an increased transverse reinforcement ratio (0.31%). Compared to previous experimental studies involving similar PC wall joints [4,20,21], the transverse reinforcement ratio represents a significant increase, ranging from 14.81% to 121.43%, exceeding the typical range for such applications. The estimated shear stress corresponding to nominal shear strength for the S-PWR joint is 0.82 MPa, which is 14.81% to 121.43% higher than those reported in previous studies. A 20 mm slit between the two wall panels is selected to minimize issues such as leakage, condensation, and insulation problems in the joint while considering construction errors.

The S-PWB is the PC wall specimen with the proposed high-tension bolted vertical joint aimed to enhance ductility compared to the conventional vertical joint. The joint details of the S-PWB are described in the previous section. Two vertical connection units with F10T-M20 standard bolts are installed to evaluate the shear capacity and performance of the S-PWB joint.

3.2. Material Properties

For each test specimen, three concrete cylinder samples, with a radius of 100 mm and a height of 200 mm, and three mortar cube samples with dimensions of 50 mm × 50 mm × 50 mm were tested. The concrete compressive strengths of S-CIPW, S-PWR, and S-PWB are 23.68 MPa, 23.96 MPa, and 23.51 MPa, respectively. The compressive strengths of the mortar cast to the vertical joints of S-PWR and S-PWB are 58.51 MPa and 53.50 MPa, respectively. The transverse reinforcements in the PC wall and vertical joint utilize 10 mm diameter rebars. The tensile test results of these rebars indicate a yield strength, ultimate strength, and yield strain of 439.0 MPa, 584.36 MPa, and 0.002, respectively.

3.3. Experimental Program

The setup for the direct shear test is illustrated in Figure 4. To prevent the horizontal slippage of the specimen during the experiment, cubes and steel plates were fixed to both ends of the specimen. A large-scale testing machine with a 5MN capacity (CCH-5000kN, Shimadzu, Kyoto, Japan) at the Intelligent Construction System Core-Support Center at Keimyung University was utilized to apply monotonic loading at a rate of 0.5 mm/min at the top of the central wall. Measurements were collected at a rate of 20 samples per sec using a static data logger (YCAM-500A, KYOWA, Tokyo, Japan). The experiment continued until a reduction of over 20% in maximum load or failure of the specimen occurred. A total of eight Linear Variable Displacement Transducers (LVDTs) were employed to measure the deformation of specimens, as shown in Figure 5. To measure the vertical displacement of the central wall in the direction of the load, two LVDTs were vertically installed in front of and behind the central wall (L3, L4). Additionally, two LVDTs were horizontally installed at the ends of the wall to measure the overall horizontal displacement of the specimen (L1, L2). Finally, to measure the shear deformation of the vertical joint, four LVDTs were installed in an "X" configuration (L5 to L8).

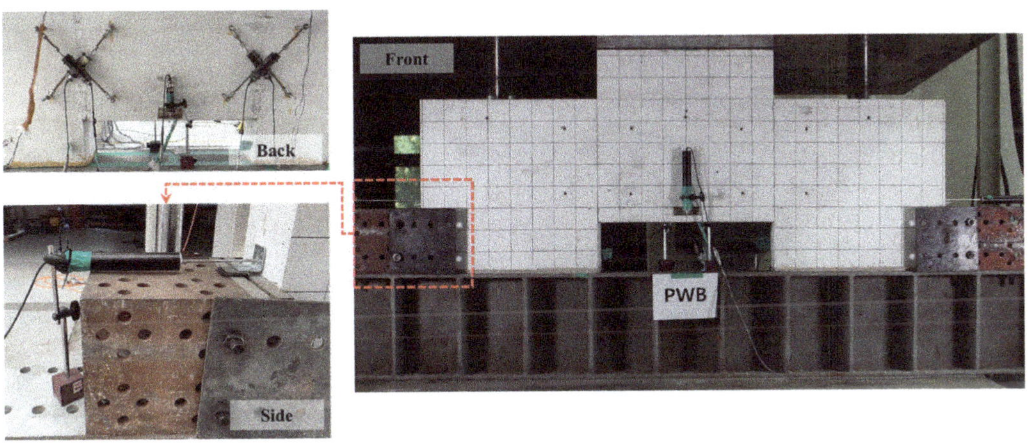

Figure 4. Experimental setup—shear tests.

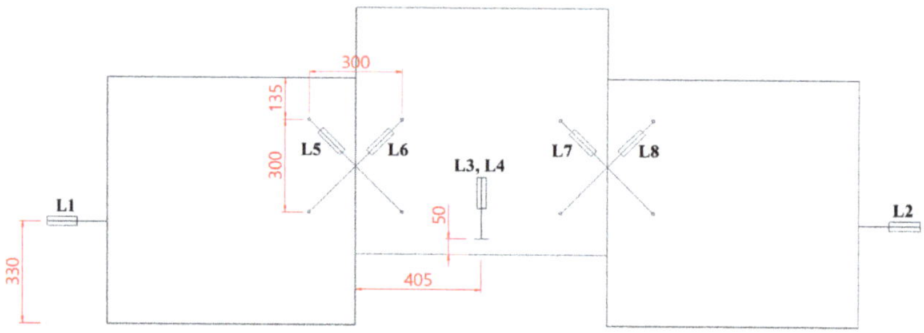

Figure 5. LVDT locations (unit: mm)—shear tests.

3.4. Experimental Results

3.4.1. Crack Patterns

The crack patterns at the failure for each specimen are presented in Figure 6. For the specimen S-CIPW, the initial cracks occurred at the connection areas and the bottom of the central wall when the load reached 390 kN. Upon reaching 980 kN, cracks developed at the bottom of both the left and right walls. Furthermore, at a load of 1200 kN, spalling began on the upper and lower sides of both walls, and diagonal cracks occurring at the vertical joints gradually extended, leading to the ultimate failure of the specimen. In the case of S-CIPW, the initial cracks extended to approximately 58% of the connection height, and shear cracks were relatively prevalent throughout the connection and the entire wall. This is attributed to the integral casting of the wall, which resulted in high strength and stiffness, allowing for the rapid attainment of high loads within a relatively short duration.

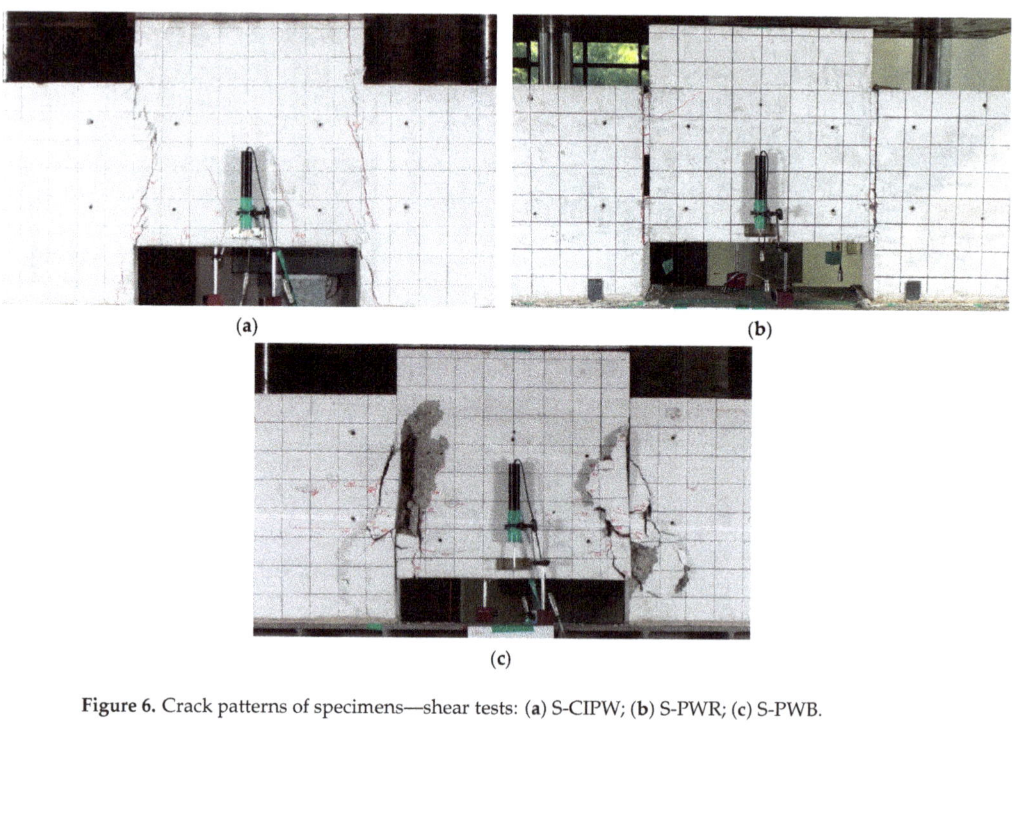

Figure 6. Crack patterns of specimens—shear tests: (**a**) S-CIPW; (**b**) S-PWR; (**c**) S-PWB.

The load on the specimen S-PWR increased sharply from 130 kN to 330 kN without any visible cracks. Cracks emerged at the upper part of the vertical joint when the load reached 400 kN and further propagated along the mortar at 550 kN. Following the attainment of the maximum load, spalling of the mortar within the joint occurred, leading to the development of significant shear cracks as they extended from the joint into the central wall. Subsequently, the experiment was terminated with the failure of the vertical joint.

The initial crack in the specimen S-PWB occurred at the vertical joint when the load reached 130 kN. Shear cracks began to develop in the upper and lower parts of the joint at 300 kN. Upon reaching a load of 530 kN, significant spalling of the concrete and mortar near the joint occurred, leading to the termination of the experiment. The specimen S-PWB exhibited a comparatively higher number of cracks than other specimens. This can be attributed to the use of high-tension bolts in the vertical joint for enhanced ductility, allowing the specimen to undergo relatively large displacements compared to other specimens.

3.4.2. Force–Displacement Relationship

The experimental results obtained from direct shear tests are summarized in Table 2, and the force–displacement relationships for the specimens are presented in Figure 7. The ultimate displacement is defined as the displacement at the point when the load is reduced by 20% after reaching the maximum load. The initial stiffnesses of S-PWR and S-PWB decreased by 3.85% and 65.38%, respectively, compared to S-CIPW. The maximum forces of those specimens also decreased by 59.06% and 64.66%, respectively. In terms of ultimate displacement, S-PWR showed a decrease of 6.54%, while S-PWB exhibited a significant increase of 410.47%. Compared to S-PWR with improved wet joints, the maximum load of S-PWB with high-tension bolted dry joints decreased by 13.67%, while its ultimate displacement increased by 446.20%. The lower maximum strength of S-PWB can be attributed to the application of dry joints. Both PC wall specimens showed incomplete monolithic structural behavior due to the presence of vertical joints, leading to lower initial stiffnesses and maximum forces compared to S-CIPW. However, considering the overall behavior, both PC wall specimens did not exhibit sudden failure after reaching the maximum force, unlike S-CIPW. S-PWB with vertical joints with high-tension bolts exhibited superior ductile behavior with a maximum displacement at failure that was approximately five times larger than that of the S-CIPW. Additionally, both S-PWR and S-PWB effectively controlled crack formation and damage at comparable load points.

Table 2. Experimental results-shear tests.

Specimen	Initial Stiffness (kN/mm)	Maximum Force (kN)	Maximum Shear Stress (MPa)	Maximum Displacement (mm)	Maximum Average Shear Strain (rad)
S-CIPW	520	1522.6	13.36	5.35	0.0022
S-PWR	500	623.3	5.47	5.00	0.0196
S-PWB	180	538.1	4.72	27.31	0.1327

3.4.3. Shear Deformation

To measure the shear deformation of the vertical joint in the PC wall specimen, "X"-shaped LVDTs were installed as described in Figure 5. Shear deformation was evaluated as the average shear strain (γ_{avg}) using Equation (1), and its configuration is illustrated in Figure 8. The force–average shear strain relationship shown in Figure 8 shows a similar trend to the force–displacement relationship shown in Figure 7. The maximum average shear strain for the specimens S-CIPW, S-PWR, and S-PWB were 0.0022, 0.0196, and 0.1327, respectively. In comparison to the shear strain observed in S-CIPW, the specimens S-PWR and S-PWB exhibit significantly higher values, with increases of 790.91% and 5931.82%, respectively. Despite the significantly higher shear strain, S-PWR and S-PWB did not exhibit abrupt failure after reaching the maximum force, indicating the effective control of crack and damage.

$$\gamma_{avg} = \frac{\left(d_1' - d_1\right) d_1 - \left(d_2' - d_2\right) d_2}{2HL} \tag{1}$$

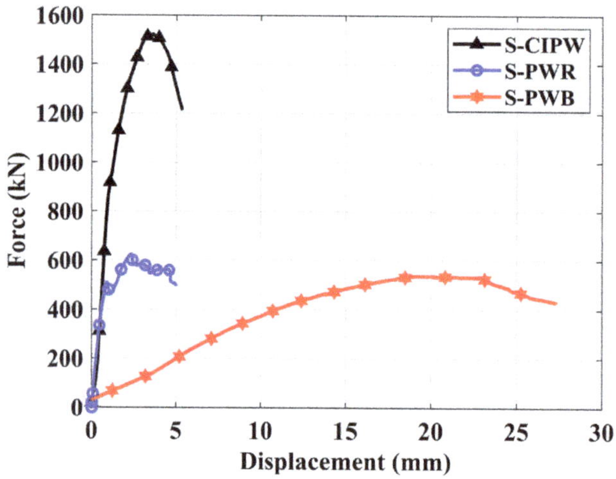

Figure 7. Force—displacement relationship—shear tests.

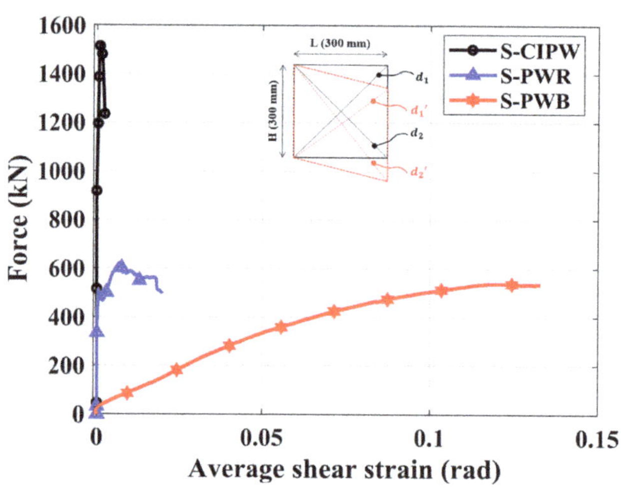

Figure 8. Force—average shear strain relationship—shear tests.

3.4.4. Strain

Significant changes in the strain of wall rebars did not occur, except for the horizontal rebars near the vertical joint. The positions and the force–strain relationships of these horizontal reinforcements are illustrated in Figure 9. The strains depicted in Figure 9 represent average values obtained from measurements taken at 200 mm from both left and right vertical joints and 285 mm above the bottom of the central wall.

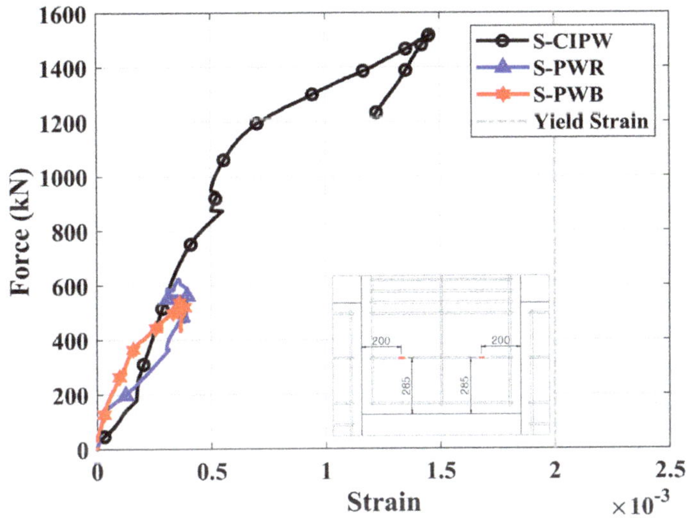

Figure 9. Force—wall strain relationship (unit: mm)—shear tests.

As shown in Figure 9, the wall rebars in all three specimens did not yield until the tests were terminated. The strains increased until reaching the maximum force and then showed a tendency to either maintain or decrease. In the case of S-CIPW, the wall strain significantly increased until reaching the maximum force, after which it rapidly decreased. The observed strain trend is deemed to be a result of the swift decrease in force after reaching the maximum force, ultimately leading to specimen failure. Furthermore, despite the specimen having the smallest ultimate displacement among the three specimens, its strain is the most significant.

The wall strain of S-PWB, which showed the largest ultimate displacement and suffered the most severe damage, was the smallest among the three specimens. It can be inferred that significant stress concentration on the high-tension bolts in the vertical joint caused substantial deformation in that area but relatively minimal deformation in the wall. Therefore, it is concluded that the proposed vertical joint effectively secures the deformation capacity of the wall itself.

4. Seismic Performance of PC Wall with the Proposed Vertical Joint

4.1. Details of Specimens

Two specimens were constructed to evaluate the seismic performance of PC walls with developed vertical joints, and static cyclic loading tests were conducted. The details of the specimens are presented in Figure 10 and Table 3. Each specimen has a configuration with a vertical joint between two wall panels. The vertical joints of the specimens used in the static cyclic loading tests are the same as those employed in the previous shear tests. The specimen C-PWR with an improved vertical wet joint serves as a reference specimen for comparing the performance of specimen C-PWB, which incorporates developed vertical dry joints.

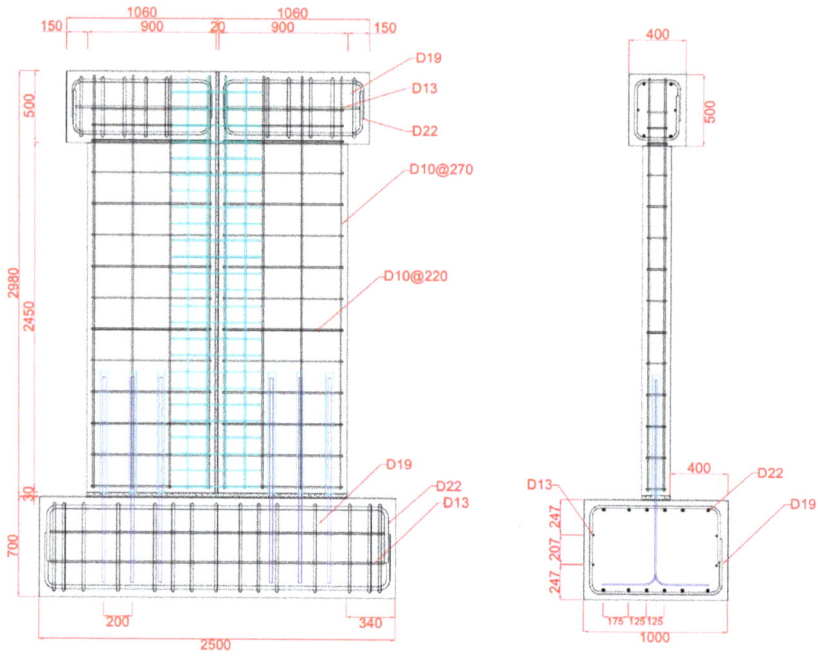

(a)

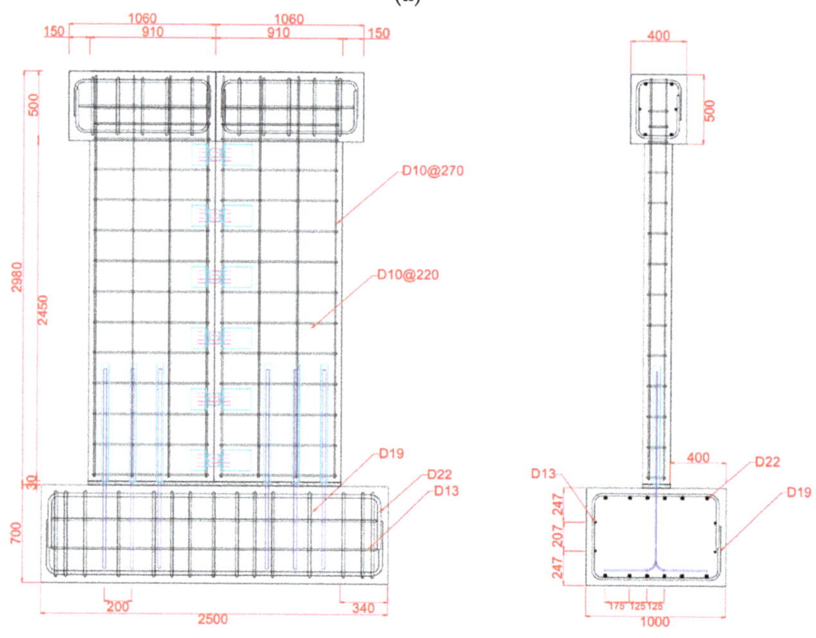

(b)

Figure 10. Elevation and section of specimens—cyclic tests (unit: mm): (**a**) C-PWR; (**b**) C-PWB.

Table 3. Specimen details—cyclic tests.

Specimen	Type of Vertical Connection	Geometric Configuration (mm)				Rebar Ratio (%)	
		Slit	Wall Panel			Vertical	Horizontal
			Width	Thickness	Height		
C-PWR	Transverse rebars	20	900	200	2450	0.32	0.34
C-PWB	High-tension bolts	-	910	200	2450	0.32	0.34

For C-PWR, a 20 mm slit is applied between the two wall panels, and reinforcements with a diameter of 10 mm are utilized at the vertical joint. C-PWB represents a PC wall with proposed high-tension bolts in the vertical joint aimed at enhancing ductility compared to the PC wall with the conventional vertical joint. The number of high-tension bolts is determined by considering the initial stiffness from the previous shear test, as shown in Table 2. By comparing the initial stiffnesses of S-CIPW and S-PWB, the shear strength and the number of high-tension bolts per unit length that satisfies 50% of the stiffness of S-CIPW are calculated. These values are then applied to the detailed design of the static cyclic loading test to determine the required number of bolts, resulting in 12 F10T-M22 standard bolts.

Additionally, to simulate the realistic structural behavior of PC walls, horizontal joints with corrugated tubes were applied between the bottom of the wall panels and the footing. Corrugated tubes are typically fabricated from corrugated alloys to enhance adhesion between the metal tube and concrete. After installing connecting rebars inside the corrugated tube, the joint is activated by casting the grout in the duct [22]. Corrugated tube joints, designed to facilitate stress transfer longitudinal direction, offer the advantage of applying uniform compressive forces to the grout, thereby inhibiting splitting failure [23].

The primary objective of this experiment is to evaluate the performance of the developed vertical joints. Therefore, to prevent failure due to the horizontal joints formed by the corrugated tubes and to ensure the integrity of the specimens, a shear friction design was implemented. The shear strength of the horizontal joints was designed to be approximately 1.7 times the shear friction strength specified in design codes for the horizontal connection in the PC wall, aiming to prevent failure caused by the horizontal joint formed by the corrugated ducts. Three corrugated tubes with a diameter of 50 mm each were installed per wall panel, and inside these ducts, a diameter of 22 mm rebars with a yield strength of 500 MPa were placed.

The clear height and thickness of the PC walls are 2450 mm and 200 mm, respectively. For the horizontal joint described previously, the thickness of the mortar pad cast between the bottom of the wall panel and the cap beam is 30 mm. The wall panel width and slit of C-PWR are 900 mm and 20 mm, while the wall panel width of C-PWB is 910 mm. Thus, the overall PC wall width, including the vertical joint and two panels, is identical to 1820 mm for both specimens.

4.2. Material Properties

For each test specimen, three concrete cylinders with a radius of 100 mm and a height of 200 mm and three mortar cube samples with dimensions of 50 mm × 50 mm × 50 mm were tested. The concrete compressive strengths of C-PWR and C-PWB are 27.44 MPa and 28.53 MPa, respectively. The compressive strengths of the mortar cast to the vertical joints of C-PWR and C-PWB are 60.34 MPa and 63.43 MPa, respectively. The tensile test results of 10 mm diameter rebars employed in PC walls and vertical joints reveal yield strength, ultimate strength, and yield strain values of 456.17 MPa, 569.53 MPa, and 0.0023, respectively. The yield strength, ultimate strength, and yield strain of 22 mm diameter rebars inside the corrugated tubes used for the horizontal joint are 552.20 MPa, 690.74 MPa, and 0.0028, respectively.

4.3. Experimental Program

The setup for the static cyclic loading test is illustrated in Figure 11. Eight dywidag thread bars were employed at the support of the specimens to securely fasten them to the reaction floor. In addition, to prevent out-of-plane deformation of the test specimens during experiments, steel rods were used to fix the space between the upper frame on the wall and the column frame. To ensure a uniform distribution of the applied axial load across the entire wall, a rigid frame was installed at the top of the specimens. A horizontal actuator with a capacity of 2000 kN was used to apply lateral forces, while two vertically installed hydraulic jacks with a capacity of 1000 kN each were responsible for applying compressive axial load. The data acquisition for these tests was performed at a rate of 1 sample per sec using a static data logger (JSM-350-32, JOOSHIN, Hanam-si, Kyeonggi-do, Republic of Korea).

Figure 11. Experimental setup—cyclic tests.

Figure 12 shows the applied lateral displacement history for the static cyclic loading tests under a constant axial load of 1092 kN, representing 10% of the ultimate axial capacity of the wall. The lateral displacement history was generated with multiple cycles at each displacement level to reflect the effect of strength degradation characteristics. The experiments were terminated either upon failure of the test specimens due to their inability to withstand the applied load or when the lateral load decreased to 80% of its maximum.

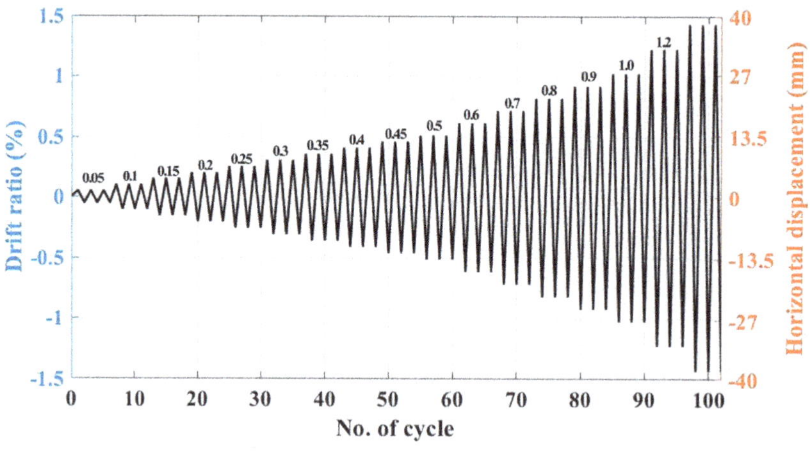

Figure 12. Loading history—cyclic tests.

LVDTs and strain gauges were utilized to measure the strains of rebars and deformation of test specimens. As shown in Figure 13a, a total of 17 LVDTs were installed. Four LVDTs were positioned horizontally (L1 to L4, L17) to measure the lateral displacement at the loading point, the rotation of the bottom cap beam, and the deformation of the mortar pad at the horizontal joint. Additionally, to monitor the occurrence of rocking of the wall panels due to horizontal joints, four LVDTs were installed vertically at each end of the wall panel (L13 to L16). "X"-shaped LVDTs were also installed to measure the shear deformation occurring on the wall panel surfaces (L5 to L12). To assess the strain of the rebars in the wall, strain gauges were installed, as depicted in Figure 13b, focusing on the plastic hinge region and the vertical joint.

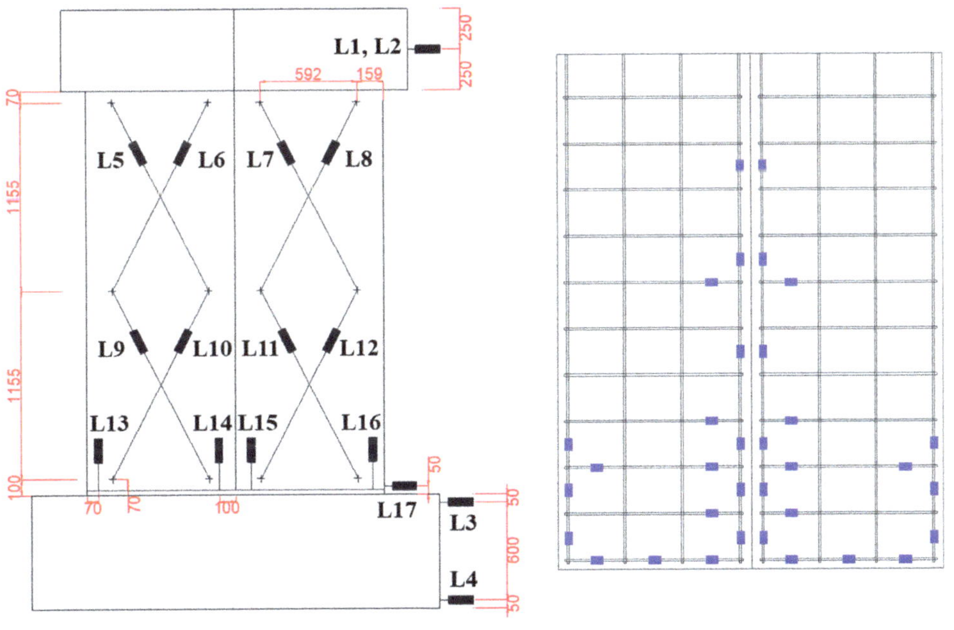

(a) (b)

Figure 13. Instrumentation (unit: mm)—cyclic tests: (**a**) LVDTs; (**b**) Strain gauges.

4.4. Experimental Results

4.4.1. Crack Patterns

Figure 14 shows the crack patterns of the specimens at the end of testing. At a drift ratio of 0.05%, the initial crack in the specimen C-PWR with the wet vertical joint occurred horizontally at the interface of the mortar pad in the horizontal joint. Subsequently, multiple cracks propagated through the wall panel and vertical joint. As the drift ratio reached 0.4%, numerous cracks occurred horizontally and diagonally at the upper part of the wall panel. At a drift ratio of 0.5%, cracks concentrated near the mortar in the vertical joint, and at a drift ratio of 0.8%, concrete spalling began at the central and lower sides of the wall panel. These cracks continued to propagate and expand until the drift ratio reached 1.0%, at which point the section with spalling experienced further failure, leading to the termination of the experiment.

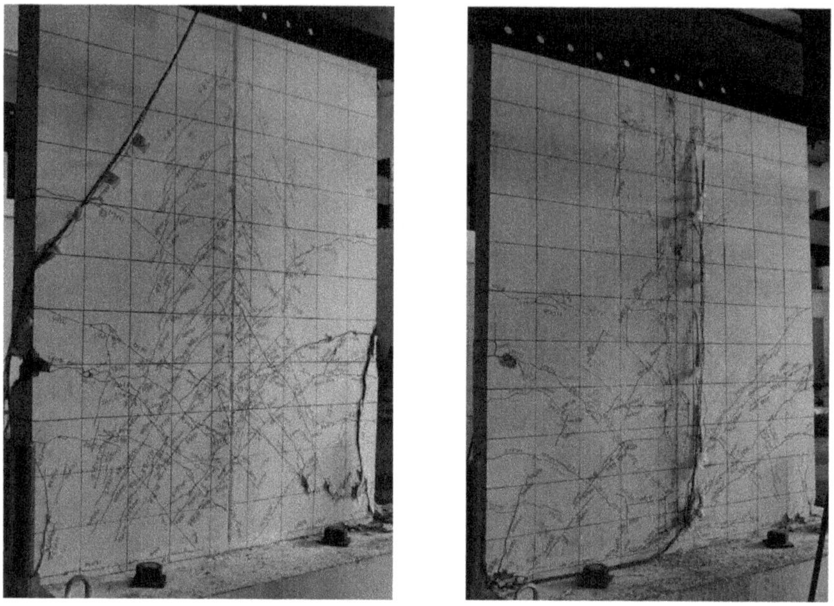

(a) (b)

Figure 14. Crack patterns of specimens—cyclic tests: (**a**) C-PWR; (**b**) C-PWB.

At a drift ratio of 0.1%, specimen C-PWB with the dry vertical joint experienced the first crack horizontally at the interface of the mortar pad between the bottom of the wall and the cap beam. Subsequently, multiple cracks developed at the lower part of the wall panel. Upon reaching a drift ratio of 0.45%, numerous cracks occurred at the interface of the vertical joint at the upper part of the wall panel, gradually expanding until concrete spalling occurred at a drift ratio of 0.8%. As the drift ratio increased, the crack width expanded, leading to the failure of the lower and central sides of the wall panel at a drift ratio of 1.1%, ultimately concluding the test.

As described above, the overall crack patterns of both specimens appeared similar. However, cracks and damage in C-PWR were more severe throughout the entire wall panel and near the vertical joint compared to those in C-PWB. Particularly, a noticeable reduction in both the number and width of cracks, along with a significant decrease in the spalled concrete of the wall panel, was observed in C-PWB. Therefore, it can be inferred that the proposed high-tension bolt vertical joint has a significant effect on controlling cracks and reducing damage in both the wall panel and the vertical joint.

4.4.2. Force–Drift Ratio Relationship

The experimental results of the specimens subjected to static cyclic loadings are summarized in Table 4. Figure 15 shows the force–drift ratio relationships of test specimens, while Figure 16 shows their envelope curves and average envelope curves. The average maximum strengths of the specimens C-PWR and C-PWB are 518.06 kN and 439.96 kN, respectively. Compared to the reference specimen C-PWR, C-PWB showed a 15.08% decrease in average maximum strength. However, the ultimate displacement at which the experiment terminated increased by 19.86% for C-PWB. Furthermore, C-PWB showed a relatively gradual reduction in force compared to C-PWR. C-PWB employs high-tension bolts alone, creating a dry connection between the PC wall panels. Conversely, C-PWR utilizes a wet joint with high-strength mortar over a larger area, effectively mimicking the behavior of a monolithic CIP wall. Thus, the observed lower strength and stiffness in C-PWB originate from these fundamental differences in connection type. Nevertheless, considering the developed specimen C-PWB aimed at enhancing ductility, its performance is deemed satisfactory, as evidenced by the observed trends of strength reduction and increased ultimate displacement rate compared to C-PWR.

Table 4. Experimental results—cyclic tests.

Specimen		Initial Stiffness (kN/mm)	F_m [1] (kN)	Average of F_m (kN)	δ_m [2] (%)	F_u [3] (kN)	δ_u [4] (%)
C-PWR	+	198.18	545.88	518.06	0.677	439.48	0.986
	−		490.24		0.557	391.45	0.988
C-PWB	+	109.33	478.03	439.96	0.674	414.28	1.184
	−		401.89		0.774	317.70	1.181

[1] F_m: max. lateral force. [2] δ_m: drift ratio at max. lateral force. [3] F_u: ultimate lateral force at failure (corresponding to 20% force reduction). [4] δ_u: ultimate drift ratio.

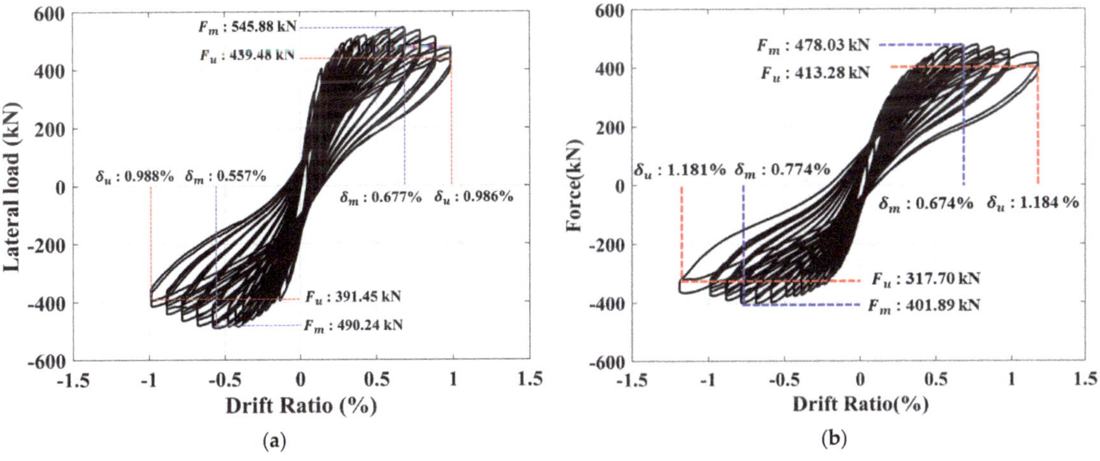

Figure 15. Force—drift ratio relationship—cyclic tests: (**a**) C-PWR; (**b**) C-PWB.

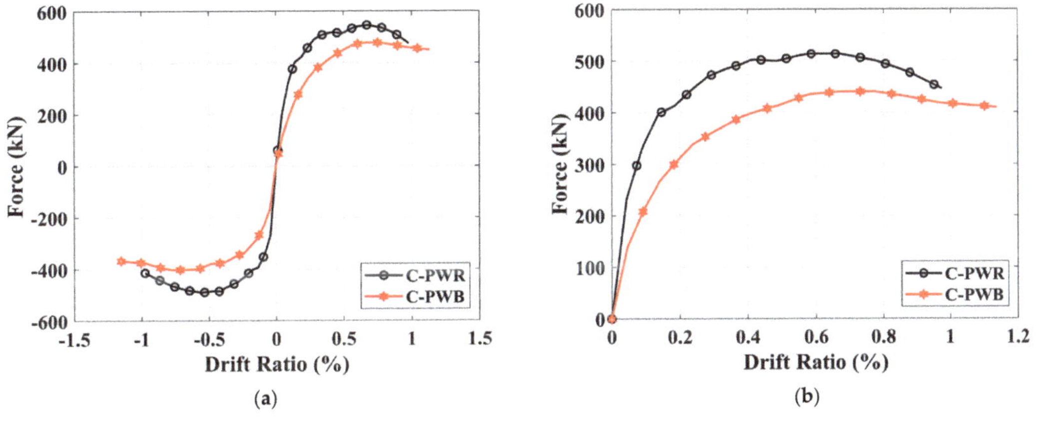

Figure 16. Envelope curve—cyclic tests: (**a**) Envelope curve; (**b**) Average envelop curve.

4.4.3. Stiffness Degradation

The relationships of stiffness degradation for both test specimens are illustrated in Figure 17. Figure 17a represents the stiffness degradation concerning the initial stiffness, while Figure 17b demonstrates the normalized stiffness degradation with respect to the initial stiffness. The initial stiffness of C-PWR and C-PWB is 198.18 kN/mm and 109.33 kN/mm, respectively, resulting in a 44.83% reduction in C-PWB compared to C-PWR. For both specimens, the most significant stiffness degradation occurred as the drift ratio increased from 0.05% to 0.1%, with reductions of 90.86% and 24.80% for C-PWR and C-PWB, respectively. Comparing the point where 50% of the initial stiffness is reduced, C-PWB exhibited a delay of 62.5% compared to C-PWR based on cycles. Similarly, at the point where 70% of the initial stiffness is reduced, C-PWB showed a 64.71% delay compared to C-PWR. In comparison to C-PWB, the substantial cracks and damages across the vertical joints and the entire wall of C-PWR, as the drift ratio increased, led to a more pronounced stiffness degradation. Therefore, it can be concluded that the application of the proposed vertical joint with high-tension bolts in this study results in a relatively lower initial stiffness but effectively delays the rate of stiffness degradation, ultimately preventing the abrupt failure of the wall.

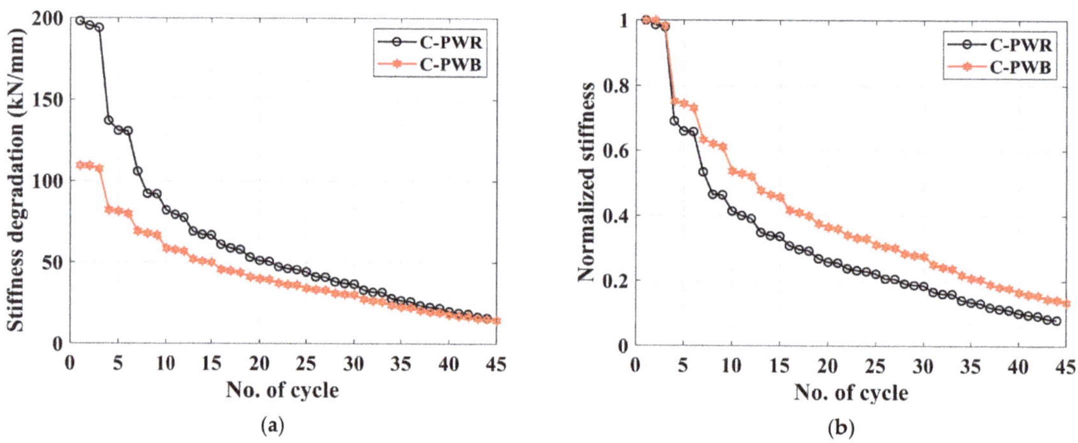

Figure 17. Stiffness degradation—cyclic tests: (**a**) Stiffness degradation per cycle; (**b**) Normalized stiffness degradation per cycle.

4.4.4. Ductility

The ductility was calculated as the ratio of the yield displacement to the ultimate displacement from the average envelop curve shown in Figure 16b. The yield displacement was defined by the energy equivalent method, and the ultimate displacement was defined as the point where the load decreased by 20% after reaching its maximum. The ductility for each specimen is detailed in Table 5.

Table 5. Ductility−cyclic tests.

Specimen	Δ_y [1] (mm)	δ_y [2] (%)	Δ_u [3] (mm)	δ_u [4] (%)	μ [5]
C-PWR	5.53	0.203	26.95	0.987	4.87
C-PWB	6.55	0.240	32.28	1.183	4.93

[1] Δ_y: Yield displacement. [2] δ_y: Yield drift ratio. [3] Δ_u: Ultimate displacement. [4] δ_u: Yield drift ratio. [5] μ: Ductility ratio.

Despite a remarkable 19.86% increase in ultimate displacement for C-PWB compared to C-PWR, its ductility ratio of 4.93 only showed a modest 1.23% increase over 4.87 for C-PWR. This seemingly counterintuitive finding can be explained by the influence of initial stiffness and peak load timing on the calculated yield displacement. C-PWB had a 44.83% lower initial stiffness and reached its peak load 17.34% later than C-PWR. These factors contributed to a delayed yield displacement for C-PWB, leading to a smaller relative increase in ductility despite its improved ultimate capacity. Additionally, the horizontal joints likely induced hinge-like behavior in the lower section of the wall, further contributing to delayed yielding and affecting overall behavior.

4.4.5. Strain Distribution

The strain distributions of vertical and horizontal rebars in the PC wall for each specimen are illustrated in Figures 18 and 19, respectively. The first yielding in vertical rebar of C-PWR occurred at a height corresponding to 23% of the total wall height when the drift ratio reached 0.3%. In contrast, the vertical rebar of C-PWB yielded first at the same height but with a drift ratio of 0.5%. Contrary to the typical trend of increasing deformation closer to the bottom of the wall, both specimens exhibited a significant increase in strain between 20% and 40% of the wall height. It could be inferred that premature damage in the mortar pad of a horizontal joint at the bottom of the wall caused the lower part of the wall to behave like a hinge, resulting in predominant yielding in the upper region. Additionally, the increased stiffness in the lower part due to the application of the corrugated tube could lead to significant deformation near the injection of the corrugated tube. Figure 18 clearly shows that the observed strains of C-PWB with a 19.86% higher ultimate drift ratio are lower than those of C-PWR. This suggests that the application of the vertical joint with high-tension bolts to the PC wall contributes to better performance in terms of deformation capacity. Furthermore, as shown in Figure 19, both specimens' horizontal reinforcement did not yield until the end of the experiment, yet their strain distribution was similar.

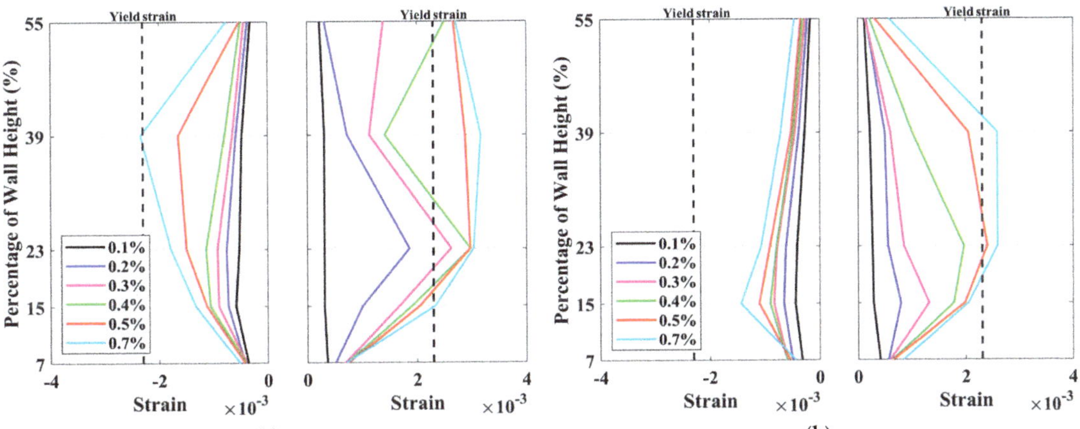

Figure 18. Strain distribution of vertical rebars—cyclic tests: (**a**) C-PWR; (**b**) C-PWB.

4.4.6. Energy-Dissipation Capacity

Figure 20 illustrates the energy dissipation per cycle, calculated by the area enclosed by the hysteresis curve, and the cumulative dissipated energy. Both specimens initially exhibited comparable energy dissipation as the loading cycles increased. A significant increase in dissipated energy occurred after 30 cycles (corresponding to a drift ratio of 0.5%). Notably, C-PWB achieved the highest energy dissipation just before failure (at a drift ratio of 1.1%). This peak dissipation coincided with the formation of numerous cracks near the wall and joint areas, accompanied by a substantial increase in crack width. In terms of cumulative energy dissipation, C-PWR consistently showed higher values than C-PWB across all cycles. Ultimately, the cumulative dissipated energy of C-PWB was 5.21% lower than that of C-PWR. This can be attributed to the relatively higher lateral stiffness and force of C-PWR, along with its transverse reinforcement ratio in the vertical joint exceeding the typical range by at least 14.81%, leading to more ductile behavior. Therefore, considering that C-PWB achieved 94.79% of the cumulative dissipated energy of C-PWR, the PC wall with the proposed vertical joint incorporating high-tension bolts could provide sufficient seismic performance in terms of energy dissipation.

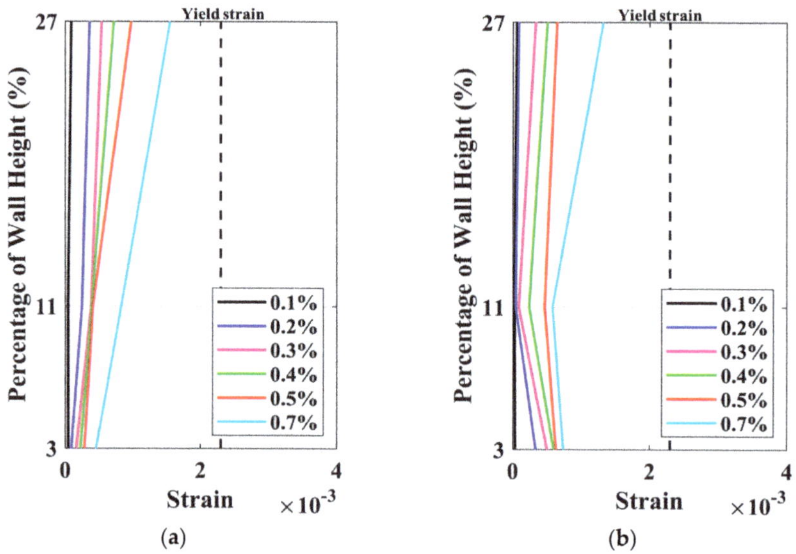

Figure 19. Strain distribution of horizontal rebars—cyclic tests: (**a**) C-PWR; (**b**) C-PWB.

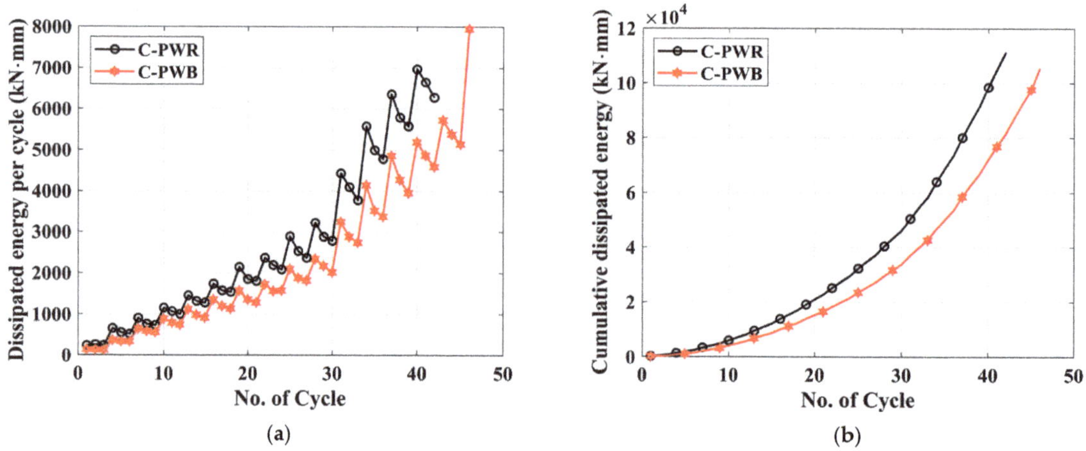

Figure 20. Comparison of energy dissipation—cyclic tests: (**a**) Dissipated energy per cycle; (**b**) Cumulative dissipated energy.

5. Conclusions

This study introduces a novel vertical joint for PC walls utilizing high-tension bolts and experimentally assesses its effectiveness in enhancing shear and seismic performance. The shear behavior of three specimens is investigated through direct shear tests, while the seismic performance of a PC wall incorporating the proposed joint is evaluated through two static cyclic tests. The most important findings are summarized below.

- Direct shear test results showed that the stiffness and strength of the PC wall with the proposed joint were lower than those of the cast-in-place concrete wall and the PC wall with the wet joint. However, the ultimate displacement increased significantly by more than 400%. In addition, the PC wall with the proposed joint did not exhibit sudden load reduction or failure, and it showed significant ductile behavior. In particular, it was found to be effective in suppressing shear cracks and damage in the PC wall.
- Static cyclic loading test results showed that the maximum strength and initial stiffness of the PC wall with the proposed joint were reduced by 15% and 44%, respectively, compared to the PC wall with wet joints. However, the ultimate drift ratio increased by about 20%. Despite the decrease in initial stiffness and strength of the PC wall with the proposed joint compared to the PC wall with wet joints, the energy-dissipation capacity was similar.

Taking into account the observations from experiments in this study, the proposed joint exhibits remarkable deformation capacity, comparable energy dissipation despite reduced initial stiffness and strength, and effective crack control in PC walls. Therefore, it can be concluded that the proposed high-tension bolted vertical joint is effective in mitigating damage and improving the ductile behavior of PC walls.

While this study focused on static loading conditions and limited material composition, further research is necessary to comprehensively assess the proposed vertical joint system. Notably, future investigations should explore its behavior under realistic earthquake loads, analyze the influence of key component material compositions, and evaluate its long-term durability under diverse environmental conditions. Investigating these aspects in future research could provide valuable insights and potentially expand the system's practical applications.

Author Contributions: Conceptualization, K.Y.M., S.J.K. and K.L.; methodology, K.Y.M., S.J.K. and K.L.; validation, K.Y.M. and S.J.K.; formal analysis, K.Y.M. and S.J.K.; investigation, K.Y.M., S.J.K. and K.L.; resources, K.Y.M. and S.J.K.; data curation, K.Y.M. and S.J.K.; writing—original draft preparation, K.Y.M.; writing—review and editing, S.J.K.; visualization, K.Y.M.; supervision, S.J.K.; project administration, S.J.K.; funding acquisition, S.J.K. All authors have read and agreed to the published version of the manuscript.

Funding: This research was supported by the Korea Basic Science Institute (the National Research Facilities and Equipment Center) grant funded by the Ministry of Education (2021R1A6C103A412).

Data Availability Statement: The data presented in this study are available on request from the corresponding author. The data are not publicly available due to privacy.

Conflicts of Interest: The authors declare no conflicts of interest.

References

1. Zhi, Q.; Xiong, X.; Yang, W.; Liu, S.; Xiong, J. Experimental study on the shear behavior of precast wall concrete joints with/without dowel reinforcement. *Materials* **2020**, *13*, 1726. [CrossRef] [PubMed]
2. Dang, L.; Liang, S.; Zhu, X.; Zhang, M.; Song, Y. Seismic performance of precast concrete wall with vertical energy-dissipating connection. *Struct. Des. Tall Spec. Build.* **2021**, *30*, e1820. [CrossRef]
3. Brunesi, E.; Peloso, S.; Pinho, R.; Nascimbene, R. Cyclic testing of a full-scale two-storey reinforced precast concrete wall-slab-wall structure. *Bull. Earthq. Eng.* **2018**, *16*, 5309–5339. [CrossRef]
4. Biswal, A.; Prasad, A.M.; Sengupta, A.K. Study of shear behavior of grouted vertical joints between precast concrete wall panels under direct shear loading. *Struct. Concr.* **2019**, *20*, 564–582. [CrossRef]
5. Zhang, C.; Li, H.; Gao, W. Development of a novel friction damped joint for damage-plasticity control of precast concrete walls. *Eng. Struct.* **2020**, *219*, 110850. [CrossRef]
6. Fib Task Group 7.3. *Seismic Design of Precast Concrete Building Structures*; Bulletin 27; International Federation for Structural Concrete (fib): Lausanne, Switzerland, 2003.
7. Li, J.; Wang, L.; Lu, Z.; Wang, Y. Experimental study of L-shaped precast RC shear walls with middle cast-in-situ joint. *Struct. Des. Tall Spec. Build.* **2018**, *27*, e1457. [CrossRef]

8. Sørensen, J.H.; Herfelt, M.A.; Hoang, L.C.; Muttoni, A. Test and lower bound modeling of keyed shear connections in RC shear walls. *Eng. Struct.* **2018**, *155*, 115–126. [CrossRef]
9. Zhao, C.; Zhang, Z.; Wang, J.; Wang, B. Numerical and theoretical analysis on the mechanical properties of improved CP-GFRP splice sleeve. *Thin-Walled Struct.* **2019**, *137*, 487–501. [CrossRef]
10. Cai, G.; Xiong, F.; Xu, Y.; Si Larbi, A.; Lu, Y.; Yoshizawa, M. A demountable connection for low-rise precast concrete structures with DfD for construction sustainability-A preliminary test under cyclic loads. *Sustainability* **2019**, *11*, 3696. [CrossRef]
11. Joergensen, H.B.; Hoang, L.C.; Hagsten, L.G. Strength of precast concrete shear joints reinforced with high-strength wire ropes. *Proc. Inst. Civ. Eng. Struct. Build.* **2017**, *170*, 168–179. [CrossRef]
12. Jorgensen, H.B.; Hoang, L.C. Load Carrying Capacity of Keyed Joints Reinforced with High Strength Wire Rope Loops. In Proceedings of the Fib Symposium: Concrete-Innovation and Design, Copenhagen, Denmark, 18–20 May 2015.
13. Pekau, O.A.; Hum, D. Seismic response of friction jointed precast panel shear walls. *PCI J.* **1991**, *36*, 56–71. [CrossRef]
14. Bora, C.; Oliva, M.G.; Nakaki, S.D.; Becker, R. Development of a precast concrete shear-wall system requiring special code acceptance. *PCI J.* **2007**, *52*, 122. [CrossRef]
15. Dal Lago, B.; Biondini, F.; Toniolo, G. Friction-based dissipative devices for precast concrete panels. *Eng. Struct.* **2017**, *147*, 356–371. [CrossRef]
16. Crisafulli, F.J.; Restrepo, J.I. Ductile steel connections for seismic resistant precast buildings. *J. Earthq. Eng.* **2003**, *7*, 541–553. [CrossRef]
17. Zhang, C.; Li, H.; Gao, W.; Li, C. Experimental and analytical investigations on new viscoelastic damped joints for seismic mitigation of structures with precast shear walls. *Struct. Control Health Monit.* **2020**, *27*, e2485. [CrossRef]
18. Christian, L.H.; Lawrence, D.R.; Chris, P.P. Behavior of Welded Plate Connections in Precast Concrete Panels under Simulated Seismic Loads. *PCI J.* **2002**, *47*, 122–133.
19. Han, Q.; Wang, D.; Zhang, Y.; Tao, W.; Zhu, Y. Experimental investigation and simplified stiffness degradation model of precast concrete shear wall with steel connectors. *Eng. Struct.* **2020**, *220*, 110943. [CrossRef]
20. Abdul-Wahb, H.S.; Sarsam, S. Strength of vertical plane joints between large, precast concrete panels. *Struct. Eng. Part A* **1988**, *66*, 211–215.
21. Sørensen, J.H.; Hoang, L.C.; Fischer, G.; Olesen, J.F. Construction-friendly ductile shear joints for precast concrete panels. In Proceedings of the International Conference on Performance-based and Life-cycle Structural Engineering, Brisbane, QLD, Australia, 9–11 December 2015.
22. Seifi, P.; Henry, R.S.; Ingham, J.M. In-plane cyclic testing of precast concrete wall panels with grouted metal duct base connections. *Eng. Struct.* **2019**, *184*, 85–98. [CrossRef]
23. Elsayed, M.; Nehdi, M.L.; Ghrib, F. Predicting behavior of grouted dowel connections using interfacial cohesive elements. *Appl. Sci.* **2019**, *9*, 2344. [CrossRef]

buildings

MDPI

Article

Post-Cracking Shear Stiffness Model of Reinforced Concrete Beams

Kaiqi Zheng [1,2], Siwen Ni [1], Yaohui Zhang [2,*], Junxuan Gu [1], Mingming Gao [1] and Yang Wei [1,*]

1 College of Civil Engineering, Nanjing Forestry University, Nanjing 210037, China; k.zheng@njfu.edu.cn (K.Z.)
2 State Key Laboratory of Mechanical Behavior and System Safety of Traffic Engineering Structures, Shijiazhuang Tiedao University, Shijiazhuang 050043, China
* Correspondence: sjzzhangyh@126.com (Y.Z.); wy78@njfu.edu.cn (Y.W.)

Abstract: Macro diagonal cracks can significantly reduce the stiffness of slender reinforced concrete (RC) beams, which results in excessive deflection compared with limitations from design specifications. To evaluate the post-cracking stiffness of slender RC beams with diagonal cracks, a shear degradation model that considers shear deformation is proposed. Based on the variable angle truss model, this study deduced the strut angle formula based on the minimum energy principle. Then, the relationship between the stirrup yielding shear stiffness and elastic shear stiffness was modeled. Finally, the calculation procedure was developed by quantifying the stiffness degradation tendency. The comparison between the experimental results of deflection and the proposed analytical method showed good agreement. Additionally, the proposed method can capture the full-range features of shear strain curves.

Keywords: reinforced concrete beam; shear crack; stiffness model; shear deformation; strut angle; variable angle truss model

1. Introduction

Diagonal cracks are commonly seen in cast-in-place and precast reinforced concrete (RC) box girder bridges, which results in additional shear deformation besides flexural deflection. Therefore, accurately computing the total deformation is a perquisite for the design of RC box girder bridges. Typically, the total deformation can be computed by adding the deformations caused by shear and flexural deformations. In order to decouple the contribution of diagonal cracking to deflection from other time-varying factors (such as concrete shrinkage and creep), it is necessary to evaluate the shear stiffness of diagonal cracked bridges. However, due to the complex influencing mechanism of shear cracks on shear stiffness degradation, quantitative and practical evaluation methods that are suitable for engineering applications are particularly needed.

From a practical perspective, current deflection models for RC beams, whether cast in situ or prefabricated, typically ignore the contribution of shear deformations [1,2]. However, some prior shear tests on slender beams revealed that the measured deflections are generally much larger than the prediction-by-code method [3–5]. One possible reason for these underestimations may be from the neglect of the contribution of shear deformations, especially the contribution of shear deformation after shear cracking [6–11].

In the face of frequently occurring excessive deflections and shear failure in large-span concrete and precast concrete bridges and buildings, it is crucial to build rational shear models. The shear stiffness of a diagonally cracked RC beam is approximately 10% to 30% of the shear stiffness of the uncracked one, depending on the web reinforcement provided [12]. This means that diagonal cracking has a much larger effect on shear stiffness than flexural cracking on bending stiffness. However, unlike the calculation of bending stiffness, which is based on the Branson's effective moment of inertia model and is accurate enough for design application [13], there is still no widely accepted shear stiffness model for cracked RC beams. Therefore, a simple, rational, and practical shear stiffness model for RC beams with shear cracks is needed.

Since the development of the truss analogy concept by Ritter and Mörsh, researchers have paid great attention to the shear deformation/stiffness of cracked beams [14]. Leonhardt and Walther (1961) [15], Kupfer (1964) [16], and Park and Paulay (1974) [12] investigated the shear stiffness of diagonally cracked beam based on the parallel-chord truss model. They also suggested different values or solving methods for the strut angle of the web strut. Vecchio and Collins (1986) [17], Bentz (2000) [18], Debernardi and Taliano (2006) [5], and Desalegne and Lubell (2012) [19] proposed or modified the MCFT (modified compression field theory) to predict the shear load-deformation response of RC beams. Pang and Hsu (1996) [20] and Zhu et al. (2001) [21] presented the shear modulus for the FA-STM (fixed-angle softened truss model) based on the smeared-crack concept. However, the iteration procedure of MCFT and FA-STM are somewhat complicated. Barzegar (1989) [22], Hansapinyo et al. (2003) [4], and Rahal (2010) [23] attempted to establish an empirical equation for the shear stiffness of beams after cracking. Their fitting parameters include concrete strain, stirrup ratio, and longitudinal reinforcement ratio.

Recently, the lower-bound theory of plasticity [9] and minimum strain energy principle [7] have been used for calculating the strut angle. Based on the minimum strain energy principle, truss models with variable strut angles were used to investigate the effective shear stiffness of RC beams with diagonal cracking [8,24,25].

For the shear stiffness degradation criteria, scholars have proposed various rules, such as linear tangent degradation [7], linear secant stiffness degradation [26], hybrid stiffness degradation [9], exponential degradation [8], adopted Branson's degradation [3], etc. However, due to the complex influencing mechanism of shear cracks on shear behavior [27–30], there is a need for shear stiffness theories that can properly reflect the gradual degradation of shear stiffness with the development of shear cracks.

This paper aims to propose an energy-based strut angle calculation method and a practical degradation rule of the effective shear stiffness of diagonally cracked RC beams. With the adoption of the minimum strain energy principle and the additional consideration of the contribution of moment, an implicit analytical calculation formula for strut angle is established. Furthermore, through a parameter analysis and comparative study with implicit expressions, a simplified practical strut angle calculation formula is obtained. Experimental validation shows that the proposed method yields good and consistent predictions of post-cracking shear stiffness and shear deformation.

2. Post-Cracking Shear Stiffness Model of Reinforced Concrete Beams

2.1. Intial Elastic Shear Stiffness K_e

Beam shear can be divided into two stages: pre-cracking stage and post-cracking stage. At the first stage, the elastic shear stiffness K_e can be calculated using elasticity theory [12], as shown in Equation (1).

$$K_e = G_c A_v = \frac{1}{2(1+\mu)} E_c A_v \quad (1)$$

where K_e is the elastic shear stiffness before cracking, G_c is the shear modulus of concrete, E_c is the modulus of elasticity of concrete, μ is the Poisson's ratio of concrete, A_v is the effective shear area ($A_v = b_w d_v$), b_w is the web width of the beam, and d_v is the effective shear depth.

In the post-cracking stage, as diagonal cracking violates the continuity of concrete, an alternative approach must be adopted for the post-cracking shear stiffness calculation.

2.2. Stirrup Yielding Shear Stiffness K_y Based on Truss Model

A truss model gives good depiction of slender beams in shear, which provides a possibility that the post-cracking shear stiffness can be depicted by the truss model. For slender beams, when the stirrups are yielded and the diagonally shear cracks are well developed, the truss model shown in Figure 1 can be used for analyses. As the observed shear cracks are generally parallel to each other in principle, the inclination of the strut angle θ_y can be assumed to be the same. The shear deformation of the truss is mainly induced by the deformation of web members δ_{web}, including the elongation of stirrups δ_s and the vertical deformation δ_c caused by the shortening of the inclined strut. The deformation of stirrups and inclined concrete struts are listed in Table 1, in which A_s, A_g, and A_{cm} are the areas of the longitudinal reinforcement, cross-section, and compression zone, respectively.

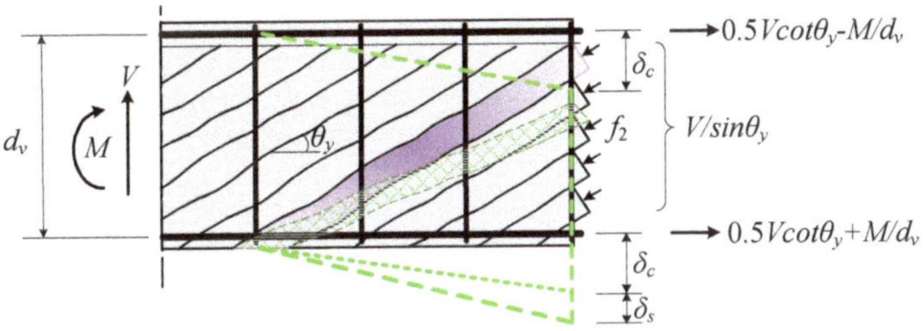

Figure 1. Truss model for shear response of diagonally cracked RC beams.

Table 1. Member deformation of the variable-angle truss model.

Inducement	Member	Force	Unit Load	Length	Rigidity	Deformation
Shear force V	Stirrups	V	1	d_v	$E_s\rho_v b_w d_v \cot\theta_y$	$\frac{V}{E_s\rho_v b_w \cot\theta_y}$
	Inclined Strut	$\frac{V}{\sin\theta_y}$	$\frac{1}{\sin\theta_y}$	$\frac{d_v}{\sin\theta_y}$	$\frac{1}{E_c b d_v \cos\theta_y}$	$\frac{V}{E_c b \sin^3\theta_y \cos\theta_y}$
	Upper chord	$\frac{V\cot\theta_y}{2}$	$\frac{\cot\theta_y}{2}$	$d_v \cot\theta_y$	$E_c A_{cm}$	$\frac{V d_v \cot^3\theta}{4E_c A_{cm}}$
	Lower chord	$\frac{V\cot\theta_y}{2}$	$\frac{\cot\theta_y}{2}$	$d_v \cot\theta_y$	$E_s A_s$	$\frac{V d_v \cot^3\theta_y}{4E_s A_s}$
Moment M	Upper chord	$-\frac{M}{d_v}$	$\frac{\cot\theta_y}{2}$	$d_v \cot\theta_y$	$E_c A_{cm}$	$\frac{-M\cot^2\theta_y}{2E_c A_{cm}}$
	Lower chord	$\frac{M}{d_v}$	$\frac{\cot\theta_y}{2}$	$d_v \cot\theta_y$	$E_s A_s$	$\frac{M\cot^2\theta_y}{2nE_c A_s}$

Equal to the shear distortion (shear deformation per unit length) induced by a unit shear force, the shear stiffness of the equivalent truss can be expressed as

$$K_y = \frac{V}{\gamma_0} = \frac{V}{\delta_{web}/(d_v \cot\theta_y)} = \frac{n\rho_v \cot^2\theta_y}{1 + n\rho_v \csc^4\theta_y} E_c A_v \tag{2}$$

where n is the ratio of the elastic modulus of the reinforcement and the concrete ($n = E_s/E_c$), and ρ_v is the stirrup ratio. Equation (2) shows that the shear stiffness is mainly affected by the material type (n, E_s, E_c), geometric properties (A_v), stirrup ratio (ρ_v), and strut angle (θ_y).

In order to evaluate the degradation of the shear stiffness of the beam after the yielding of stirrups, it is necessary to establish the relationship between the stirrup yielding shear stiffness K_y and the initial shear stiffness K_e. Combine Equations (1) and (2), the ratio of K_y to K_e can be expressed as

$$\xi_y = \frac{K_y}{K_e} = \frac{2n(1+\mu)\rho_v \cot^2\theta_y}{1 + n\rho_v \csc^4\theta_y} \tag{3}$$

where the ratio factor ξ_y is defined as the degradation coefficient of shear stiffness. Apparently, the parameters that have an effect on the shear stiffness degradation are the stirrup ratio ρ_v and the strut angle θ_y. For a specified specimen, n, μ, and ρ_v are all known; if

θ_y is determined, the degradation coefficient can be easily calculated from Equation (3). Similarly, the yielding shear stiffness can be obtained from Equation (2).

Considering that the stirrup ratio ρ_v of practically used RC beams is generally between 1.0% and 2.0%, $n = E_s/E_c \approx 6$ and the Poisson's ratio of the concrete $\mu \approx 0.2$. For θ_y between 25° and 45°, the shear stiffness degradation coefficient is between 0.12 and 0.29. That is to say, for concrete beams with severe web shear cracks, the yielding shear stiffness can be reduced by more than 70% compared with the elastic shear stiffness.

2.3. Shear Stiffness Degradation Rules for Partially Shear-Cracked RC Beams

Because the transition mechanism from an elastic shear stiffness to a stirrup yielding shear stiffness is affected by many factors, such as reinforcement ratio, section size, concrete strength, etc., it is unrealistic to give an accurate and quantitative expression to describe this degradation process.

Scholars have developed some empirical or simplified degradation rules to depict the degraded shear stiffness of partially shear-cracked RC beams, such as linear secant stiffness degradation [30], linear tangent stiffness degradation [7], mixed degradation [9], exponential degradation [8], regression fitting degradation [10], adopted Branson's degradation [3], etc. Below are four representative effective shear stiffness models,

$$K_{eff,1} = K_y + \frac{V - V_{cr}}{V_y - V_{cr}}(K_e - K_y) \in [K_y,\ K_e] \tag{4}$$

$$K_{eff,2} = \frac{V}{\frac{V_{cr}}{K_e} + \frac{V-V_{cr}}{V_y-V_{cr}}\left(\frac{V_y}{K_y} - \frac{V_{cr}}{K_e}\right)} \in [K_y,\ K_e] \tag{5}$$

$$K_{eff,3} = K_{eff,1} + K_{eff,2} \tag{6}$$

$$K_{eff,4} = K_y + \left(\frac{V_y - V}{V_y - V_{cr}}\right)^3 (K_e - K_y) \in [K_y,\ K_e] \tag{7}$$

where $K_{eff,1}$, $K_{eff,2}$, $K_{eff,3}$, and $K_{eff,4}$ are the effective shear stiffness considering linear secant degradation, linear tangent degradation, mix degradation, and adopted Branson's degradation, respectively; V_{cr} and V_y are the diagonal cracking shear force and stirrup yielding shear force, which can be analytically calculated using Equations (8) and (9).

$$V_{cr} = 0.17\sqrt{f'_c} b_w d_v \tag{8}$$

$$V_y = V_{cr} + \rho_v f_{yv} b_w d_v \cot\theta_y \tag{9}$$

After determining the key parameters of the yield shear stiffness and the degradation rule of the shear stiffness, the effective shear stiffness of a partially shear-cracked RC beam can be calculated according to the following steps:

a. Before shear cracking, the elastic shear stiffness K_e can be calculated using Equation (1) and used for the calculation of elastic shear deformation;
b. The cracking shear force V_{cr} and stirrup yielding force V_y of the critical section are calculated using Equations (8) and (9);
c. After shear cracking, the yielding shear stiffness K_y can be obtained from Formula (2), and which the shear stiffness degradation coefficient ξ_y can be calculated using Equation (3);
d. The shear increment V-V_{cr} under the shear load V at different loading levels is calculated and substituted into the selected formula from Equations (4)–(7) to calculate the effective shear stiffness K_{eff} of the RC beam with shear cracks;
e. The average shear strain or shear deformation of the corresponding beam segment is calculated according to the effective shear stiffness K_{eff} under a specific shear load.

3. Determination of Strut Angle, θ_y

3.1. Strut Angle Based on Minimum Energy Principle

It is believed that the angle θ_y will occur at an inclination that requires the minimum potential energy [16]. The strain energy stored in the truss members can be determined using the virtual work principle. Assume that a unit shear force is applied to the truss, the internal force and deformation of each member (including web members and chord members) of the truss can be obtained (See Figure 1 and Table 1). The total vertical deformation of the truss is the sum of the members' deformations listed in Table 1, that is:

$$\delta = \frac{V(1 + n\rho_v \csc^4\theta_y)}{n\rho_v E_c b \cot\theta_y} + \frac{V d_v \cot^3\theta_y \left(1 + n\frac{A_s}{A_{cm}}\right)}{4nE_c A_s} + \frac{M\cot^2\theta_y\left(1 - \frac{nA_s}{A_{cm}}\right)}{2nE_c A_s} \tag{10}$$

The strain energy stored in the truss is equal to the external shear work (EWD) applied to the truss, which is the distortion angle of the truss induced by a unit load. The distortion angle can be obtained by dividing the vertical deformation of the truss by its horizontal length:

$$EWD = \gamma = \frac{\delta}{d_v \cot\theta_y} = \frac{V(1 + n\rho_v \csc^4\theta_y)}{n\rho_v E_c A_v \cot^2\theta_y} + \frac{V\cot^2\theta_y\left(1 + n\frac{A_s}{A_{cm}}\right)}{4nE_c A_s} + \frac{M\cot\theta_y\left(1 - \frac{nA_s}{A_{cm}}\right)}{2nE_c d_v A_s} \tag{11}$$

Based on the minimum strain energy principle, the rational strut angle leads to a minimum energy. By differentiating Equation (11) with respect to θ_y and equating it to zero:

$$\frac{d(EWD)}{d\theta_y} = \frac{d\gamma}{d\theta_y} = 0 \tag{12}$$

Through simplification, Equation (12) becomes

$$\left(4n\rho_v + \frac{\rho_v}{\rho_s}R_1\right)\cot^4\theta_y + \left(\lambda_{MV}\frac{\rho_v}{\rho_s}R_2\right)\cot^3\theta_y = 4 + 4n\rho_v \tag{13}$$

where $\left(1 + n\rho_s\frac{A_g}{A_{cm}}\right)\frac{A_v}{A_g} = R_1$, $\left(1 - n\rho_s\frac{A_g}{A_{cm}}\right)\frac{A_v}{A_g} = R_2$, and $\frac{M}{Vd_v} = \lambda_{MV}$, in which ρ_s (=A_s/A_g) is the longitudinal reinforcement ratio. The analytical solution of Equation (13) is too complex for expression. So, a trial-and-error procedure is used for the θ_y calculation.

The factors R_1, R_2 in Equation (13) remain constant for a specified section. For the rectangular section, the effective shear depth d_v is the maximum of {0.72h, 0.9d} [14]; therefore, A_v/A_g can be assumed to be 0.72 for simplicity; the depth of the compression zone c can be simply assumed to be 0.35d [31]; therefore, A_{cm}/A_g is equal to 0.35. For the I-section, A_v/A_g is equal to A_{web}/A_g, and A_{cm}/A_g is still simply assumed to be 0.35.

The factor λ_{MV} in Equation (13), which can be regard as the shear span-to-depth ratio, has a larger influence on shear deformation than the stirrup ratio ρ_v during the pre-cracking stage but less influence during the post-cracking stage [3,6]. This interesting phenomenon is further discussed in the next subsection and in the experimental validation.

3.2. Simplification Formula for Strut Angle

As is mentioned above, the shear-to-depth ratio λ_{MV} has little effect on the post-cracking shear deformation. To validate this supposition, first assume that the influence of moment is neglected, that is, $\lambda_{MV} = 0$. Then, Equation (13) can be rearranged and worked out:

$$\theta_y = \arctan\left[\left(\frac{1 + \frac{R_1}{4n\rho_s}}{1 + \frac{1}{n\rho_v}}\right)^{0.25}\right] \tag{14}$$

The strut angles calculated using Equations (13) and (14) are compared to investigate the influence of λ_{MV} (see Figure 2).

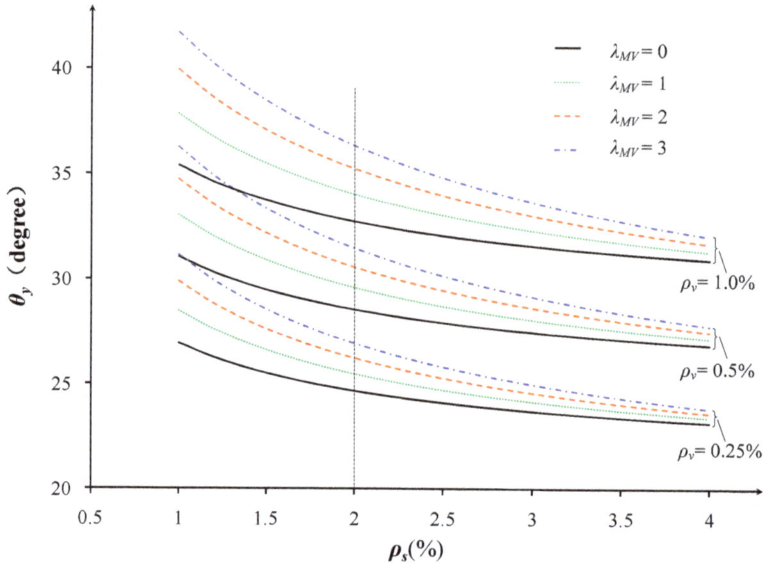

Figure 2. The influence of λ_{MV} on angle θ_y.

As is shown in Figure 2, the factor λ_{MV} has an effect on the strut angle θ_y, but slightly so, especially for a beam with a higher longitudinal reinforcement ratio ρ_s. For example, when the longitudinal reinforcement ratio $\rho_s > 2.0\%$, the calculated strut angle of Equation (13) (taking into account the influence of λ_{MV}) is approximately 1.0 to 1.1 times of the calculated angle obtained using Equation (14) (assuming $\lambda_{MV} = 0$). The relative difference is very small and can be ignored.

Particularly, when we set $A_v = 0.72\ A_g$ and $A_{cm} = 0.35\ A_g$, the value of R_1 is approximately equal to 1 for cross-sections with ordinary longitudinal reinforcement arrangements. Therefore, Equation (14) can be simplified to:

$$\theta_y = \arctan\left[\left(\frac{1+\frac{1}{4n\rho_s}}{1+\frac{1}{n\rho_v}}\right)^{0.25}\right] \tag{15}$$

4. Experimental Validation

4.1. Shear Deformation Test in the Literature

To evaluate the proposed shear stiffness model, three series of shear test data were used for validation, in which the shear deformation was directly measured, and its corresponding equivalent shear stiffness was deduced. A total of 23 lattices/zones of 10 beams are used for comparison. It should be noted that many RC beam test data in which only the total deformation was measured cannot be used for validation and, therefore, were excluded.

The authors conducted shear deformation tests on six large thin-webbed RC beams. Self-tailored strain measuring lattices were installed on one side of the beams, and each measuring lattice was composed of five mechanical dial gauges. The mean shear strain of each lattice was obtained through a strain analysis. The detailed procedures are illustrated in reference [6].

Debernardi and Taliano [5] reported an experiment comprising six RC I-section beams with a thin web. Square lattices made up of transducers were arranged to measure the mean curvature and the mean shear strain, from which the shear deformations can be decoupled from flexural deformation. Fifteen lattices in four specimens (TR1, TR2, TR3, and TR6) with detailed shear strain curves were selected for validation. The specimens were simply supported, subjected to two symmetrical loads (TR1 and TR2), a symmetrical load (TR3), or a non-symmetrical load (TR6).

Hansapinyo et al. [4] tested four RC beams with web reinforcement to investigate the shear deformation after diagonal cracking. Electronic transducers were also used to measure the shear strains of each panel based on the rosette concept. All of the four specimens were selected for validation. The main parameters of all the specimens are listed in Tables 2 and 3.

Table 2. Details of specimens.

Resources	Specimen ID	f'_c (MPa)	E_c (GPa)	d_v (mm)	b_w (mm)	f_{yv} (MPa)	ρ_v (%)	ρ_s (%)	V_{cr} (kN)	V_y (kN)
Zheng [6]	C1	39.0	29.4	684	100	327	0.5	4.8	150	240
	C2	36.0	28.2	684	100	327	0.4	4.8	160	240
Debernardi [5]	TR1	22.0	22.0	300	100	570	0.5	0.74	40	160 *
	TR2	22.0	22.0	300	100	570	0.5	1.34	50	200
	TR3	20.0	21.0	300	100	570	0.5	0.74	40	160 *
	TR6	33.5	27.2	300	100	570	0.5	1.34	55	240
Hansapinyo [4]	S1	33.0	27.0	350	150	370	0.47	4.26	73.6	180
	S2	33.0	27.0	320	150	370	0.47	4.26	64.1	170
	S3	33.0	27.0	320	150	370	0.47	2.13	61.3	160
	S4	33.0	27.0	320	150	370	0.31	2.13	61.6	130

"*"—The yielding shear force V_y was calculated using Equation (9), as the specimen failed in flexure before the stirrups yielding in shear span.

Table 3. Shear-to-span depth ratio of each measuring zone.

Resources	Specimen ID	Zone	a/h or $M/(Vh)$
Zheng [6]	C1, C2	G3	0.5
		G4	0
Debernardi [5]	TR1, TR2	A, E	1.67
		B, D	2.5
	TR3	A, E	2.08
		B, D	2.92
	TR6	A	2
		G	4
		F	4.83
Hansapinyo [4]	S1, S2, S4	-	2.6
	S3	-	3.5

4.2. Comparison of Strut Angle and Degradation Coefficient

The stirrup yielding strut angles θ_y, θ_P, θ_H were calculated using the proposed method, Pan et al. [7], and He et al. [9], respectively. The stirrup yielding strut angles θ_y, θ_P, θ_H and corresponding calculated degradation coefficient ξ_y, ξ_P, ξ_H are listed in Table 4. Overall, the average ratio of the predicted values to the observed degradation coefficient ξ_{exp} and its coefficient of variation (CV) are 0.95 and 0.19, 0.76 and 0.25, and 0.95 and 0.20 for the proposed method, Pan's method, and He's method, respectively.

As is shown in Table 4, Pan's method obtain a higher prediction of strut angle and lower prediction of degradation coefficient than the other two methods, and it has the largest value of CV. The other two methods obtain a relatively good prediction of shear stiffness degradation. However, He's prediction method ignores the influence of the longitudinal reinforcement ratio and may cause larger deviations for specimens with a low longitudinal reinforcement ratio.

Table 4. Strut inclination θ_y and degradation coefficient ξ_y at stirrup yielding status.

Resource	Specimen ID	θ_y (Degree)	θ_P (Degree)	θ_H (Degree)	ξ_y	ξ_P	ξ_H	ξ_{exp}	ξ_y/ξ_{exp}	ξ_P/ξ_{exp}	ξ_H/ξ_{exp}
Zheng [6]	C1	26.3	31.3	25.1	0.182	0.156	0.186	0.190	0.96	0.69	0.97
	C2	25.2	30.0	24.1	0.169	0.145	0.173	0.178	0.95	0.71	0.98
Debernardi [5]	TR1 *	33.9	42.3	35.0	0.164	0.107	0.156	-	-	-	-
	TR2	31.1	38.5	35.0	0.183	0.132	0.156	0.225	0.81	0.59	0.69
	TR3 *	34.0	42.4	36.1	0.169	0.112	0.154	-	-	-	-
	TR6	30.8	38.4	30.8	0.162	0.113	0.162	0.253	0.64	0.45	0.64
Hansapinyo [4]	S1	26.4	31.5	26.8	0.179	0.152	0.178	0.155	1.15	0.98	1.15
	S2	26.4	31.5	26.8	0.179	0.152	0.178	0.143	1.25	1.06	1.24
	S3	28.5	35.2	26.8	0.170	0.128	0.178	0.178	0.96	0.72	1.00
	S4	26.1	32.5	23.8	0.142	0.106	0.152	0.168	0.85	0.63	0.90
								Average	0.95	0.76	0.95
								CV	0.19	0.25	0.20

"*"—Stirrups did not yield before specimen failure.

4.3. Comparison of Effective Shear Stiffness and Shear Strain

The measured and calculated effective shear stiffness reduction factor ξ and shear strain of 23 zones of 10 beams are shown in Figures 3–7. All the shear force–shear strain curves have a distinct turn point, which means the shear stiffness decreases abruptly after the first diagonal cracking. Behind the turn point, the curve stays nearly linear before the stirrup yielding. Based on the experimental results, here, the linear tangent degradation model, which depicts the main features of shear strain curves of slender RC beams, was adopted to evaluate the post-cracking shear stiffness and was used for comparison.

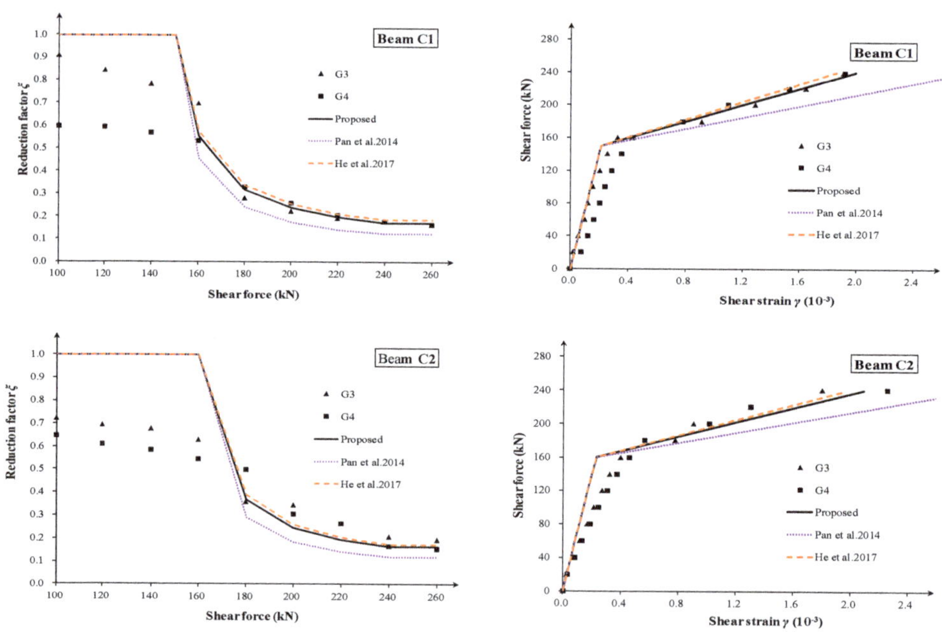

Figure 3. Comparison of measured and calculated effective shear stiffness and shear strain of Beams C1, C2 [6,7,9].

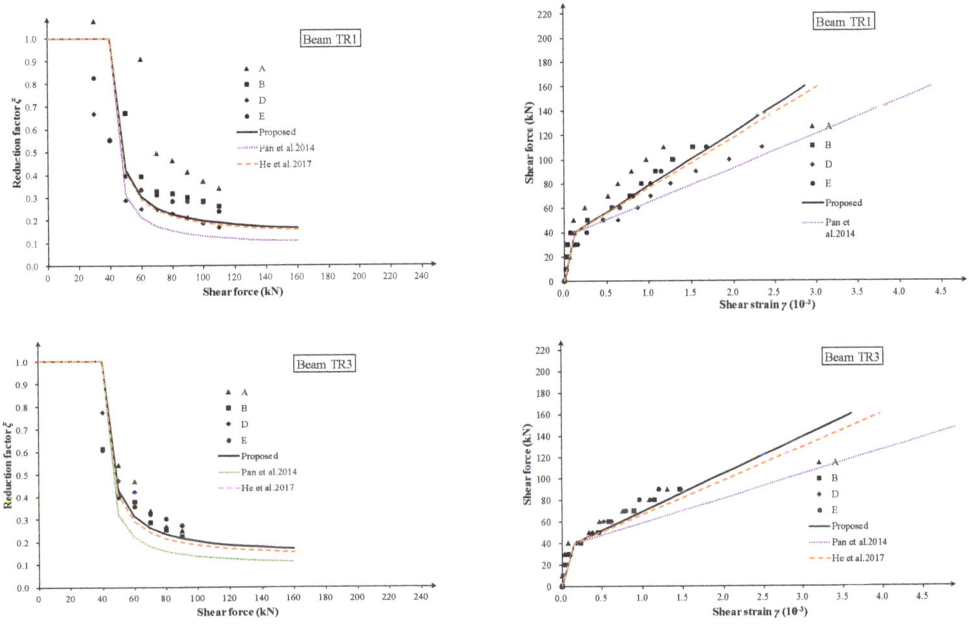

Figure 4. Comparison of measured and calculated effective shear stiffness and shear strain of Beams TR1, TR3 [5,7,9].

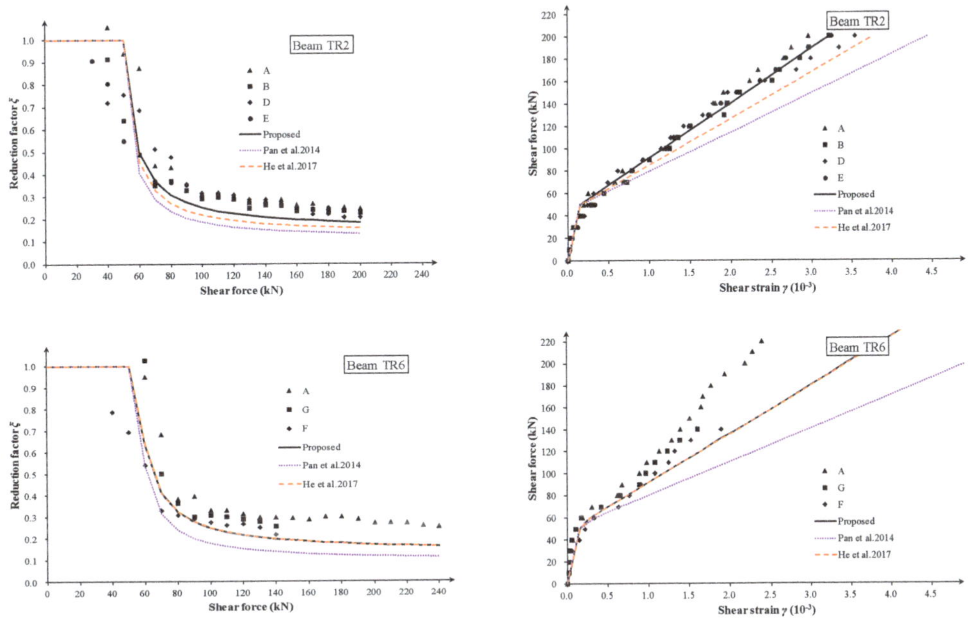

Figure 5. Comparison of measured and calculated effective shear stiffness and shear strain of Beams TR2, TR6 [5,7,9].

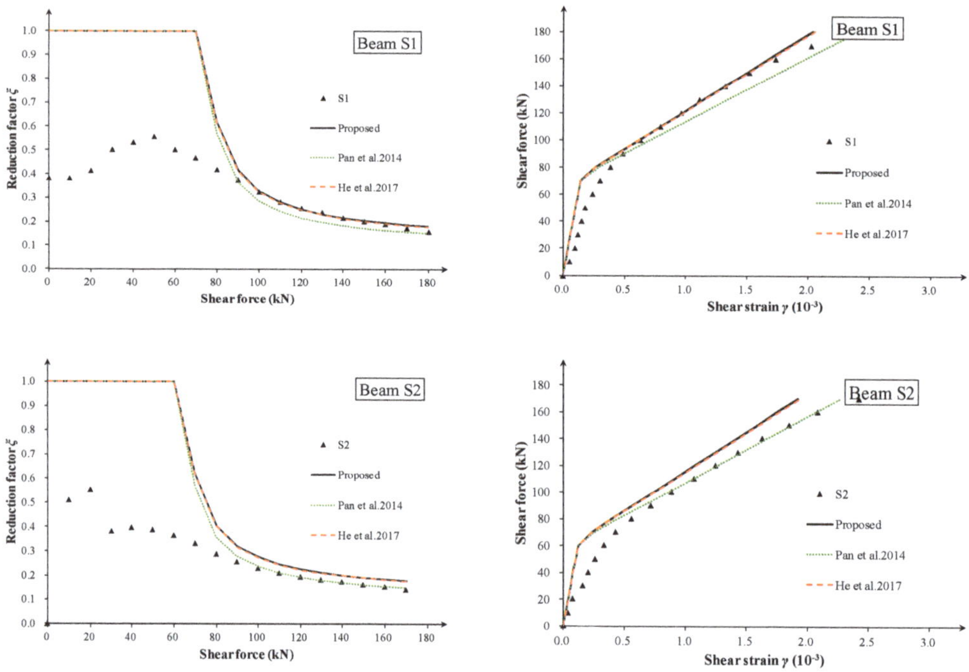

Figure 6. Comparison of measured and calculated effective shear stiffness and shear strain of Beams S1, S2 [4,7,9].

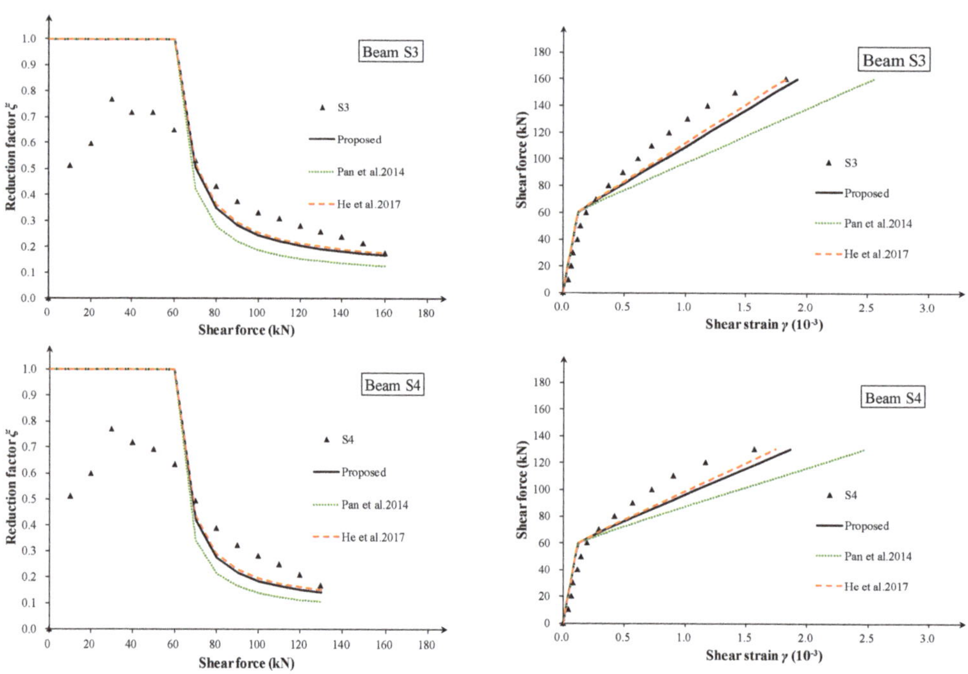

Figure 7. Comparison of measured and calculated effective shear stiffness and shear strain of Beams S3, S4 [4,7,9].

As is known, the shear span-to-depth ratio λ_{MV} has a significant effect on shear strain before cracking. However, as is observed from the tests, the influence becomes very small for RC beams at the post-cracking stage (See A to H zones with different λ_{MV} values of specimens TR1 to TR6 in Figures 4 and 5).

Three methods for determining the strut angle are used for comparison: the proposed method, method by Pan et al. [7], and method by He et al. [9]. As is shown in Figures 3–7, all of the three methods give a rational evaluation of effective shear stiffness, while Pan's method gives a more conservative result than the two other methods. Compared to He's method, the proposed method accounts for the influence of ρ_s and gives a better prediction for beams with a lower ρ_s (Specimen TR2 in Figure 5). Based on the assumption that the longitudinal reinforcement will not yield before stirrup yielding, He's method tends to give a similar prediction of effective shear stiffness with the proposed method for RC beams, with $(\rho_s f_{ys})/(\rho_v f_{yv}) > 3$.

From Figures 3–7, it can be seen that the proposed shear stiffness degradation model simulates the degradation process of shear stiffness very well and provides better stiffness prediction results in most cases. Although the stiffness prediction results of specimen S2 are slightly larger (about 20%), the shear stiffness degradation law still conforms well. It should be noted that for specimen S2, the deviation between the predicted value calculated using the analytical method in the original literature [4] and the measured shear strain is also the largest, which may be due to the shear stiffness anomaly caused by the specimen's own defects.

For thin-webbed beams, a linear tangent stiffness degradation tends to give safe predictions. Therefore, the authors suggest that linear tangent degradation criteria are used for the concrete box girder bridge's shear stiffness evaluation in engineering practice.

In addition, although there is a significant difference in the shear span-to-depth ratio between different beam segments of the same specimen in the experiment (such as TR1~TR6), there is no clear connection between the measured shear stiffness degradation coefficient and the shear span-to-depth ratio. The stiffness degradation curves of each beam segment are also relatively close, indicating that the shear span ratio has little effect on the post-cracking shear stiffness degradation, and its impact on the shear stiffness and strain after cracking can be ignored.

Finally, the proposed shear stiffness model for RC beams is only validated using 23 beam segments. More shear deformation data are needed for further validation. The digital image correlation (DIC) technique based on shear deformation measurements may be widely used in a future study [32–35].

5. Conclusions

To evaluate the effective shear stiffness of RC beams with diagonal cracks, this paper proposed a post-cracking shear stiffness degradation model. Based on the variable truss model, and using the minimum strain energy principle, the strut angle was analytically deduced. Then, the linear tangent stiffness degradation rules were found to be more suitable for thin-webbed concrete beams. With experimental validation, the following conclusions can be drawn:

1. Based on the variable-angle truss model, the relationship between the stirrup yielding shear stiffness and elastic shear stiffness was established. Then, a practical linear tangent shear stiffness degradation model, which depicts the main features of shear strain curves of slender RC beams at the post-cracking stage, was adopted to evaluate the effective shear stiffness.
2. The strut angle θ_y, which was found as a function of the stirrup ratio, longitudinal reinforcement ratio, and elasticity modulus ratio, was determined using the minimum energy principle. Compared with other two methods in the literature, the proposed angle equation tended to give a moderate prediction of strut angles and degradation coefficients for varying beam parameters.

3. A turning point occurs in the shear strain curves corresponding to the first diagonal crack. Behind the turn point, the tangent slope of the curve remains nearly constant before the stirrup yielding, especially for thin-webbed beams. Additionally, the shear span-to-depth ratio λ_{MV} has little effect on the shear deformation of slender RC beams at the post-cracking stage.
4. The analytical prediction was compared with the shear strain data of 23 zones. The results showed that the proposed method gives a good and consistent prediction of the effective shear stiffness and shear strain. The proposed degradation model can be used for the post-cracking shear stiffness evaluation of shear-cracked RC beams.
5. In practice, the procedure introduced in Section 2.3 can be used for a quick shear stiffness evaluation of concrete box girder bridges in service, on which performing experimental tests may be impractical and costly.

Author Contributions: Conceptualization, K.Z.; Data curation, S.N. and J.G.; Funding acquisition, K.Z.; Investigation, K.Z. and S.N.; Methodology, K.Z.; Project administration, Y.Z. and Y.W.; Resources, Y.W.; Software, J.G. and M.G.; Supervision, Y.Z. and Y.W.; Validation, S.N., J.G. and M.G.; Writing—original draft, K.Z.; Writing—review and editing, Y.Z. and Y.W. All authors have read and agreed to the published version of the manuscript.

Funding: This research was funded by the National Natural Science Foundation of China (52108150), China Postdoctoral Science Foundation (2021M691606), State Key Laboratory of Mechanical Behavior and System Safety of Traffic Engineering Structures (KF2020-17), S&T Program of Hebei (215676145H), and Nanjing Forestry University Undergraduate Innovation Training Program (202310298024Z).

Data Availability Statement: Data are contained within the article.

Acknowledgments: The authors greatly appreciate the support from State Key Laboratory of High-Performance Civil Engineering Materials. Also, the authors sincerely appreciate the help from Xingxin Zou and Dongzhi Guan in improving the paper quality.

Conflicts of Interest: The authors declare no conflict of interest.

References

1. *ACI318-19*; Building Code Requirements for Structural Concrete and Commentary. ACI: Farmington Hills, MI, USA, 2019.
2. AASHTO. *LRFD Bridge Design Specifications*, 9th ed.; AASHTO: Washington, DC, USA, 2020.
3. Zheng, K.; Zhou, S.; Zhang, Y.; Wei, Y.; Wang, J.; Wang, Y.; Qin, X. Simplified evaluation of shear stiffness degradation of diagonally cracked reinforced concrete beams. *Materials* **2023**, *16*, 4752. [CrossRef] [PubMed]
4. Hansapinyo, C.; Pimanmas, A.; Maekawa, K.; Chaisomphob, T. Proposed model of shear deformation of reinforced concrete beam after diagonal cracking. *J. Div. Mater. Concr. Struct. Pavements* **2003**, *725*, 305–319. [CrossRef] [PubMed]
5. Debernardi, P.G.; Taliano, M. Shear deformation in reinforced concrete beams with thin web. *Mag. Concr. Res.* **2006**, *58*, 157–171. [CrossRef]
6. Zheng, K.; Kuwornu, M.; Liu, Z. Shear test of variable depth rc beams with inflection point. *MATEC Web Conf.* **2019**, *275*, 02003. [CrossRef]
7. Pan, Z.; Bing, L.; Lu, Z. Effective shear stiffness of diagonally cracked reinforced concrete beams. *Eng. Struct.* **2014**, *59*, 95–103. [CrossRef]
8. Wang, T.; Dai, J.G.; Zheng, J.J. Multi-angle truss model for predicting the shear deformation of rc beams with low span-effective depth ratios. *Eng. Struct.* **2015**, *91*, 85–95. [CrossRef]
9. He, Z.; Liu, Z.; Ma, Z.J. Shear deformations of RC beams under service loads. *J. Struct. Eng.* **2017**, *143*, 04016153. [CrossRef]
10. Huang, Z.; Tu, Y.; Meng, S.; Ohlsson, U.; Taljsten, B.; Elfgren, L. A practical method for predicting shear deformation of reinforced concrete beams. *Eng. Struct.* **2020**, *206*, 110116. [CrossRef]
11. Liu, J.; Jia, Y.; Zhang, G.; Wang, J. Effect of diagonal cracks on shear stiffness of pre-stressed concrete beam. *Int. J. Struct. Integr.* **2018**, *9*, 414–428. [CrossRef]
12. Park, R.; Paulay, P. *Reinforcement Concrete Structures*; John Wiley & Sons: Christchurch, New Zealand, 1975; pp. 315–319.
13. Branson, D.E. *Deformation of Concrete Structures*; McGraw-Hill Companies: Toronto, ON, Canada, 1977; pp. 545–546.
14. *ACI 445-R99*; Committee Recent Approaches to Shear Design of Structural Concrete. American Concrete Institute: Farmington Hills, MI, USA, 2009.
15. Leonhardt, F.; Walther, R. *The Stuttgart Shear Tests*; Cement and Concrete Association: London, UK, 1964; Volume 111, pp. 1–134.
16. Kupfer, H. Generalization of Morsch's truss analogy using the principle of minimum strain energy. *CEB Bull.* **1964**, *40*, 44–57.

17. Vecchio, F.J.; Collins, M.P. The modified compression field theory for reinforced concrete elements subjected to shear. *ACI Struct. J.* **1986**, *83*, 219–231.
18. Bentz, E.C. *Sectional Analysis of Reinforced Concrete Members*; University of Toronto: Toronto, ON, Canada, 2000.
19. Desalegne, A.S.; Lubell, A.S. Consideration of shear deformations for slender concrete beams. *ACI Spec. Publ.* **2012**, *SP284-15*, 1–18.
20. Pang, X.B.; Hsu, T.T.C. Fixed-angle softened-truss model for reinforced concrete. *ACI Struct. J.* **1996**, *93*, 197–207.
21. Zhu, R.; Hsu, T.T.; Lee, J.Y. Rational shear modulus for smeared-crack analysis of reinforced concrete. *ACI Struct. J.* **2001**, *98*, 443–450.
22. Barzegar, F. Analysis of RC membrane elements with anisotropic reinforcement. *J. Struct. Eng.* **1989**, *115*, 647–665. [CrossRef]
23. Rahal, K.N. Post-cracking shear modulus of reinforced concrete membrane elements. *Eng. Struct.* **2010**, *32*, 218–225. [CrossRef]
24. Kim, J.H.; Mander, J.B. Influence of transverse reinforcement on elastic shear stiffness of cracked concrete elements. *Eng. Struct.* **2007**, *29*, 1798–1807. [CrossRef]
25. Li, B.; Tran CT, N. Reinforced concrete beam analysis supplementing concrete contribution in truss models. *Eng. Struct.* **2008**, *30*, 3285–3294. [CrossRef]
26. Zheng, K.; Liu, Z.; Zhang, Y.; Qin, S. Practical evaluation method for shear stiffness of concrete beam bridges with web diagonal cracks. *Bridge Constr.* **2015**, *45*, 46–51. (In Chinese)
27. Kuo, W.W.; Hsu, T.T.; Hwang, S.J. Shear Strength of Reinforced Concrete Beams. *ACI Struct. J.* **2014**, *111*, 809–818. [CrossRef]
28. Słowik, M. Shear failure mechanism in concrete beams. *Procedia Mater. Sci.* **2014**, *3*, 1977–1982. [CrossRef]
29. Armaghani, D.J.; Hatzigeorgiou, G.D.; Karamani, C.; Skentou, A.; Zoumpoulaki, I.; Asteris, P.G. Soft computing-based techniques for concrete beams shear strength. *Procedia Struct. Integr.* **2019**, *17*, 924–933. [CrossRef]
30. Wang, Y.D.; Yang, S.; Han, M. Experimental study of section enlargement with reinforced concrete to increase shear capacity for damaged reinforced concrete beams. *Appl. Mech. Mater.* **2013**, *256*, 1148–1153. [CrossRef]
31. Muttoni, A.; Fernández Ruiz, M. Shear strength of members without transverse reinforcement as function of critical shear crack width. *ACI Struct. J.* **2008**, *105*, 163–172.
32. Duan, M.; Zou, X.; Bao, Y.; Li, G.; Chen, Y.; Li, Z. Experimental investigation of headed studs in steel-ultra-high performance concrete (UHPC) composite sections. *Eng. Struct.* **2022**, *270*, 114875. [CrossRef]
33. Zhang, F.; Wang, C.; Liu, J.; Zou, X.; Sneed, L.H.; Bao, Y.; Wang, L. Prediction of FRP-concrete interfacial bond strength based on machine learning. *Eng. Struct.* **2023**, *74*, 115156. [CrossRef]
34. Huang, Z.; Tu, Y.; Meng, S.; Sabau, C.; Popescu, C.; Sas, G. Experimental study on shear deformation of reinforced concrete beams using digital image correlation. *Eng. Struct.* **2019**, *181*, 670–698. [CrossRef]
35. Chen, S.; Wei, Y.; Ding, M.; Zhao, K.; Zheng, K. Combinatorial design and flexural behavior of laminated bamboo-timber composite beams. *Thin-Walled Struct.* **2022**, *181*, 109993. [CrossRef]

buildings

MDPI

Article

Effect of Entrance Frame on Crack Development around Prefabricated Subway Station Openings

Zhenze Mo [1], Shuaike Feng [2], Dongzhi Guan [2,*] and Zhengxing Guo [2]

[1] Xicheng-CRRC (Wuxi) Urban Rail Transit Engineering Co., Ltd., Wuxi 214432, China
[2] School of Civil Engineering, Southeast University, Nanjing 211189, China
* Correspondence: guandongzhi@seu.edu.cn

Abstract: The openings at the sidewalls of subway station entrances generally reduce the localized load-bearing capacity of the sidewalls and lead to concentrated stress around the openings. In this study, to strengthen the sidewalls with openings in a newly-developed prefabricated subway station, a prefabricated steel-reinforced concrete (SRC) frame around the entrance was developed. To further investigate the effect of the developed entrance frame on the mechanical behavior of the sidewalls, a monotonic static test and finite element analysis were performed on a 1/2 scale station entrance substructure, including the proposed entrance frame and the adjacent top slab, bottom slab, and sidewalls. It was found that the developed entrance frame could effectively prevent stress concentration in the adjacent sidewall region. The most severe crack development was concentrated at the corner of the opening, which could be attributed to the torsional moment at the SRC beam end. The ratio of the torque shared by the beam to the total bending moment of the slab end varied from 21.2% to 26.8% in the elastic stage of all cases. In addition, both the improvement in the torsional bearing capacity of the SRC beam and the out-of-plane flexural capacity of the SRC column could positively contribute to controlling the crack development around the opening.

Keywords: prefabricated subway stations; monotone static test; entrance frame; development of cracks; stress distribution

1. Introduction

A prefabricated subway station that comprises assembled technology applied to the construction of the station conforms to the sustainable construction development trend. Unlike the construction of traditional subway stations, which involves unavoidable carbon emissions and serious resource consumption, assembled station construction technology is not only energy-saving and environmentally friendly but also has advantages such as short construction periods, reliable product quality, and a high degree of industrialization [1,2]. The assembly technology was first researched and developed in the construction of the Changchun subway station in China. The proposed prefabricated subway station structure comprises prefabricated elements such as bottom plates, sidewalls, and top arches. A novel grouted mortise–tenon joint was developed to connect the aforementioned prefabricated components. Furthermore, Yang et al. [3–5] experimentally assessed the bearing capacity characteristics and flexural stiffness of the developed joints and found that the proposed joints could effectively transfer the internal forces of prefabricated elements. Additionally, theoretical formulas were derived to calculate the ultimate flexural resistance and bending stiffness of the novel joints. Tao et al. [1,6] conducted a shaking table test to evaluate the seismic response of a prefabricated station structure and found that the mortise–tenon joint exhibited a higher bearing capacity than that of the prefabricated components, guaranteeing excellent integrity and mechanical properties of the station structure under seismic loads. Ding et al. [7] performed a numerical simulation of the single-ring structure of a Changchun prefabricated subway station and traditional cast-in situ subway station. Comprehensive analysis indicated that the single-ring structure of the Changchun station exhibited more favorable deformation resistance and mechanical properties than those of the traditional station. The popularization and application of assembly technology in the construction of subway stations have resulted in many novel connected methods and precast element forms being extensively investigated and employed in a variety of station structures. Liu et al. [8] studied the mechanical behavior of a precast concrete beam–slab–column interior joint comprising many connected methods and optimized the thickness of semi-precast slabs in the Jinanqiao prefabricated subway station. Du et al. [9] subsequently presented an experimental program on precast sidewalls with grouted sleeve connectors, which provided a foundation for predicting the mechanical properties of grouted splices used in prefabricated subway stations. Additionally, a novel wall–beam–strut joint connected using welded steel plates was adopted in the Guangzhou subway station, and the bearing capacity of the proposed joint was experimentally investigated [10]. Thus, the connected and assembled forms in subway prefabricated stations were implemented with reference to above-ground prefabricated structures, and the assessment of the mechanical behavior was performed through experimental and numerical investigations.

A subway station entrance is the passageway for to enter and exit the station meant for the evacuation and transfer of passengers. Therefore, it is necessary to know about the structural behavior and the design method of an entrance structure. Wang et al. [11] investigated structural behavior of a subway station entrance structure under three different buried depths. The optimal sectional dimensions of the deep buried entrance structure were proposed according to the analysis results. Monika [12] reported and discussed the measured displacement of the entrance hall to the metro station. Ou [13] compared the analysis method of an entrance structure in a subway station using a 2D plane model and 3D space models, and the results showed that the corners of an entrance structure should be enhanced. Liu et al. [14] analyzed the pressure distribution of the horizontal entrance of a subway station, which was caused by airflow. Yang and Lin [15] introduced a prefabricated ring frame structure of an entrance and exit in a prefabricated underground metro station structure. Choi et al. [16] presented the project-specific design approach of the station structure, which could be followed for the entrance structure. Most investigations on the mechanical performance of entrance structures in subway stations focus on conventional monolithic entrance structures. Studies on the prefabricated entrance structures are lacking, especially in the progress of promoting prefabricated underground stations.

The design of the entrance inevitably requires large openings to be cut into structural sidewalls, which directly reduces the localized load-bearing capacity of the sidewalls, and this should be rigorously estimated [17]. The mechanical performance of sidewalls with openings in subway stations has been investigated less intensively. Previous designers commonly referred to the recommendations provided by current standards for above-ground structures to evaluate the effects of cut-out openings on the behavior of station sidewalls. Standards such as EN 1992-1-1 [18], AS 3600 [19], ACI 318 [20], and CAN/CSA-A23.3 [21] all provide calculation assumptions to evaluate the mechanical behavior of the reinforced concrete (RC) walls with openings. However, as the load pattern applied to underground sidewalls may be different from that applied to above-ground structural walls, the feasibility of the current design recommendations requires further assessment.

RC structural walls in the above-ground structures can be considered as compression members that are designed to carry in-plane vertical loads [22]. Although a small eccentricity may be present, the failure pattern of the wall is still dominated by compressive forces [23–27]. The load applied to RC structural walls can present another type of the in-plane lateral seismic forces. Most previous studies have focused on the seismic performance of structural walls [28–30].

In contrast, sidewalls in subway stations are generally subjected to negligible in-plane lateral loads [31]. They are typically designed to carry in-plane axial loads and lateral out-of-plane loads such as moments from the top slab ends. Note that the out-of-plane behavior of RC walls with openings has also received little attention in above-ground structures. Hence, extensive experimental tests may be required to evaluate the out-of-plane mechanical performance of sidewalls with openings.

In general, for all subway stations, the area of the entrance openings introduced in the sidewalls is less than a tenth of the total continuous sidewall area. At the structural level, the entrance opening generally does not significantly affect the ultimate strength of integral sidewalls [18,19]. However, entrance openings in the sidewalls inevitably affect the load transfer paths and cause a redistribution of sidewall stresses. This further results in stress concentration around the opening, which encourages the development of cracks around the opening. Compared with above-ground buildings, the crack width of the components in underground structures must be strictly controlled to satisfy waterproofing requirements [32]. In such cases, the vicinity of the opening where stress concentrations and cracks generally occur first should be strengthened [33].

To strengthen RC walls with openings, additional reinforcing bars or steel plates are always embedded around the openings [34], and some advanced composites on the surface of the cut-outs can be used as externally bonded reinforcements [35]. Most previous studies on the effectiveness of reinforcement details around the edges of the openings have focused on above-ground structures. Therefore, the effects of the above reinforcement methods on the static behavior of sidewalls should be re-evaluated under the specific load pattern carried by the subway station. In addition, the application of these methods in prefabricated subway stations has been limited, owing to complex fabrication and high costs. To improve the efficiency of on-site construction, a steel-reinforced concrete (SRC) frame around the entrance was developed for a prefabricated subway station. The developed entrance frame structure can be used to compensate for the effects of entrance openings on the localized bearing capacity of the continuous sidewalls and to prevent stress concentrations around the openings.

Figure 1 shows the connection of the proposed entrance frame and its adjacent members. To facilitate component transportation, the precast column and beam in the SRC frame can be separately manufactured and connected onsite. As illustrated in Figure 1a, H-steels partially embedded into the bottom slab of the entrance story are applied to connect the precast SRC column to the bottom slab. Furthermore, a cantilever H-steel beam is integrated into the precast SRC column during production to connect the end of the exposed H-steel beam. As shown in Figure 1b, partial precast sidewalls are used in the prefabricated subway station. Horizontal U-type rebars are arranged longitudinally along the column overlapped with the transverse reinforcements of the sidewalls. Similarly, vertical U-type rebars protruding from the bottom slab are utilized to overlap the longitudinal rebars of the sidewalls. After the entrance frame and its adjacent precast sidewalls are placed at the designed positions, on-site concrete is poured into the reinforcement overlapping regions and the prepared region of the semi-sidewall. Subsequently, a partial precast top slab is placed on the corbel and connected to the entrance frame and sidewalls. As shown in Figure 1c, U-type rebars are formed by extending the longitudinal rebars of the sidewalls and partial precast top slab. The connection type between the sidewalls and top slab can be also regarded as overlapping with the U-type rebars. Finally, on-site concrete is poured into the upper region of the partial top slab to improve structural integrity, as shown in Figure 1d.

Due to the influence of the proposed entrance frame on the mechanical behavior of its adjacent sidewalls being ambiguous, the study of the mechanical behavior of the proposed frame and its adjacent sidewalls under the common stress state is an important prerequisite for promoting the application of the proposed frame in prefabricated subway stations. To experimentally investigate the overall behavior, cracking behavior, and stress distribution of the frame and adjacent members under vertical static loads, in this study, a 1/2 scale station entrance substructure comprising the developed frame, the adjacent top slab, the bottom slab, and sidewalls was constructed and tested under monotonically increasing vertical loads on the top slab. The effect of the frame on the mechanical behavior of its adjacent sidewalls was analyzed. Additionally, a numerical parametric study was conducted using the FE model to further investigate the influence of the developed frame with different configuration parameters on the crack development around the opening. A comparative analysis based on experimental and numerical tests revealed the crack development mechanism of the openings at the sidewalls of subway station entrances.

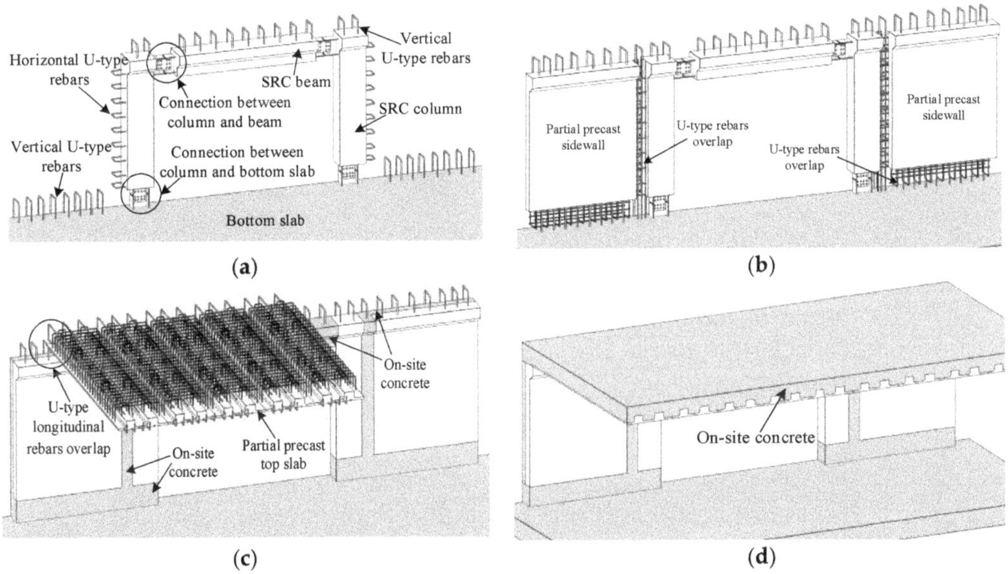

Figure 1. Developed entrance system and assembly process: (**a**) entrance frame in assembly, (**b**) sidewalls connected to entrance frame, (**c**) top slab connected to frame and sidewalls, and (**d**) assembled entrance system.

2. Experimental Program

2.1. Representative Subway Station

An actual engineering case under construction in China was selected as the prototype structure for this study. The subway station comprises a two-story, three-span structure, as shown in Figure 2. The total height of the story where the entrance system is located is 6600 mm, of which 800 mm is the thickness of the laminated top slab, and 5800 mm is the height of the precast sidewalls. The entrance opening dimensions are 4800 (height) $\times$ 7000 (width) mm, and the thickness of the laminated sidewalls is 700 mm, wherein the cast-in-place and precast layers are 350 mm. The height of the entrance frame is 5800 mm, and the distance between the column centers is 7800 mm. The cross-sectional dimensions of the columns are 800 $\times$ 700 mm, and the height and width of the beam are 1000 and 700 mm, respectively. Furthermore, the thickness of the entrance frame is identical to that of the adjacent sidewalls. One entrance substructure (Figure 2) is taken out from the top floor of the prototype building. The total height of the substructure was uniquely determined by the height of the story wherein the entrance system was located. The overhang length of the assembled top slab was determined based on the bending moment diagram of the top story of a subway station under uniform vertical loads (Figure 3). A single standardized precast sidewall panel was retained on both sides of the entrance frame, and the width of each sidewall panel was 3000 mm. In addition, the boundary conditions of the substructure are completely consistent with the actual station structure.

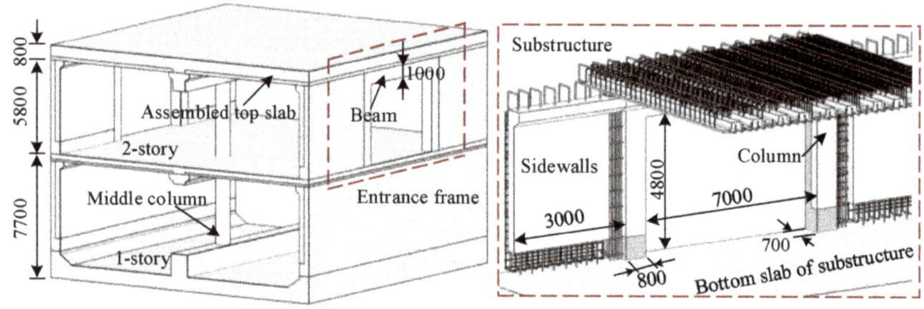

Figure 2. Schematic illustration of the full-scale subway station (units: mm).

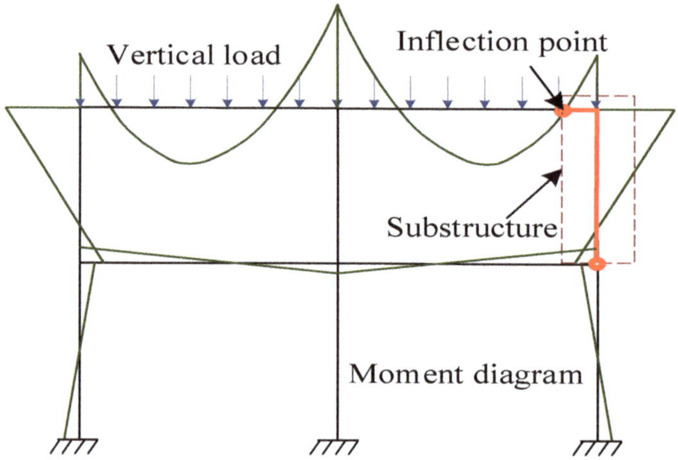

Figure 3. Simplified model of a cross-section of the subway station.

2.2. Specimen Details

As the constraints of the available experimental manipulation space, the substructure was scaled down by a factor of 2 to serve as the test specimen. Table 1 summarizes the parameters of the elements in the specimen. Scaled-down specimens are commonly used for studying the effects of cut-out openings on the strength of RC walls [17,22,26]. Previous studies have determined that it is feasible to test scaled-down RC walls with cut-out openings to further understand the mechanical properties of the structures. The design of the substructure considered the similarity between the scaled-down specimen and the original structure. The ratio of the geometric dimensions of each member in the substructure model to those of the corresponding member in the actual station structure was maintained at 1/2, and the substructure model was completely consistent with the material used in the actual station structure. Therefore, according to the scalability theory for the static loading test on a structure of civil engineering, the load-bearing mechanism and force-transferring path of the test scale model can be regarded as the same as those of the prototype structure for the case in this paper. As a consequence, the substructure model specimen could exhibit the same failure and crack distribution mode as the actual structure specimen, so as to achieve the experimental purpose of studying the static performance of the sidewall at the opening position. Figure 4 shows the dimensions and reinforcement details of the test specimen. The reinforcement ratio of each member in the station specimen was almost identical to that of the representative station, except for the differences in the bottom slab. In this study, the effect of the bottom slab on the bearing capacity of the entrance system was ignored. Hence, to ensure the restraint on the bottom ends of the columns and sidewalls and to prevent bottom slab failure before substructure failure during the testing process, the thickness and reinforcement ratio of the bottom slab were strengthened. In addition, the steel ratios of the SRC column and SRC beams were approximately the same as those in the representative station.

Table 1. Parameters of the test specimen.

Members	Section Size (mm)	ρ_r (%)	ρ_s (%)	$f_{cu,p}$ (MPa)	$f_{c,p}$ (MPa)	$f_{cu,c}$ (MPa)	$f_{c,c}$ (MPa)	N_{ut} (kN)	V_{ut} (kN)	M_{ut} (kN·m)
SRC column	400 × 350	3.74	5.90	55.6	35.8	—	—	7516.9	—	541.9
SRC beam	500 × 350	1.58	4.86	55.6	35.8	—	—	—	2203.7	2250.9
Sidewall	1500 × 350	1.07	—	60.1	38.6	49.6	32.1	15,535.1	—	365.3
Top slab	7300 × 400	1.28	—	60.1	38.6	40.9	27.3	—	3252.8	3417.5

Note: ρ_r is the longitudinal reinforcement ratio of components; ρ_s is the steel ratio; $f_{cu,p}$ is the cube compressive strength of precast concrete; $f_{cu,c}$ is the cube compressive strength of cast-in situ concrete; $f_{c,p}$ is the prism compressive strength of precast concrete; $f_{c,c}$ is the prism compressive strength of cast-in situ concrete; N_{ut} is the theoretical value of the axial compressive bearing capacity of the member; V_{ut} is the theoretical value of the shear resistant capacity of the member; M_{ut} is the theoretical value of the ultimate flexural capacity of the member.

As shown in Figure 4, the cross-sectional dimensions of the scaled-down SRC column and beam were 400 × 350 mm and 500 × 350 mm, respectively. The sectional dimensions of the H-steel in the column were 250 mm × 200 mm × 12 mm × 14 mm (height × width × web thickness × flange thickness, respectively), and those of the H-steel in the beam were 350 mm × 200 mm × 9 mm × 14 mm, respectively. The cantilever H-steel beam integrated into the SRC column comprised sectional dimensions that were identical to those inside the beam. A bolting and welding mixed connection was utilized to connect the SRC column and SRC beam. This type of connection has already been successfully used by the authors in [36,37]. Thirty-four pieces of 14 mm diameter HRB400 longitudinal rebar were distributed around the cross-section of the SRC column, which had an approximate longitudinal reinforcement ratio of 3.74%. Grade HRB400 stirrups with a nominal diameter of 10 mm were arranged at a spacing of 100 mm along the column height. This arrangement accounted for a transverse reinforcement ratio of 0.96% in the columns. Simultaneously, the eighteen pieces of 14 mm diameter HRB400 longitudinal rebars were arranged around the SRC beam cross-section, and 10 mm diameter HRB400 stirrups were distributed at a longitudinal spacing of 150 mm along the beam span. The longitudinal reinforcement and stirrup ratios of the SRC beam were 1.58% and 0.30%, respectively.

The cross-sectional dimensions of the scaled-down single standard sidewall were 1500 mm × 350 mm, and the sectional dimensions of the scaled-down top slap were 7300 mm × 400 mm. The total height of the sidewall was 3300 mm, and the horizontal distance between the cantilevered and the fixed ends of the top slab was 2350 mm. As illustrated in Figure 4, grade HRB400 deformed 18 mm-diameter rebars with 150 mm spacing in the longitudinal direction were uniformly placed in the sidewall and the top slab to function as longitudinal reinforcement. The longitudinal reinforcement ratios of the sidewalls and top slab resulting from this configuration were 1.07% and 1.28%, respectively. Furthermore, the outer-side reinforcement of the sidewalls (i.e., tensile reinforcement) was arranged double-layered within a vertical range of 1500 mm from the bottom slab. This design can be attributed to the flexural moment at the bottom end of the sidewalls being significantly less than that at the top end under actual vertical loads (Figure 3). This could be utilized to guarantee that the expected flexural failure is controlled in the top region instead of the bottom end of the sidewall. Grade C50 concrete was used as the precast concrete of the entrance frame, sidewalls, and slabs. Grade C40 concrete was used for the cast-in situ concrete of the beam–column connection region, sidewalls, and top slab. In addition, the contacting surfaces of both the semi-precast top slab and sidewalls were roughened based on the recommendations in the concrete design specification [38] to transfer shearing force.

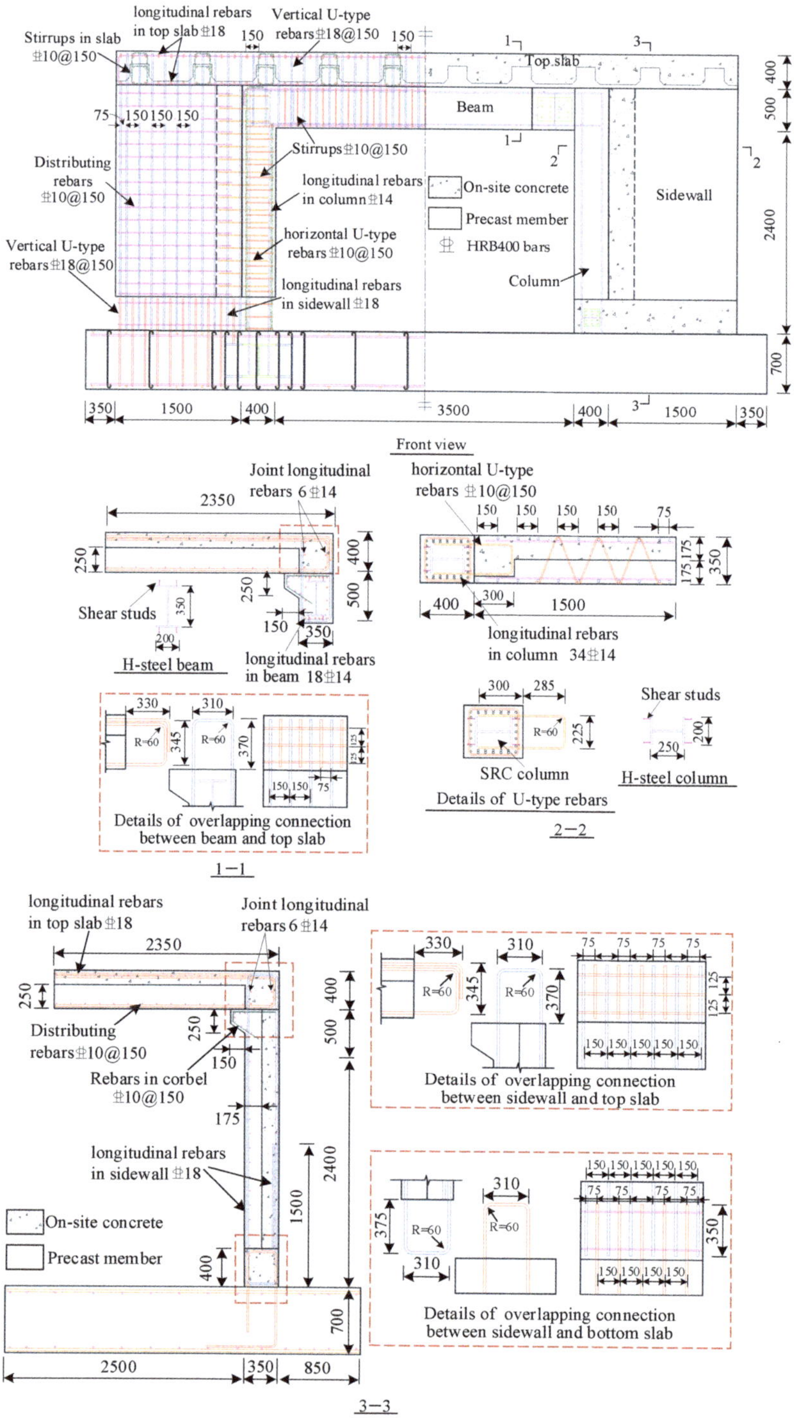

Figure 4. Dimensions and reinforcement details of the test specimen (units: mm).

Detailed U-type reinforcement overlapping connections are shown in Figure 4, in which it is observed that the design parameters of the top slab–sidewall connection and top slab–frame connections are similar. The total length of the horizontal and vertical overlaps of the adjacent U-type rebar was 655 mm. The center-to-center distance between adjacent U-type rebars in the longitudinal direction was 75 mm. In general, the beam–column knee joints with U type reinforcement overlap have been commonly used in above-ground frame structures. Numerous experimental studies have verified the behavior of the knee joints and provided a basis for the current design methodology. As discussed in previous studies [39,40], joints with U-type reinforcements exhibited better mechanical behavior than those with standard 90 degree hook details. The developed knee joints for the top slab were designed based on the recommendations provided in the above literature, and the overlapping length of the lap splices met the Chinese code requirements [38]. For the connection between the sidewalls and the bottom slab, the bond strength evaluation of the U-type rebars was performed considering the bond strength of 90 degree hook-ended bars. Therefore, the U-type reinforcement overlapping connection can be considered as a general non-contact lap splice, and the connection parameters can be derived from both the current Chinese code [38] and the ACI building code [20]. The overlapping length of the non-contact splice was 350 mm, and the center-to-center distance between adjacent U-type rebars in the longitudinal direction was 75 mm. Simultaneously, the Chinese code recommendations [38] could be utilized to determine the overlapping length between the horizontal U-type rebars from the SRC columns and the distribution of the rebars along the sidewalls.

The construction process of the test specimens is illustrated in Figure 5. A partial precast top slab and sidewalls were prefabricated in a factory. For convenience, the bottom slab was manufactured directly at the designated test position. The H-steels of the entrance frame were installed in their designed positions before the concrete of the base plate was poured. The semi-sidewalls were then connected to a steel frame when the bottom slab was cured. Subsequently, the precast concrete encasing the steel frame and the on-site concrete of the sidewalls were poured in batches. Finally, the precast top slab was connected to the assembled frame and sidewalls. The precast part of the concrete in the specimen was cured under indoor conditions, and the cast-in-place part of the concrete was cured under outdoor conditions. During the curing process, attention was paid to the maintenance of the concrete surface temperature, and no obvious cracking of the concrete surface occurred.

2.3. Materials

To determine the characteristic value of cubic concrete compressive strength, six cubic specimens with dimensions of 150 mm × 150 mm × 150 mm were made for each batch of concrete. After curing for 28 days under identical conditions as the specimens, compression tests were performed on these cube specimens to determine the real concrete strength according to the standard GB 50204 [41], and the averaged results were calculated. The differences between the maximum and average results and between the minimum and average results of the cubic compressive strength of the six cubic specimens did not exceed 15% of the average result. According to the relevant provisions in standard GB 50204 [41], the average value could be taken as the cube compressive strength characteristic strength of this group of specimens. Based on the recommendations in standard GB 50010 [38], the prism compressive strength of concrete can be determined by the characteristic value of its cube compressive strength, which can be further used to calculate the theoretical bearing capacity of each member in the substructure. Both the cubic compressive strength of concrete ($f_{cu,p}$, $f_{cu,c}$) and the prism compressive strength of concrete ($f_{c,p}$, $f_{c,c}$) are shown in Table 1. The reinforcement and steel plate coupons were made from the corresponding batches of specimens. Based on the recommendations of standard GB/T 228.1 [42], coupon tensile tests were conducted on three coupons of each steel component and rebar to measure their material properties, and the averaged test results are listed in Table 2.

Figure 5. Construction process of the test specimen.

Table 2. Material properties of steel components.

Elements	t or d (mm)	f_y (MPa)	f_u (MPa)	E_s (GPa)	$\mathcal{E}_y$ (με)
Flange of H-steel	14	410.5	556.6	202	2032
Web of H-steel	12	407.6	551.2	201	2028
	9	411.5	552.0	201	2047
Reinforcement	18	425.0	640.2	201	2114
	14	440.1	630.3	201	2189
	10	455.0	550.0	201	2263

Note: t is the thickness of steel components; d is the diameter of reinforcements; f_y is the yield strength of steel components and reinforcements; f_u is the ultimate strength of steel components and reinforcements; E_s is the modulus of elasticity; $\mathcal{E}_y$ is the yield strain of steel components and reinforcements.

2.4. Bearing Capacity of Components

The ultimate bearing capacity of each component can be utilized as a reference for formulating the loading protocol. The out-of-plane flexural capacity and axial compressive bearing capacity of the SRC columns were calculated according to the calculation assumptions in standards JGJ 138 [43] and EN 1994-1-1 [44]. The out-of-plane flexural capacity and axial compressive bearing capacity of the precast sidewalls were determined based on the recommendations of EN 1992-1-1 [18] and GB 50010 [38]. In addition, standards EN 1992-1-1 [18] and GB 50010 [38] can be used to evaluate the ultimate flexural capacity and shear resistance capacity of the precast top slab. The ultimate bearing capacity of each component in the substructure was calculated by using the measured strength values of steel shown in Table 2 and the calculated prism compressive strength of concrete shown in Table 1. The theoretical calculations of the bearing capacity of each member in the substructure are presented in Table 1. As observed, the flexural resistance of the top slab was higher than the total flexural resistance of the sidewalls and SRC columns. Furthermore, the calculated values of the shear resistance of the top slab indicated that bending failure of the laminated slab occurred before shear failure. Therefore, the typical flexural damage in the sidewalls was expected to be the primary failure mode of the test specimen.

2.5. Test Setup and Loading Protocol

A general view of the test setup is shown in Figure 6. Five high-strength finishing rolling rebars with a nominal diameter of 40 mm were used to connect the hydraulic jacks and top slab. When the concrete of the bottom slab was poured, the bottom ends of these high-strength rebars were vertically embedded in the bottom slab. Five hydraulic jacks, each with a maximum capacity of 1000 kN, were networked together to enable a single operator to apply a uniformly distributed load through the loading steel beam arranged along the longitudinal length of the top slab. The horizontal distance between the loading point and the center of the sidewall was 1935 mm, as shown in Figure 6a. The hydraulic fluid for these jacks was supplied by a power steering pump for controlling the total test loads by setting the working pressures. To measure the induced force, compression load cells were placed between each hydraulic jack and the loading steel beam. The loading process was initiated for a monotonous static downward load, with a total load increment of 44 kN, corresponding to a bending moment increment of 85 kN·m at the end of the top slab. A load-holding period of 5 min was maintained to observe the cracks, record the phenomena, and test the data. The displacement and strain information of the specimen during the loading process can be monitored by the field data acquisition instrument, as shown in Figure 6b. The arrangement of the measuring points of the specimen is reported in detail in Section 2.6. For safety, the test was terminated when all the longitudinal reinforcements in the tension zone of the sidewalls entered the yield stage, and the specimen had not yet reached the final failure.

2.6. Test Instrumentation

The response quantities of interest consisted of the crack patterns, overall deformation, and strain distribution of the sidewalls, entrance frame, and top slab. A concrete automatic digital crack width tester was used to monitor crack development from the initiation of cracks. Figure 6a illustrates the distribution of linear variable deformation transducers (LVDTs). Three LVDTs with a capacity of 300 mm (V1, V2, and V3) were installed at the bottom of the loading end of the top slab to measure the vertical deformation. Nine LVDTs (numbered H1–H9) with a capacity of 100 mm were installed horizontally at the backside of the sidewalls and entrance frame to respectively monitor their out-of-plane displacements.

Figure 7 presents the arrangement of the strain gauges for each member of the specimen. To measure the tensile strain distribution along the longitudinal direction of the sidewall, 10 general-purpose strain gauges (i.e., S1 to S10) were arranged longitudinally on the tensile reinforcement along cross-section 4 at a vertical distance of 800 mm from the beam bottom, as shown in Figure 7a. Simultaneously, 13 strain gauges (i.e., T1 to T13) were attached to the upper-layer reinforcement at the top slab end section to monitor the longitudinal strain distribution between the entrance opening and sidewalls, as shown in Figure 7b. To measure the stress state of the SRC column, strain gauges CR1 and CR2 were arranged on the compressive and tensile reinforcements of the column, respectively, and strain gauges F1 to F4 were placed on the column flange at cross-section 4.

To monitor the stress distribution around the opening, eight strain rosettes (each consisting of three 60 mm gauges) were attached to the concrete surface on the rear end of the frame (i.e., the tension side) at the locations shown in Figure 7a. Strain rosettes CW1 to CW4 were arranged vertically from the bottom of the beam to the top of the slab, well aligned with the corner of the opening. Strain rosettes CW5 to CW8 were arranged vertically from the bottom of the beam to the top of the slab, which were aligned to the opening centroid. Moreover, six strain rosettes (i.e., SW1 to SW6), each with three general gauges, were used to measure the stress distribution between the supports and midspan of the steel beam, and their arrangements are illustrated in Figure 7a. In addition, the strain development of U-type rebars in the connection zone was measured for better understanding of the joint behavior. As shown in Figure 7c, strain gauges TJ1 to TJ3 and ST1 to ST3 were used to monitor the U-type rebars in the assembly joint between the sidewall and top slab, while strain gauges SB1 to SB6 and BB1 to BB4 were used to measure the U-type rebars in the assembly joint between the sidewall and bottom slab.

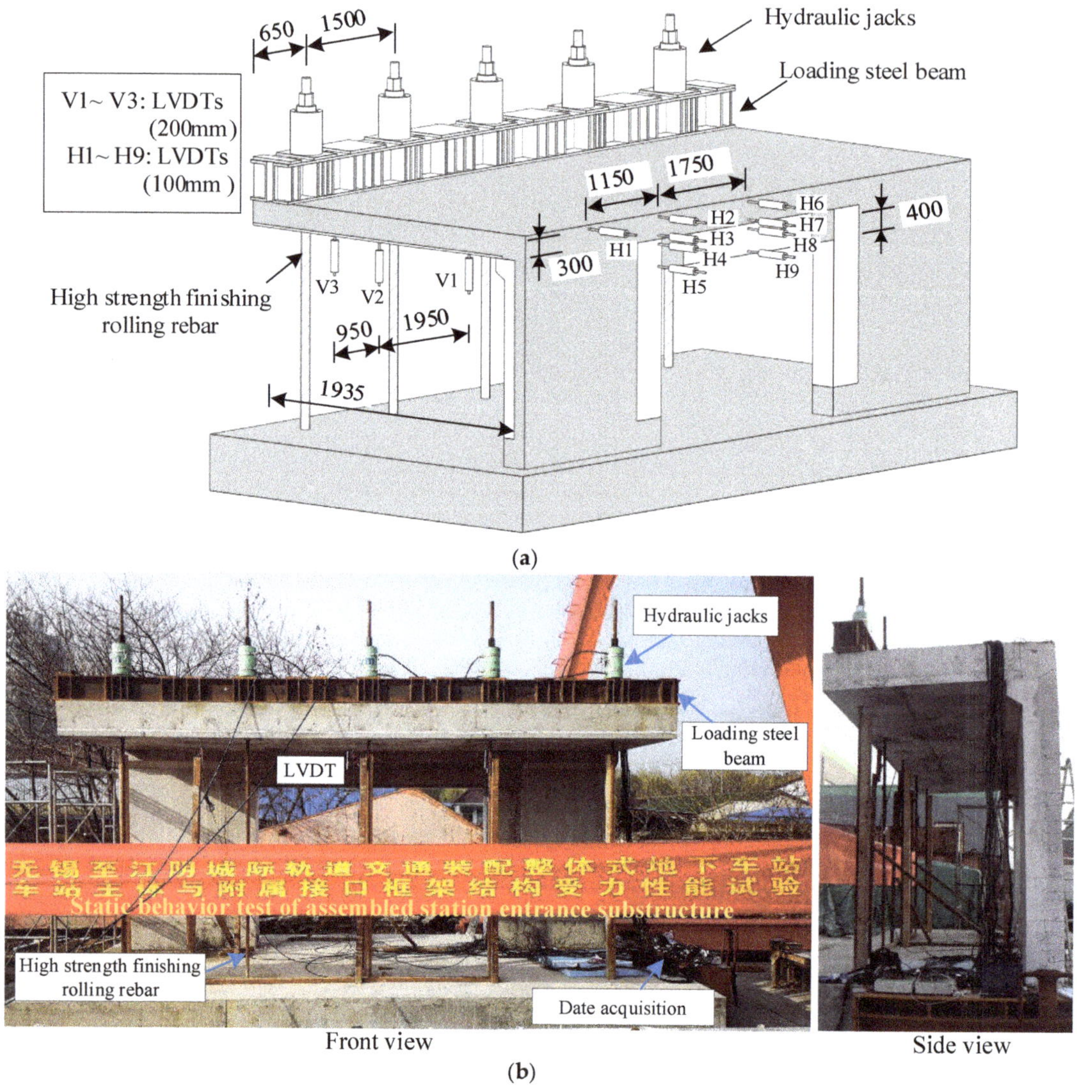

Figure 6. General overview of a test setup (units: mm): (**a**) schematic of the test setup and (**b**) photograph of the test setup.

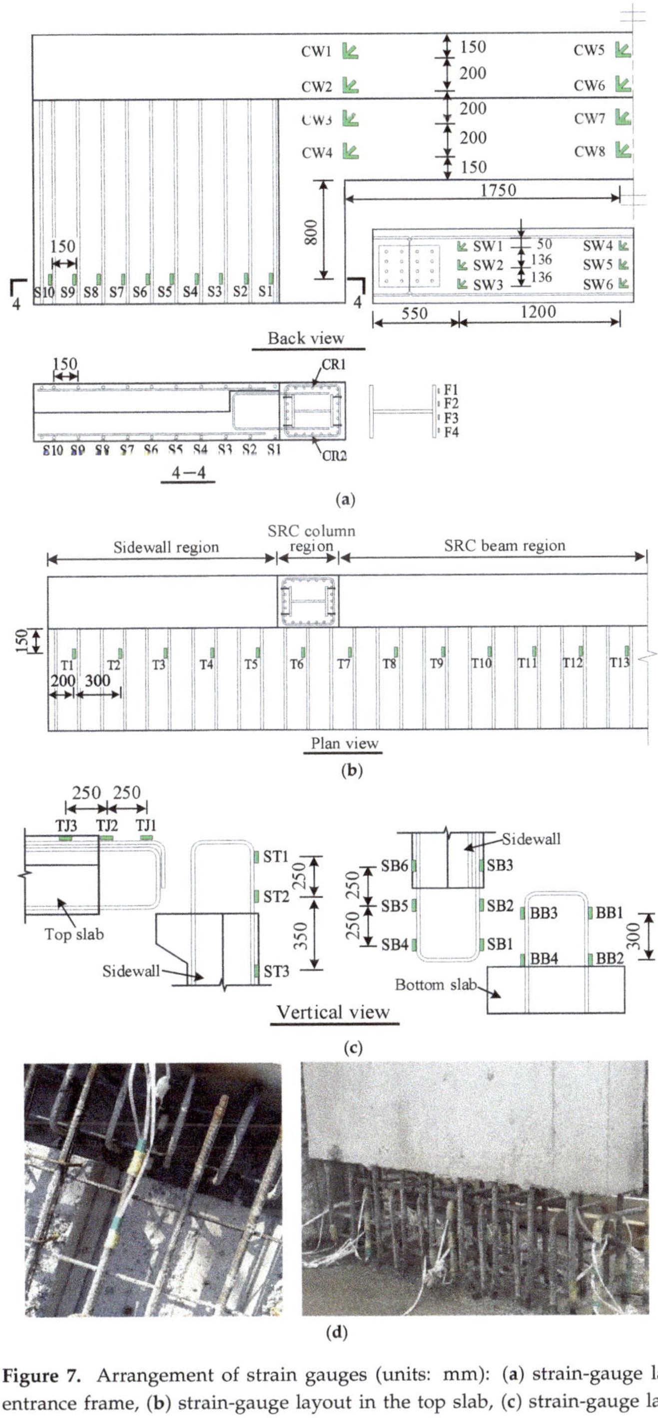

Figure 7. Arrangement of strain gauges (units: mm): (**a**) strain-gauge layout in sidewall and entrance frame, (**b**) strain-gauge layout in the top slab, (**c**) strain-gauge layout in the U-type rebar overlapping connections, and (**d**) photograph of strain gauge arrangement at U-type rebar overlapping connections.

3. Experimental Results

3.1. Overall Behavior and Crack Pattern

The total applied load measured by the compression load cell versus the horizontal and vertical displacements measured by the LVDTs is shown in Figure 8. Here, the vertical displacement at the loading end of the top slab (TSVD) was measured using LVDTs V1–V3. It can be seen that the incremental difference in the TSVD along the longitudinal direction of the station was negligible. Furthermore, the out-of-plane horizontal displacements at the top of the sidewall (STHD) and the top of the entrance frame (ETHD) were monitored using LVDTs H1 and H6, respectively, and the measurements indicated an insignificant difference between the two displacement increments. As the formal loading was terminated when all tensile rebars of the sidewalls entered the yield stage, the applied load increased almost linearly, with extremely limited inelastic deformations. The total applied loads and displacements corresponding to each stage of the test specimen during the loading process are shown in Figure 8.

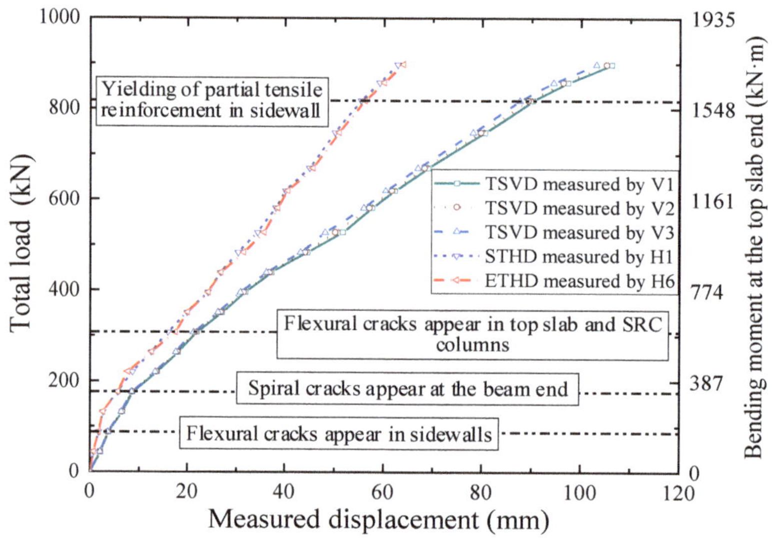

Figure 8. Load displacement curves for the specimen.

The crack pattern of the test specimen at the yield stage of all tensile reinforcements in the sidewalls is illustrated in Figure 9. As shown in the back view, the first flexural crack was observed on the tension side of the sidewalls when the applied load reached 88 kN, corresponding to a bending moment of 170.3 kN·m at the end of the top slab. The load level at this stage was approximately $0.1P_y$. Here, P_y is the maximum load corresponding to the load at which all tensile reinforcements in the sidewalls yielded. Subsequently, the number of flexural cracks along the vertical direction of the sidewalls gradually increased, and the crack width increased as the applied load increased. As the applied load increased, the horizontal cracks appearing on the sidewalls further increased in the SRC column section. This further indicates the effectiveness of the connection between the SRC column and sidewalls in transferring stress.

When the applied load of 176 kN was measured, corresponding to a bending moment of 340.6 kN·m at the end of the top slab, diagonal cracks were initiated at both ends of the SRC beam corresponding to the corners of the opening. The load level at this stage was approximately $0.2P_y$. With a further increase in the applied load, these diagonal cracks gradually propagated to the fixed end of the top slab. The angle between the diagonal cracks and the beam axis ranged from 35° to 55°. As shown in the front view in Figure 9, diagonal cracks could also be observed at the beam ends corresponding to the corners of the opening. Interestingly, the propagation direction of the front diagonal crack was opposite to that of the diagonal cracks on the rear side. It is therefore deduced that the diagonal cracks appearing on the front and back of the beam ends may be credited to the torsional deformation of the SRC beam.

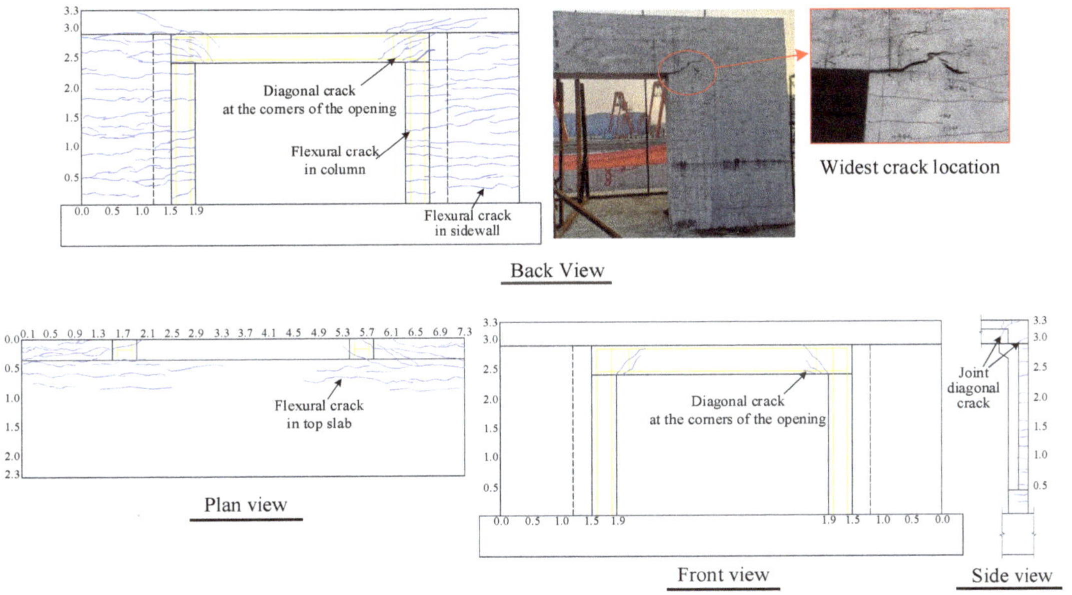

Figure 9. Crack patterns of the specimen (units: m).

When the horizontal cracks in the sidewalls spread over the entire column section, flexural cracks were also observed in the top slab at an applied load of 308 kN (approximately $0.35P_y$). As shown in the plan view in Figure 9, the propagation of flexural cracks was concentrated at the fixed end of the slab within the range of the sidewalls, and only a limited number of cracks grew in the opening region throughout the test. Upon further increasing the applied load to 396 kN (approximately $0.45P_y$), a few diagonal cracks initially appeared at the joint between the sidewalls and top slab, as shown in the side view in Figure 9. Interestingly, the joint diagonal cracks did not propagate to the core region throughout the loading process, and the maximum crack width was 0.3 mm. In addition, no splitting cracks occurred at the location of the U-type rebar, implying that no bond slippage occurred between the U-type rebar and concrete throughout the test.

When the applied load increased to 818.5 kN, corresponding to a bending moment of 1583.8 kN·m at the end of the top slab, partial tensile reinforcement in sidewalls entered the yield stage. The load level at this stage was approximately $0.9P_y$. In addition, a horizontal crack with a width of 2.5 mm and length of 150 mm appeared in the right column section that intersected with the bottom of the SRC beam, as shown in the back view in Figure 9. Subsequently, this crack propagated obliquely to the top of the sidewall at an angle of approximately 45° and continued to cause the concrete cover to spall at the top of the sidewall. As the load increased further, this crack became the widest crack, finally attaining its maximum width (3.5 mm). As the reinforcement in the sidewall began to yield, the load increased nonlinearly until it reached a maximum of 897.5 kN (P_y) when the test was terminated.

3.2. *Crack Width Development*

Figure 10 shows a comparison of the crack width development at the opening corners, SRC columns, top slab end, and sidewalls of the specimen. As mentioned in [17,33], the presence of cut-outs in traditional concrete walls encourages cracks to appear first at the corners of the opening. However, owing to the entrance frame setting in the specimen, cracks in the sidewalls appeared before those at the corners of the opening. At the initial loading stage, the width of the flexural crack in the sidewall was the largest compared to other members. When the total applied load was increased to 582 kN, the crack width at the corners of the opening gradually exceeded that of the sidewalls. The extremely rigorous crack width control requirement in underground structures is meant for the expectation of providing excellent resistance to the penetration of water and chemicals. Notably, the development of the crack width at the corners of the opening was more abrupt than that in other members. Therefore, more attention must be paid to the reason for the formation and width control of cracks at the opening corners. For the top slab and sidewalls, only the maximum crack widths during each monotonic loading process are shown in Figure 10. In actuality, the width of the horizontal flexural cracks in both the sidewall and top slab gradually increased from the opening to the outside edges, thereby directly demonstrating the lag effect of the entrance frame on the stress development of longitudinal rebars in adjacent members.

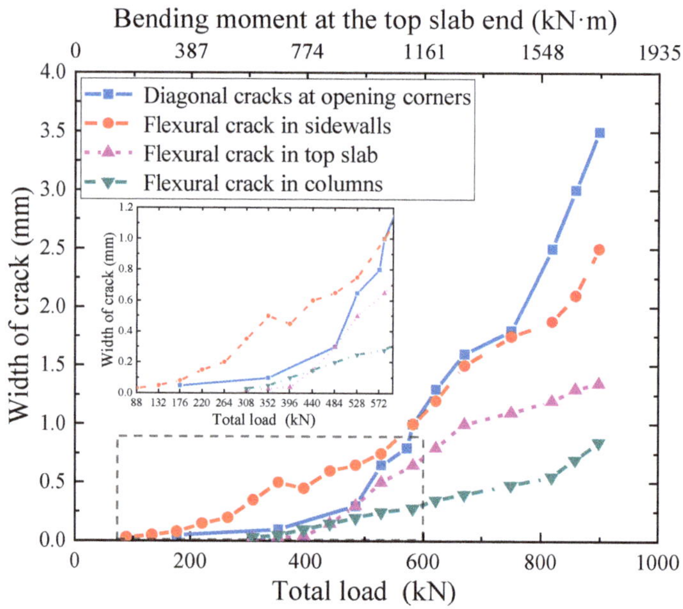

Figure 10. Load–crack width curves of the specimen.

3.3. *Strains of the U-Type Reinforcement*

Figure 11a illustrates the relationships between the applied load and the strains of the U-type reinforcement in the connection between the sidewalls and the bottom slab. Gauges SB6 and SB3 were attached to the longitudinal reinforcement outside the overlapping region of the U-type rebar. Their arrangement was utilized to directly realize the development of the longitudinal strain in the sidewall. Furthermore, at each load level, the longitudinal strain value of the U-type reinforcement decreased gradually from the sidewall to the core region of the joint. This was attributed primarily to the presence of bond stress between the overlapping rebars and surrounding concrete. In such cases, the longitudinal strain of the overlapping U-type rebar maintained elasticity throughout the test.

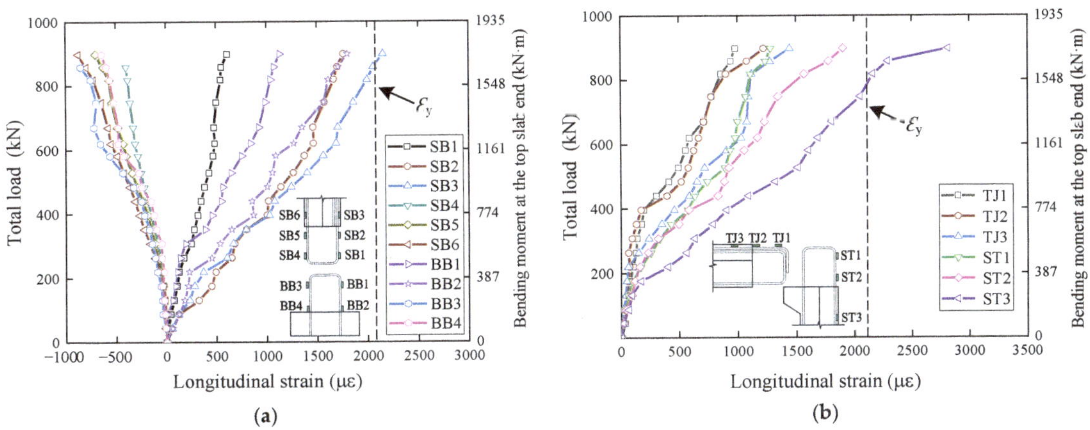

Figure 11. Strains of the U-type reinforcement at the joint: (**a**) connection of sidewalls and bottom slab and (**b**) connection of sidewalls and top slab.

Figure 11b further shows the development of the tensile strain of U-type rebars in the connection between the sidewalls and top slab. Similar to the lap joint at the bottom of the sidewall, the transferring stress attained between the overlapping U-type rebars in the connection was also primarily determined by the bond stress around rebars. Therefore, the minimum longitudinal reinforcement strain appeared at the ends of the U-type rebars extending from the respective members, where the gauges TJ1 and ST1 were located. As a result of the larger flexural resistance at the end of the top slab, only the tensile reinforcement at the top end of the sidewall entered the yield stage, and during the testing process, no overlapping U-type rebars were yielded. Moreover, as discussed in Section 3.1, no bond-splitting cracks occurred in the top and bottom joints of the sidewall. It was therefore deduced that there was no bond stress degradation between the U-type reinforcement and the concrete in the proposed connection type, and the manner of overlapping U-type reinforcements could effectively guarantee the stress transfer between the sidewalls and top and bottom slabs.

3.4. Strains of the Sidewall Longitudinal Reinforcement

The longitudinal reinforcement strain distribution along the sidewall cross-section is shown in Figure 12, which can be used to further analyze the effects of the entrance frame on the stress distribution within the sidewall. As observed, the longitudinal reinforcement in the edge of the sidewall, further from the entrance frame, attained yield strength first during the testing process. Subsequently, the tensile reinforcement in the sidewall region close to the frame began to gradually yield. It could be seen that the presence of the entrance frame had a lag effect on the stress development of the longitudinal reinforcement in its adjacent local regions. This further demonstrated that the developed entrance frame could effectively prevent stress concentration on both sides of the opening and could facilitate crack width control in the sidewalls.

Figure 13 demonstrates the development of longitudinal strains in the SRC column under different load levels. The distribution of the longitudinal steel strains along the H-steel flange width is illustrated in Figure 13a, and the applied load versus longitudinal reinforcement strain curves are given in Figure 13b. Evidently, the SRC column and sidewalls jointly resisted the out-of-plane flexural moment from the top slab end. The distance between the neutral axis and the centerline of the column was approximately 55 mm. Both the longitudinal reinforcement and the H-steel in the column remained in the elastic stage throughout the test. This phenomenon could be attributed to only a limited sidewall section size being selected in the substructure design. However, the longitudinal section of the sidewall is much larger than that of the SRC column in actual subway stations. In such cases, the flexural resistance of the column has a negligible effect on the bearing capacity of the sidewalls. The experimental results showed that the yielding of all tensile rebars in the sidewalls occurred earlier than concrete crushing in the compression zone. Therefore, it is deduced that although the sidewalls were also subjected to vertical loads, the failure was primarily controlled by the out-of-plane bending moment.

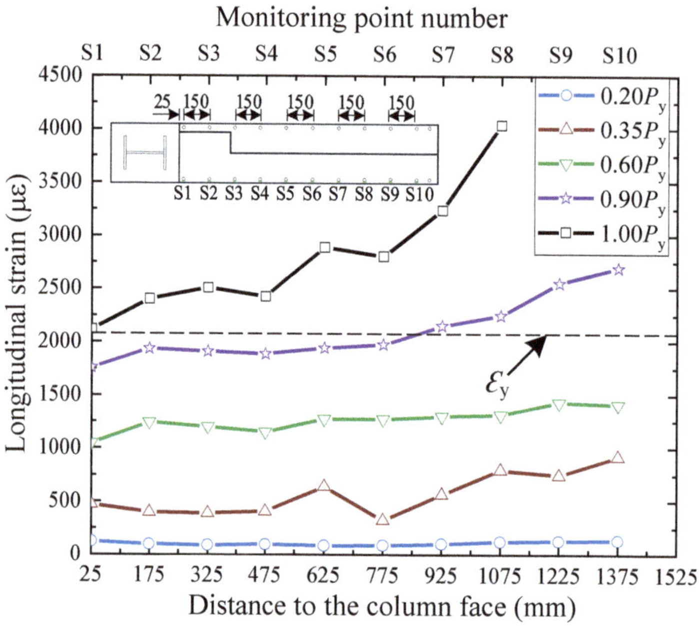

Figure 12. Strain distribution along the sidewall cross-section.

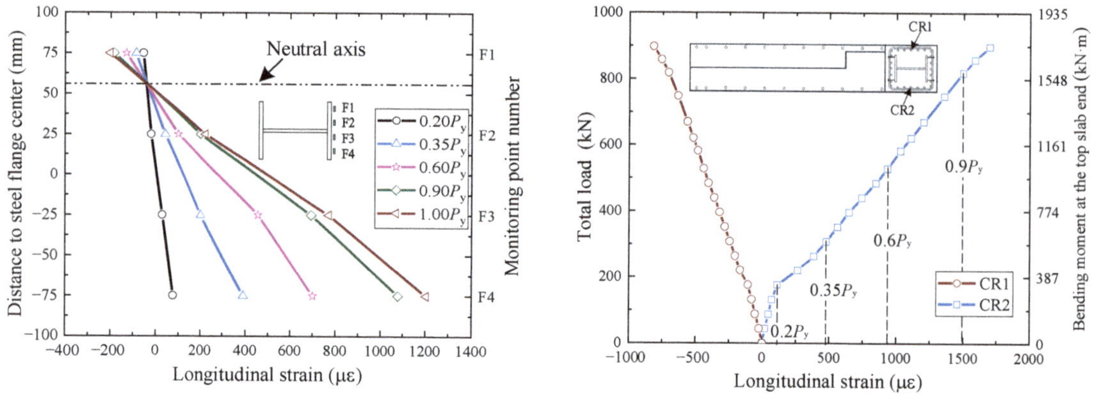

(a) (b)

Figure 13. Strains development in the SRC column: (**a**) typical strain distribution along the flange and (**b**) load–strain curves of the longitudinal rebars.

3.5. Strains of the Top Slab Longitudinal Reinforcement

Figure 14 shows the strain distribution of the longitudinal reinforcement at the end of the top slab from the outer edge of the sidewall to the opening. Note that the longitudinal rebars at the slab end still maintained elasticity, even when all tensile reinforcements in the sidewalls yielded. Furthermore, the distribution of the reinforcement strain values at the slab end generally exhibited a decreasing trend from the sidewall to the opening, and it only marginally increased in the range of the SRC column. The strain distribution further illustrated the transmission mechanism of the bending moment of the slab end to the sidewall and entrance frame. It was therefore deduced that the out-of-plane moment from the slab end was primarily carried by the sidewalls, and that an extremely limited moment load was transferred to the SRC beam above the opening. However, the limited bending moment borne by an SRC beam must receive attention. This may be the primary reason for torsional deformation of the frame beam. In addition, the out-of-plane bending moment carried by the beam was further transmitted to the SRC column through the beam–column connection. As a result of this process, diagonal cracks may be generated at the corners of the opening, that is, on the front and rear of the beam ends.

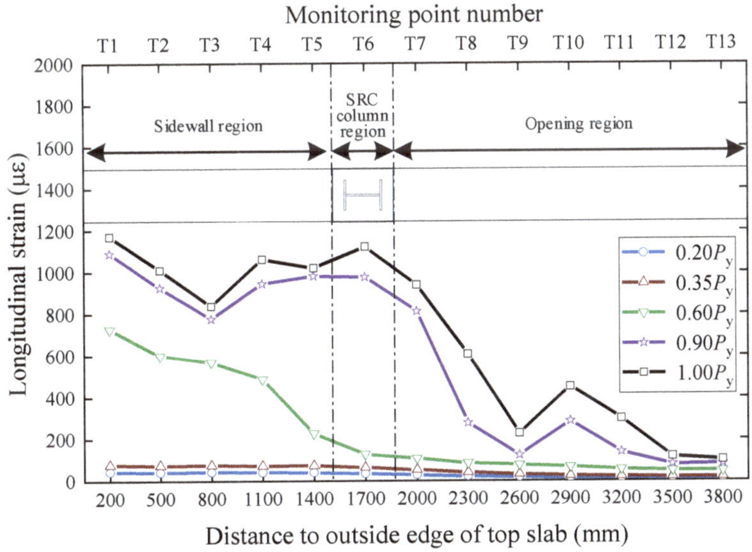

Figure 14. Strain distribution of the longitudinal reinforcement at the top slab end.

3.6. Stressed State of the SRC Beam

To measure the out-of-plane torsional deformation of the SRC beam during the testing process, a series of horizontal LVDTs (H4, H5, H8, and H9) was installed at the rear side of the beam, as shown in Figure 6a. Here, LVDTs H4 and H5 were almost aligned to the SRC beam end section, and LVDTs H8 and H9 to the midspan section of the SRC beam. The torsion angle (θ) of the midspan section of the SRC beam can be calculated as follows:

$$\theta = \arctan\frac{\delta_{H8} - \delta_{H9}}{400} - \arctan\frac{\delta_{H4} - \delta_{H5}}{400} \quad (1)$$

where δ_{Hi} is the reading of LVDT numbers H4, H5, H8, and H9 during the test. During the final loading stage, the experimental value of θ was evaluated to be 0.007 rad. The captured θ value directly demonstrated an extremely slight torsional deformation of the beam.

To further illustrate the stressed state of the SRC beam under different load levels, the principal stress along the beam depth was monitored using strain rosettes attached to the concrete surface and the H-steel web at the locations shown in Figure 7a. In the previous stage when the diagonal cracks appeared, the principal stress of the concrete obtained from strain rosettes CW1 to CW8 is shown in Figure 15a. It can be seen that the maximum principal stress of the concrete at the corner region of the opening increased continuously from the top of the slab to the bottom of the beam. This directly resulted in the different development of the diagonal crack width along the beam depth. In addition, the angle between the major principal stress direction and the beam axis is also shown in Figure 15a. The principal tensile stress direction at the beam end was at 46–66° to the beam axis. However, the angles between the principal tensile stress direction at the beam midspan section and beam axis were drastically reduced to 5–18°. Furthermore, the principal stress values of the beam midspan section were almost zero. Figure 15b further illustrates the variation in the principal stress of the H-steel between the midspan and supports of the beam at the last loading level, where the stress was measured from strain rosettes SW1 to SW6. Similar to the concrete stress case, the major principal stress of the H-steel decreased significantly from the end to the midspan of the beam, and the corresponding directions were also gradually oriented toward the beam axis.

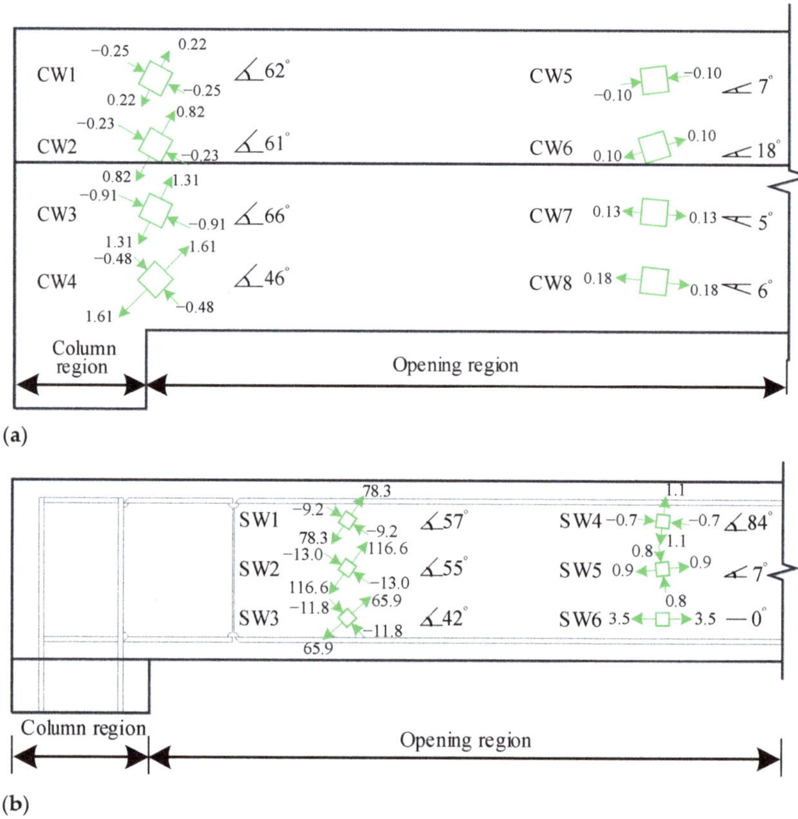

Figure 15. Stress distribution of the SRC beam (units: N/mm^2): (**a**) stress distribution of the concrete before cracking and (**b**) stress distribution of H-steel in the last loading stage.

The reason for this significant difference can be attributed to the variation in the torsional moment along the beam length. As described in Section 3.5, after the bending moment was transmitted from the top slab to the SRC beam, the load could then be further transferred in the torsional moment from the beam to the columns through beam-to-column connections. Therefore, the torsional moment was primarily concentrated at the beam ends, which could be considered the primary cause of stress development at the beam ends. Owing to the constraint of the column on the torsional deformation of the beam ends, the principal stress direction was not at an angle of 45° or 135° to the beam axis. It had a negligible effect on stress development owing to the lower torsional moment at the midspan location. Therefore, it was deduced that the development of diagonal cracks at the open corners was primarily dominated by the torsional moment at the beam ends.

4. Finite Element Analysis

To further examine the effect of the proposed entrance frame on the static behavior of adjacent members under vertical loading and provide brief design criteria for the frame, ABAQUS software was utilized to construct a three-dimensional (3D) finite element (FE) model of the substructure. Nonlinear inelastic analysis was conducted using the ABAQUS/implicit static general procedure. The FE model is illustrated in Figure 16.

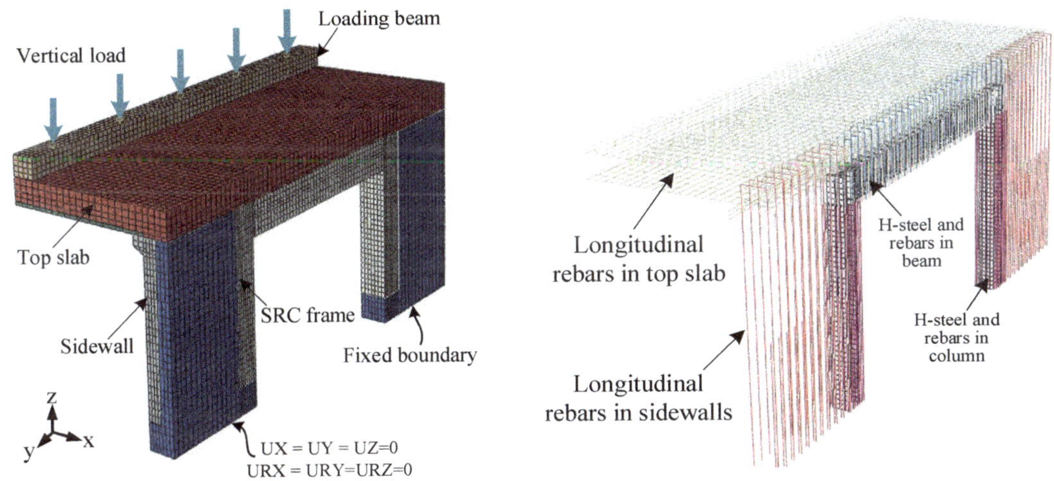

Figure 16. Meshed model and the loading methods.

4.1. FE Model

In the model, the cross-sectional parameters and reinforcement details of each member in the model were identical to those of the test specimen. The concrete and H-steel were discretized using the eight-node linear 3D solid elements with hourglass control (C3D8R). All reinforcing bars were modeled by the three-dimensional two-node linear truss elements (T3D2). To ensure accurate prediction of the static behavior and deformability of the specimen at a reasonable computational cost, a sensitivity study was conducted by discretizing continuous solid elements in different regions using a variety of nominal element sizes. The meshing of the model is also shown in Figure 16. Here, the nominal concrete element sizes in the thickness direction of the sidewalls and top slab were 50 mm and 100 mm, respectively. In the longitudinal direction of the station, the nominal concrete member sizes of the sidewalls and top slab were both 100 mm. The concrete and H-steel of SRC columns were discretized in three dimensions using an element size of 50 mm. The three-dimensional mesh size of concrete and H-beam in the longitudinal 500 mm region at both ends of the SRC beam was 50 m, while the mesh size of the remaining regions was 100 mm. In addition, 100 mm truss elements were utilized to discretize the reinforcement. To simulate the interactions of the interfaces between cast-in-place concrete and precast concrete in the laminated slabs, the "hard contact" and the "penalty–friction" algorithms from the default interaction library of ABAQUS were adopted [45]. The friction coefficient between the interfaces of different concrete layers was set as 0.6, which was determined based on the results of a previous study [46]. Embedded technology was used to couple the degrees of freedom between concrete and steel rebar elements and between concrete and H-steel elements, since no bond slippage occurred during the testing process.

The boundaries and loading conditions of the model were applied based on the test results. Because of the constant restraint of the bottom slab to the sidewalls and entrance frame and the effectiveness of the bottom joints of the sidewall throughout the test, fixed boundary conditions were uniformly formulated to simulate the constraints of the bottom slab on the sidewalls and frame. Here, the degrees of freedom in the X-, Y-, and Z-directions at the bottom ends of the sidewalls and frame were restricted, respectively. To reproduce a uniformly distributed load at the top slab end, a loading beam model with a flexural stiffness similar to that of the experimental steel beam was developed. The modeled beam was tied at the end of the top slab using the constraint feature in ABAQUS [47]. A vertical load was directly applied at the top of the modeled beam and controlled by the loading protocol used in the test.

The rebar and H-steel in the model were treated as the elastic–plastic materials, and the material characteristics in compression and tension were similar. The multiple linear kinematic hardening model [48] can be used to describe the mechanical behavior of steel. The parameters of the steel model were determined based on the material test results, as summarized in Table 2. The stress–strain curves of the steel are depicted in Figure 17a. The axial compressive and tensile plasticity damage model [49] can be used to analyze the mechanical performance of concrete. The critical parameters of the model were obtained from the material test results, as listed in Table 1. The compressive and tensile stress–strain curves of the concrete were drawn based on the method proposed in GB 50010 [38], as shown in Figure 17b,c. In the concrete damage plasticity (CDP) model, the dilation angle is 30°, the eccentricity is 0.1, the viscosity parameter is 0.005, and the ratio between the concrete strength in terms of biaxial and uniaxial strength is 1.16. The concrete damage factors of different components in the CDP model can be determined according to the recommendations provided in the ABAQUS standard user's manual [47], as shown in Equations (2) and (3).

$$D_c = 1 - \frac{\sigma_c E_c^{-1}}{\varepsilon_c^{in}(1-\eta_c) + \sigma_c E_c^{-1}} \tag{2}$$

$$D_t = 1 - \frac{\sigma_t E_c^{-1}}{\varepsilon_t^{in}(1-\eta_t) + \sigma_t E_c^{-1}} \tag{3}$$

where D_c and D_t are the tensile and compressive damage factors of materials in the CDP model, respectively; η_c is the ratio of compressive damage plastic strain to inelastic strain, which is 0.7; η_t is the ratio of tensile damage plastic strain to inelastic strain, which is 0.1; ε_t^{in} is inelastic tensile strain; ε_c^{in} is inelastic compressive strain; E_c is the elastic modulus of concrete; and σ_c and σ_t represent the compressive stress and tensile stress of concrete, respectively.

4.2. Validation of the FE Model

The developed FE model was validated by comparing the numerical results with the test results in terms of the load–deformation responses and crack patterns. A comparison of the load–deformation diagrams obtained from the FE analysis and test is presented in Figure 18. Although the displacements calculated by the FE model were marginally less than the experimental results after the cracking of the top slab concrete, the predicted displacements under the P_y load were basically consistent with the experimental results. Generally, the load–deformation diagrams obtained from the FE analysis resembled well the experimental results.

The maximum principal inelastic strain patterns of the concrete materials from the FE analysis could be used to simulate the distribution of cracks in the specimen. Figure 19 displays the principal inelastic strain distributions of the concrete materials under the P_y load. As shown in Figure 19, the most critical plastic damage occurred at the SRC beam ends and the adjacent top slab and SRC columns, coinciding with the diagonal crack-concentration regions of the specimens during the test. The principal strain on the sidewalls gradually decreased from the outside edge to the entrance frame. This trend was in favorable agreement with the development of the longitudinal reinforcement strain and concrete crack width along the sidewall cross-section. The comparison confirmed that the developed FE model could accurately show the development of cracks around the opening.

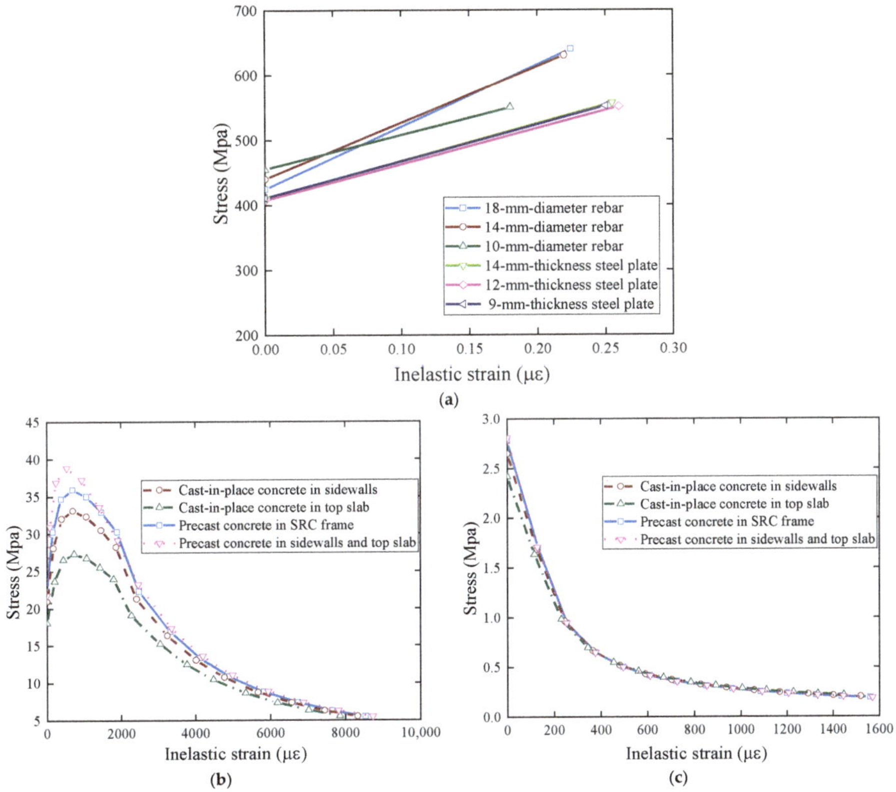

Figure 17. Stress–strain curves of steel and concrete: (**a**) steel constitutive curve, (**b**) concrete compressive stress–strain curve, and (**c**) concrete tensile stress–strain curve.

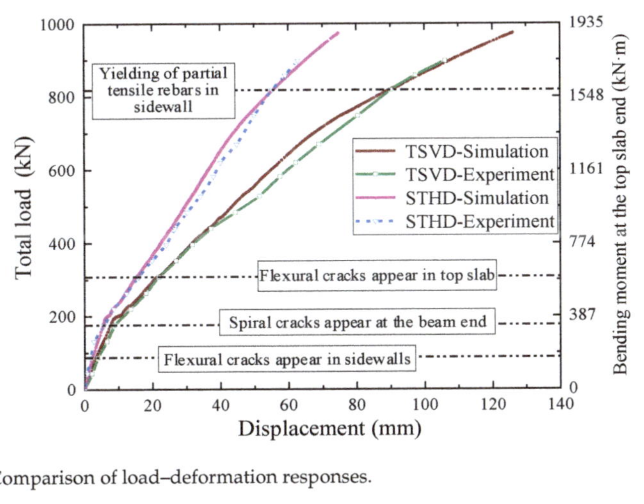

Figure 18. Comparison of load–deformation responses.

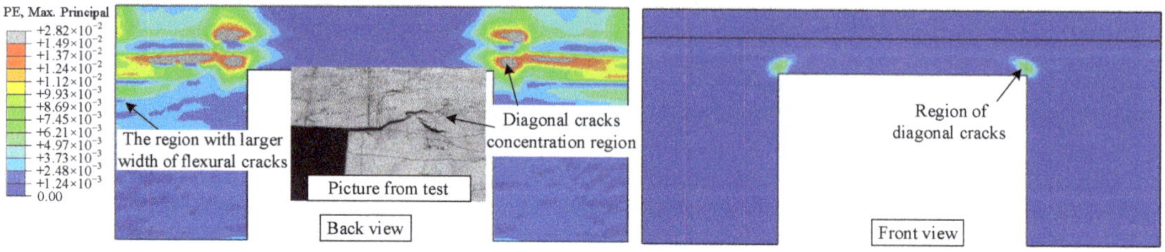

Figure 19. Maximum principal inelastic strain patterns from FE analysis.

4.3. Parametric Study

A parametric study was conducted using the validated FE model to further investigate the effects of the entrance frame with different configurations on the development of cracks around the opening. The parameters were the steel ratio of the SRC frame and cross-sectional height of the column and beam. Table 3 summarizes the details of all cases in this parametric study. For convenience, these cases are denoted as E-X. Case E-1 employed a validated FE model with a configuration similar to that of the test specimen, which could be treated as the baseline configuration for the other cases. Compared to E-1, the steel ratio of the SRC beam in E-2 was zero, and that of the SRC column in E-3 was zero as well. Both cases were designed to investigate the effects of the steel ratio of the entrance frame. In addition, the cross-sectional height of the SRC beam increased to 800 mm in Case E-4, and the cross-sectional width of the SRC column increased to 550 mm in Case E-5. An increase in the cross-sectional heights of the SRC beam and column directly improved the bearing capacity and out-of-plane stiffness. For Cases E-4 and E-5, although the cross-sectional heights of the SRC column or beam were varied, their longitudinal reinforcement ratios and steel ratios for all frame sections were still similar to those in Case EJ-1.

Table 3. Details and results of different cases in the parametric study.

Name	SRC Beam Section				SRC Column Section				M_y (kN·m)	T_y (kN·m)	k_1 (%)	k_y (%)
	Height (mm)	Width (mm)	ρ_s (%)	ρ_r (%)	Height (mm)	Width (mm)	ρ_s (%)	ρ_r (%)				
E-1	500	350	4.86	1.58	400	350	5.90	3.74	1807.7	264.5	21.2	29.3
E-2	500	350	0.00	1.58	400	350	5.90	3.74	1694.6	174.8	21.4	20.6
E-3	500	350	4.86	1.58	400	350	0.00	3.74	1504.7	172.2	21.5	22.9
E-4	800	350	4.86	1.58	400	350	5.90	3.74	1498.0	261.3	26.3	34.9
E-5	500	350	4.86	1.58	400	550	5.90	3.74	3798.6	799.8	26.8	42.1

4.4. Effect on Beam End Torques

Table 3 shows the data when all the tensile reinforcements in the sidewalls yielded, that is, reached the P_y load, the bending moment at the top slab end (M_y), and the torsional moment at the beam end (T_y) in each case. It can be seen that the varied steel ratios and cross-sectional heights of the SRC column and beam contributed to the yield resistance of specimens. Nevertheless, as discussed in Section 3.4, the dimensions of the sidewall in the actual project were significantly larger than those of the entrance frame. Therefore, the variations in the bearing capacity and out-of-plane stiffness of the frame contributed insignificantly to the overall mechanical performance of the sidewall systems. A proportionality coefficient (k) can be utilized to further illustrate the effect of configuration parameters on the proportion of the load carried by the beams above the opening. k can be defined as the ratio of the summation of the torques (T) at both ends of the beam to the corresponding bending moment (M) at the top slab end, i.e., it can be calculated as $2T/M$. Here, the coefficient k_1 was obtained when the concrete in the FE model was not cracked, and the coefficient k_y was obtained under the P_y load. It should be noted that T

was obtained by integrating points over the entire height of the section from the bottom of the SRC beam to the top of the roof in the FE analysis.

In the elastic stage, the coefficient k_1 values varied from 21.2% to 26.8%, as shown in Table 3. The difference between the coefficient k_1 values of Cases E-1 to E-3 with different steel ratios were negligible, which indicates that the variation in the steel ratio of the entrance frame had negligible effects on the load distribution coefficient of the SRC beams and its adjacent members. For Cases E-4 and E-5, the coefficient k_1 values were 24.1% and 26.4% higher than those of Case E-1, respectively. This implied that both the improvement in the torsional stiffness of the SRC beam and in the out-of-plane flexural stiffness of the SRC column could significantly increase T at the beam ends in the elastic stage. In general, the steel ratio exhibited a marginal contribution to the frame stiffness. This resulted in a negligible effect on the SRC frame sharing of the out-of-plane moments from the top slab end. Consequently, the variation in the steel ratio of the SRC frame contributed limitedly to T at the beam ends under identical loading levels.

As can be seen from Table 3, the coefficient k_y values ranged from 20.6% to 42.1% when all tensile reinforcements in the sidewalls of the respective models entered the yield stage. For E-1, E-3 E-4, and E-5, the coefficient k_y values were 38.2%, 6.5%, 32.6%, and 57.1% higher than those of k_1, respectively. However, the coefficient k_y of E-2 was 3.7% lower than k_1. The change in coefficients implies a redistribution of the internal force between the SRC beams and its adjacent members, which can be attributed to the constant variation in the torsional stiffness of the SRC beam and in the flexural stiffness of the vertical members during the loading process due to concrete cracking and reinforcement yielding in the members. Owing to the rapid development of cracks at the beam ends of E-2, the torsional stiffness degradation of its beam was more pronounced than that of other cases. Therefore, its proportionality coefficient presented an overall decreasing trend with increasing load. For other cases, the increase in k_y could be explained by the development of flexural cracks in the sidewalls exacerbating the degradation of its flexural stiffness, and changes of the configuration parameters restraining the cracks of beam ends. This also indicates that the steel ratio of the SRC beam had a significant effect on the control of crack development at the beam ends.

4.5. Effect on Inelastic Strain Distribution of Concrete

Figure 20 further shows the maximum principal inelastic strain distributions of the concrete materials under the P_y load for Cases E-2 to E-5. Similar to E-1 (Figure 19), the maximum plastic damage of concrete materials in E-2 was also concentrated at the SRC beam ends, and the damage region gradually expanded to the top slab and sidewalls. Note that the plastic damage area of E-2 was significantly greater than that of E-1, which can be explained by the lower torsional capacity of the normal concrete beam than that of the SRC beam. In contrast to E-1 and E-2, the location of the initial plastic damage at the beam ends of E-3 was almost aligned to the central section of the SRC column. This phenomenon can be attributed to the absence of H-steel in the column, which significantly reduced the resistance of the column to the torsional moment transmitted from the beam end. It is therefore deduced that the variations in the flexural resistance of the column had a significant effect on the development of the cracks at the open corners.

Cases E-4 and E-5 effectively limited the development of concrete inelastic strain at the beam ends. For E-4, although the increment in the cross-sectional heights of the SRC beam directly increased the torsional moment shared by the beam, the torsional bearing capacity improved simultaneously as well. In addition, the higher beam section effectively reduced the torsional shear stress transmitted to the column, which could prevent the column concrete from entering the plasticity prematurely. For E-5, the increment in the cross-sectional height of the SRC column significantly improved the out-of-plane flexural capacity and its restraint on the torsional deformation of the beam ends. Therefore, this method was effective in preventing column concrete from entering plasticity and limiting further propagation of the beam end torsional shear stress in the top slab. The aforementioned results prove that both the improvement in the torsional bearing capacity of the beam and out-of-plane flexural capacity of the column positively influenced the control of crack development at the open corners.

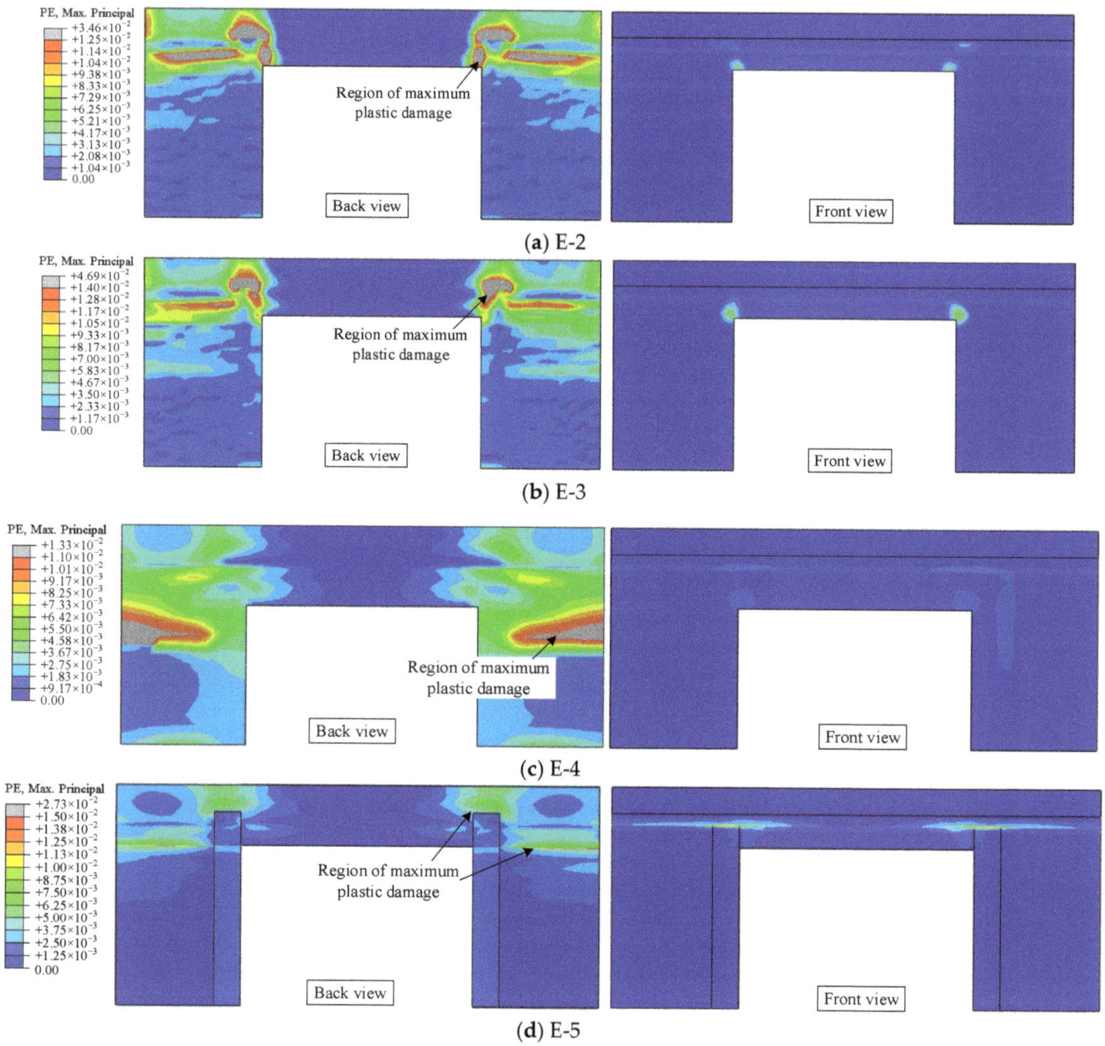

Figure 20. Maximum principal inelastic strain patterns of the concrete for Cases E-2 to E-5.

5. Conclusions

In this study, a novel type of precast entrance frame was proposed to compensate for the effects of entrance openings on the localized bearing capacity of the continuous sidewalls and prevent stress concentrations around the openings. To further investigate the effect of the developed entrance frame on the mechanical behavior of its adjacent sidewalls, a monotonic static test and finite element analysis were performed on a 1/2 scale station entrance substructure. The overall static behavior, cracking behavior, and stress distribution of the substructure were reported. The main findings can be summarized based on the experimental and numerical results as follows:

(1) The U-type rebar overlapping connection between the developed frame and adjacent members exhibited satisfactory mechanical performance throughout the test, implying that this connection type could effectively guarantee stress transfer between precast members.
(2) The developed frame influenced the stress distribution of the adjacent members. Furthermore, it could effectively prevent stress concentration in the adjacent sidewall region and facilitate crack-width control in the sidewalls.
(3) The out-of-plane moment transmitted from the top slab end to the SRC beam was concentrated at the beam ends in the form of torsional moment. This directly resulted in a stress concentration at the corner region of the opening, which could be regarded as the primary cause of the serious development of cracks in this region.
(4) Parametric studies were performed to investigate the influence of the steel ratio of the SRC frame and cross-sectional dimensions in the frame on the proportion of the load carried by the beam above the opening. The results showed that both the improvement in the torsional stiffness of the SRC beam and the out-of-plane flexural stiffness of the SRC column could significantly increase the torsional moment at the beam ends under identical rooftop vertical loads.
(5) Parametric studies also investigated the influence mechanism of the developed frame with different configuration parameters on the crack development around the opening. The results showed that the improvement in the torsional bearing capacity of the SRC beam as well as the out-of-plane flexural capacity of the SRC column could contribute positively to the control of crack development around the opening.

Note that the number of test samples in this study was relatively limited due to the limitations of testing costs. In addition, due to the limitation of available experimental operation space, the test specimen was designed as a scale substructure. To address this research gap, further finite element studies can be carried out based on the validated FE model to simulate the mechanical behavior of the overall station structure with an entrance opening, and to investigate the effects of other variables not investigated in the present study.

Author Contributions: Conceptualization, Z.M. and D.G.; methodology, S.F. and D.G.; software, S.F.; validation, S.F., Z.M. and D.G.; riting—original draft preparation, S.F.; writ-ing—review and editing, D.G.; visualization, S.F.; supervision, Z.M. and Z.G.; project administration, Z.G.; funding acquisition, Z.M. and Z.G. All authors have read and agreed to the published version of the manuscript.

Funding: This research is supported by the National Natural Science Foundation of China (52278154) and the Fundamental Research Funds for the Central Universities (2242021R10031).

Data Availability Statement: Some or all data, models, or codes generated or used during this study are available from the corresponding author by request.

Acknowledgments: Thanks for the supported by the National Natural Science Foundation of China (52278154) and the Fundamental Research Funds for the Central Universities (2242021R10031).

Conflicts of Interest: The authors declare no conflict of interest.

References

1. Tao, L.; Ding, P.; Shi, C.; Wu, X.; Wu, S.; Li, S. Shaking table test on seismic response characteristics of prefabricated subway station structure. *Tunn. Undergr. Space Technol.* **2019**, *91*, 102994. [CrossRef]
2. Liu, M.; Jia, S.; Liu, X. Evaluation of mitigation potential of GHG emissions from the construction of prefabricated subway station. *J. Clean. Prod.* **2019**, *236*, 117700. [CrossRef]
3. Yang, X.; Lin, F.; Huang, M. Research on flexural bearing capability of long grouted single mortise-tenon joints for prefabricated metro station structures. *China Civ. Eng. J.* **2020**, *53*, 111–118. (In Chinese)
4. Yang, X.; Lin, F.; Huang, M. Experimental research on bending bearing capability of grouted double mortise-tenon joint for pre-fabricated metro station structure. *Adv. Civ. Eng.* **2021**, *2021*, 6635409.
5. Yang, X.; Lin, F.; Huang, M. Experimental Analysis of Bending Stiffness Characteristics of Grouted Double Mortise-Tenon Joint for Prefabricated Metro Station Structure. *Adv. Civ. Eng.* **2021**, *2021*, 9958436. [CrossRef]

6. Tao, L.; Ding, P.; Yang, X.; Lin, P.; Shi, C.; Bao, Y.; Wei, P.; Zhao, J. Comparative study of the seismic performance of prefabricated and cast-in-place subway station structures by shaking table test. *Tunn. Undergr. Space Technol.* **2020**, *105*, 103583. [CrossRef]
7. Ding, P.; Tao, L.; Yang, X.; Zhao, J.; Shi, C. Three-dimensional dynamic response analysis of a single-ring structure in a prefabricated subway station. *Sustain. Cities Soc.* **2019**, *45*, 271–286. [CrossRef]
8. Liu, H.; Wang, Z.; Du, X.; Shen, G.Q. The seismic behaviour of precast concrete interior joints with different connection methods in assembled monolithic subway station. *Eng. Struct.* **2021**, *232*, 111799. [CrossRef]
9. Du, X.L.; Liu, H.T.; Lu, D.C.; Xu, C.S.; Luo, F.R.; Li, S.M. Study on seismic performance of sidewall joints in assembled monolithic subway station. *China Civ. Eng. J.* **2017**, *50*, 38–47.
10. Liu, T.J.; Lu, J.D.; Wang, D.; Liu, H.Y. Experimental investigation of the mechanical behaviour of wall–beam–strut joints for prefab-ricated underground. *Int. J. Concr. Struct. Mater.* **2021**, *15*, 63–81. [CrossRef]
11. Wang, Z.H.; Ouyang, Y.P.; Hong, L. Effect of deep buried tunnel on structural behavior of subway station entrance structure. *Modern Urban Transit.* **2020**, *10*, 92–95. (In Chinese)
12. Monika, M.C. A study of displacements of structures in the vicinity of deep excavation. *Arch. Civ. Mech. Eng.* **2019**, *19*, 547–556.
13. Ming, O. Internal force calculation and comparative analysis of subway entrance structure. *Eng. Constr.* **2021**, *53*, 43–49. (In Chinese)
14. Liu, Y.A.; Hu, Y.X.; Xiao, Y.M.; Chen, J.L.; Huang, H.T. Effects of different types of entrances on natural ventilation in a subway station. *Tunn. Undergr. Space Technol.* **2020**, *105*, 103578. [CrossRef]
15. Yang, X.R.; Lin, F. Prefabrication technology for underground metro station structure. *Tunn. Undergr. Space Technol.* **2021**, *108*, 103717. [CrossRef]
16. Choi, S.J.; Hansmire, W.H.; Horvath, Z.; Herranz, C. Seismic design approach for underground structures of los angeles metro regional connector transit corridor project. *Lifelines* **2022**, *2021*, 308–321.
17. Mohammed, B.S.; Ean, L.; Malek, M. One way RC wall panels with openings strengthened with CFRP. *Constr. Build. Mater.* **2013**, *40*, 575–583. [CrossRef]
18. *EN 1992-1-1*; Eurocode 2: Design of Concrete Structures—Part 1-1: General Rules and Rules for Buildings. British Standards Institution: London, UK, 2004.
19. *AS 3600*; Concrete Structures. Standards Australia: Sydney, Australia, 2009.
20. *ACI 318-14*; Building Code Requirements for Structural Concrete. Am Concrete Institute: Farmington Hills, MI, USA, 2014.
21. *CAN/CSA-A23.3*; Design of Concrete Structures. Canadian Standards Association: Mississauga, ON, Canada, 2004.
22. Popescu, C.; Sas, G.; Sabău, C.; Blanksvärd, T. Effect of cut-out openings on the axial strength of concrete walls. *J. Struct. Eng.* **2016**, *142*, 04016100. [CrossRef]
23. Robinson, G.P.; Palmeri, A.; Austin, S.A. Design methodologies for one way spanning eccentrically loaded minimally or centrally reinforced pre-cast RC panels. *Eng. Struct.* **2013**, *56*, 1945–1956. [CrossRef]
24. Popescu, C.; Sas, G.; Blanksvärd, T.; Täljsten, B. Concrete walls weakened by openings as compression members: A review. *Eng. Struct.* **2015**, *56*, 172–190. [CrossRef]
25. Mattock, A.H.; Kriz, L.B.; Hognestad, E. Rectangular concrete stress distribution in ultimate strength design. *ACI Struct. J.* **1961**, *57*, 875–928.
26. Saheb, S.M.; Desayi, P. Ultimate Strength of RC Wall Panels with Openings. *J. Struct. Eng.-Asce.* **1990**, *116*, 1565–1578. [CrossRef]
27. Ganesan, N.; Indira, P.; Santhakumar, A. Prediction of ultimate strength of reinforced geopolymer concrete wall panels in one-way action. *Constr. Build. Mater.* **2013**, *48*, 91–97. [CrossRef]
28. Dazio, A.; Beyer, K.; Bachmann, H. Quasi-static cyclic tests and plastic hinge analysis of RC structural walls. *Eng. Struct.* **2009**, *31*, 1556–1571. [CrossRef]
29. Zhi, Q.; Guo, Z.; Xiao, Q.; Yuan, F.; Song, J. Quasi-static test and strut-and-tie modeling of precast concrete shear walls with grouted lap-spliced connections. *Constr. Build. Mater.* **2017**, *150*, 190–203. [CrossRef]
30. Wu, D.; Liang, S.; Shen, M.; Guo, Z.; Zhu, X.; Sun, C. Experimental estimation of seismic properties of new precast shear wall spatial structure model. *Eng. Struct.* **2019**, *183*, 319–339. [CrossRef]
31. *GB50011-2010*; Code for Seismic Design of Buildings. China Architecture and Building Press: Beijing, China, 2015.
32. Gong, C.; Ding, W.; Mosalam, K.M. Performance-based design of joint waterproofing of segmental tunnel linings using hybrid computational/experimental procedures. *Tunn. Undergr. Space Technol.* **2020**, *96*, 103172. [CrossRef]
33. Mohammed, B.S.; Abu Bakar, B.H.; Choong, K.K. The Effects of Opening on the Structural Behavior of Masonry Wall Subjected to Compressive Loading—Strain Variation. *Open Civ. Eng. J.* **2009**, *3*, 62–73. [CrossRef]
34. Casadei, P.; Ibell, T.; Nanni, A. Experimental results of one-way slabs with openings strengthened with CFRP laminates. In Proceedings of the Fiber-Reinforced Polymer Reinforcement for Concrete Structures, Singapore, 8–10 July 2003; FRPRCS-6. pp. 1097–1106. [CrossRef]
35. Hollaway, L.C. A review of the present and future utilisation of FRP composites in the civil infrastructure with reference to their important in-service properties. *Constr. Build. Mater.* **2010**, *24*, 2419–2445. [CrossRef]
36. Feng, S.; Guan, D.; Guo, Z.; Liu, Z.; Li, G.; Gong, C. Seismic performance of assembly joints between HSPC beams and concrete-encased CFST columns. *J. Constr. Steel Res.* **2021**, *180*, 106572. [CrossRef]
37. Feng, S.; Guan, D.; Ni, L.; Lin, Y.; Liu, Z.; Guo, Z.; Li, G. Experimental study on seismic behavior of joints connecting precast H-steel reinforced concrete beams and concrete-filled steel tube columns. *J. Build. Eng.* **2022**, *45*, 103444. [CrossRef]

38. *GB 50010-2010;* Code for Design of Concrete Structures. China Building Industry Press: Beijing, China, 2010.
39. Megget, L.M. The seismic behaviour of small reinforced concrete beam-column knee joints. *Bull. N. Z. Soc. Earthq. Eng.* **1998**, *31*, 215–245. [CrossRef]
40. Megget, L.M. The Seismic Design and Performance of Reinforced Concrete Beam-Column Knee Joints in Buildings. *Earthq. Spectra* **2003**, *19*, 863–895. [CrossRef]
41. *GB50204-2015;* Code for Acceptance of Constructional Quality of Concrete Structures. Ministry of Housing and Urban-Rural Development of the People's Republic of China: Beijing, China, 2015.
42. *GB/T 228.1-2010;* Metallic Materials: Tensile Testing: Part 1: Method of Test at Room Temperature, General Administration of Quality Supervision. Inspection and Quarantine of the People's Republic of China: Beijing, China, 2011.
43. *JGJ 138-2015;* Code for Design of Composite Structures. China Building Industry Press: Beijing, China, 2016.
44. *EN 1994-1-1;* Eurocode 4: Design of Composite Steel and Concrete Structures—Part 1. 1: General Rules and Rules for Buildings. British Standards Institution: London, UK, 2006.
45. Kabir, I.; Lee, C.; Rana, M.M.; Zhang, Y. Flexural behaviour of ECC-LWC encased slender high strength steel composite beams. *J. Constr. Steel Res.* **2020**, *173*, 106253. [CrossRef]
46. Tassios, T.P.; Vintzēleou, E.N. Concrete-to-concrete friction. *J. Struct. Eng.* **1987**, *113*, 832–849. [CrossRef]
47. Smith, M. *ABAQUS/standard User's Manual, version 6.14;* Dassault Systemes Simulia Corp: Providence, RI, USA, 2014.
48. Yu, Y.; Chen, Z.; Wang, X. Effect of column flange flexibility on WF-beam to rectangular CFT column connections. *J. Constr. Steel Res.* **2015**, *106*, 184–197. [CrossRef]
49. Yu, J.; Zhang, W.; Tang, Z.; Guo, X.; Pospíšil, S. Seismic behavior of precast concrete beam-column joints with steel strand inserts under cyclic loading. *Eng. Struct.* **2020**, *216*, 110766. [CrossRef]

buildings

MDPI

Article

Prediction of the Shear Resistance of Headed Studs Embedded in Precast Steel–Concrete Structures Based on an Interpretable Machine Learning Method

Feng Zhang, Chenxin Wang, Xingxing Zou *, Yang Wei, Dongdong Chen, Qiudong Wang and Libin Wang

College of Civil Engineering, Nanjing Forestry University, Nanjing 210037, China
* Correspondence: civilzou@njfu.edu.cn; Tel.: +86-15586188579

Abstract: Headed shear studs are an essential interfacial connection for precast steel–concrete structures to ensure composite action; hence, the accurate prediction of the shear capacity of headed studs is of pivotal significance. This study first established a worldwide dataset with 428 push-out tests of headed shear studs embedded in concrete with varied strengths from 26 MPa to 200 MPa. Five advanced machine learning (ML) models and three widely used equations from design codes were comparatively employed to predict the shear resistance of the headed studs. Considering the inevitable data variation caused by material properties and load testing, the isolated forest algorithm was first used to detect the anomaly of data in the dataset. Then, the five ML models were established and trained, which exhibited higher prediction accuracy than three existing design codes that were widely used in the world. Compared with the equations from AASHTO (the one that has the best prediction accuracy among design specifications), the gradient boosting decision tree (GBDT) model showed an 80% lower root mean square error, 308% higher coefficient of determination, and 86% lower mean absolute percent error. Lastly, individual conditional expectation plots and partial dependence plots showed the relationship between the individual parameters and the predicted target based on the GBDT model. The results showed that the elastic modulus of concrete, the tensile strength of the studs, and the length–diameter ratio of the studs influenced most of the shear capacity of shear studs. Additionally, the effect of the length–diameter ratio has an upper limit which depends on the strength of the studs and concrete.

Keywords: headed shear studs; steel–concrete composite structures; machine learning (ML); isolation forest; partial dependence plot

1. Introduction

Steel–concrete composite structures are efficient structural members that fully utilize the compressive strength of concrete and the tensile strength of steel in most loading scenarios [1]. Therefore, steel–concrete composite structures have been widely used in enormous buildings and bridges [2]. As steel parts are usually installed first as supports and form systems for concrete [3], steel–concrete composite structures show great potential for precast structures. For steel–concrete composite structures, the interfacial shear connection is a prerequisite to ensure the composite action between steel and concrete. Headed studs are one of the most popular and cost-effective interfacial shear connections for composite structures due to their fast installation, equal shear strength and stiffness in all directions normal to the axis of the stud, and little obstruction to the slab reinforcement ([4,5]).

Since the 1950s, numerous push-out tests have been carried out to investigate the mechanical performance of headed studs in steel–concrete composite structures under different loads and effects ([6–8]). Viest [6] identified three failure modes in push-out tests, including steel failures, concrete failures, and mixed failures that included the failure of both materials, and proposed the first equation for calculating the resistance of headed studs. Then, Ollgaard et al. [7] proposed an empirical equation for the shear resistance of headed studs based on the experimental tests on 48 push-out tests with headed studs embedded in both normal-weight and lightweight concrete. The equation later became the basis of the calculation equations for the design of headed stud shear connectors in many international design codes ([9–11]). In 1981, the first design code for steel–concrete composite structures adopted a design equation for the resistance of headed studs suggested by CEB-ECCS-FIP-IABSE [12]. In 2004, Lee and Shim [13] conducted static and fatigue tests on large-diameter headed stud connectors and found that the equations in Eurocode-4 [9] and AASHTO LRFD [10] underestimated the static and fatigue performances of large-size studs to a large extent. The following research focused on modifying the coefficients in existing design equations to achieve higher prediction accuracy [14] or developing new equations by fitting push-out tests ([15,16]). To summarize, traditional prediction equations provided limited accuracy based on parameters with physical meaning. Nevertheless, current design codes are based on limited test data and a lack of reliability beyond the range of the data. Additionally, ultra-high-performance concrete (UHPC) and high-performance concrete (HPC) have been proposed and applied to precast steel–concrete structures as a more durable and efficient solution in lieu of conventional concrete. However, studies have shown that using traditional equations, which were derived based on the experimental data of normal concrete specimens, to predict the shear capacity of headed studs in UHPC and HPC is questionable because the strength of normal concrete is much lower than that of UHPC and HPC ([16,17]). As a more efficient and accurate method to derive equations compared with physics-based equations in most cases, data-driven machine learning (ML) models have been widely used in engineering research in recent years, which can get rid of time-consuming and costly experimental tests.

ML has shown extremely high accuracy in many studies in predicting the interfacial behavior between steel and concrete, and there are many ML algorithms that have shown to be promising in modeling different civil engineering problems, namely, artificial neural network (ANN), multivariate adaptive regression splines, Gaussian process regression, minimax probability machine regression, random forest (RF), support vector machine (SVM), and least squares support vector machine (LS-SVM) available in the literature to develop models ([18–38]). Esteghamati et al. [19] presented a framework to develop generalizable surrogate models to predict seismic vulnerability and environmental impacts of a cluster of buildings at a particular location and performed sensitivity assessments to evaluate the most important parameters and study assumptions. Avci-Karatas [21] presented regression methodologies to predict the resistance of headed studs by using the concepts of minimax probability machine regression and extreme ML methods. Wang et al. [25] predicted the stud shear stiffness and achieved automatic hyperparameter optimization with an auto-tuning deep forest. Mahjoubi et al. [26] presented a logic-guided neural network to predict the interfacial properties of steel–concrete composites. Additionally, accurate predictions made by ML models can serve as supplementary data, expanding the amount of data that can be exploited by empirical equations ([27–29]). However, most of the existing research focused on developing high-performance ML models or comparing the performance differences of several algorithms ([30–33]), which is limited by the datasets that need to extend the quantity and be updated in time, and they are still a kind of "black box" although these ML models have very high prediction accuracy. How to give explanatory analysis to these ML models with superior performance is very important, which is related to the better application of ML in engineering. Cakiroglu et al. [34] developed ensemble ML models to predict the axial compression capacity of rectangular concrete-filled steel tubular columns and the SHAP model for interpreting the ensemble learning models indicates that the side length of the cross-section has the highest impact on the predicted compressive load capacity. To quantify the contribution of each feature to the predictive ability of the model through feature importance is effective and widely used ([35,36]), but it cannot explain the complex influence of features and features on output variables. Several techniques have been developed to study the interpretability of ML models and visualize them, such as partial dependence plot (PDP) and individual conditional expectation (ICE) [37]. Setvati et al. [38] predicted the stud resistance based on six trained ML models and visualized relationships between the input variables within the predicted resistance, but the study is based on a dataset with 242 push-out specimens, which is rather few for training ML models, and no anomaly detection was performed on the dataset like most previous studies, although it is crucial in developing ML models. In comparison, this study is dedicated to establishing the largest, to the authors' knowledge, dataset and to using isolation forests for anomaly detection to ensure prediction accuracy.

This study presents an interpretable ML method to predict the shear resistance of headed studs in steel–concrete structures. Five main contributions are made in this study: (1) a significantly larger database than used by many international design codes with 428 push-out test specimens was established; (2) an isolation forest algorithm is used to determine the presence of any anomaly in the dataset; (3) the performance of five advanced ML algorithms, namely ANN, SVM, RF, decision tree (DT), and gradient boosting decision tree (GBDT), was derived and compared; (4) three international design codes, namely GB50017–2017, Eurocode-4, and AASHTO LRFD Bridge Design Codes, were re-evaluated base on the established dataset; (5) an interpretable analysis of the GBDT model was performed by visualizing and evaluating the relationships between the characteristic parameters. The results from the present work demonstrate that interpretable ML models have a certain guiding significance and can be used as potential candidates in engineering design.

2. Physics-Based Equations

2.1. Eurocode-4

Eurocode-4 predicts the shear capacity of the headed studs based on two failure modes, namely, concrete failure and steel failure. On the basis of the research of Ollgaard et al. [7], Eurocode-4 [9] revised the equation through more push test data, together with additional rules when $3 \leq \frac{h}{D} \leq 4$ [17]:

$$P_{stud} = min\begin{cases} \frac{0.29\alpha D\sqrt{f_c{}'E_c}}{\gamma}(\text{concrete failure}) \\ \frac{0.8f_u\pi D^2/4}{\gamma}(\text{steel failure}) \end{cases} \quad (1)$$

where P_{stud} is the diameter of the headed stud; $f_c{}'$ is the cylindrical compressive strength; γ is the partial factor for the design shear resistance of a headed stud, which is recommended in Eurocode-4 to have a value of 1.25; D is the diameter of the stud shank;E_c is the concrete elastic modulus; f_u is the tensile strength of the headed studs; α is a parameter related to the geometric shape of the headed stud, and is given by

$$\alpha = \begin{cases} 0.2\left(\frac{h}{D}+1\right) & \text{for } 3 \leq \frac{h}{D} \leq 4 \\ 1 & \text{for } \frac{h}{D} > 4 \end{cases} \quad (2)$$

where h is the nominal height of the stud.

2.2. Chinese GB50017–2017 Code

According to the Chinese GB50017–2017 code [11], the capacity of headed studs themselves are the upper limit to determine the bearing capacity of the connector, and the cubic compressive strength f_{cu} is used to replace the cylindrical compressive strength $f_c{}'$. The design code is as follows:

$$P_{stud} = 0.43A_s\sqrt{f_{cu}E_c} \leq 0.7A_sf_u \quad (3)$$

where A_s is the cross-sectional area of the headed studs; f_{cu} is the cubic compressive strength.

2.3. AASHTO LRFD Bridge Design Codes

In the AASHTO LRFD Bridge Design Codes [10], the resistance of the headed studs is calculated as follows:

$$P_{stud} = \varphi 0.5 A_s \sqrt{E_c f_c{}'} \leq \varphi A_s f_u \tag{4}$$

where φ is the resistance coefficient of the stud, which is generally taken as 0.85 in steel–concrete composite constructions [39]. The AASHTO used the stud performance parameters to determine the upper limit of the resistance of the headed studs. The diameter of the stud, the compressive strength, and the elastic modulus of the concrete are selected as the decisive factors for the headed studs' resistance.

3. Data

3.1. Dataset of Headed Studs Embedded in Concrete Push-Out Test Specimens

At present, beam tests and push-out tests are the most common methods to test the resistance of headed studs. Among them, the beam tests agree better with the actual loading situations but are also much more complicated than push-out tests. More importantly, the materials are more expensive, and the beam tests take much longer than push-out tests. The push-out tests are simpler and easier to implement and more convenient to carry out. As some research showed that the resistance of the headed studs obtained by the push-out test is lower than that of the beam test, which is more favored in code development [40]. Therefore, the international design codes for the resistance of shear connectors are proposed based on the push-out test results. Figure 1 shows the push-out tests; the diameter of the headed part and the shank are denoted by d_k and D, respectively. The depth of the headed part and the total length of the stud is denoted by k and h, respectively.

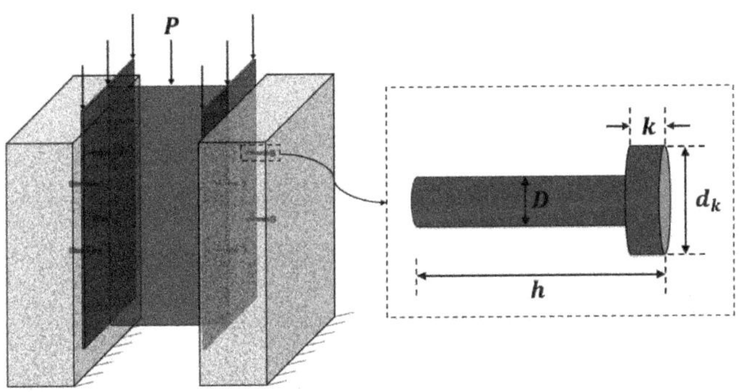

Figure 1. Push-out test.

This study collects a total of 428 specimens from 37 references ([14,16,38,41–74]) according to the following criteria:

1. The test is a push-out test and uses two symmetrical concrete slabs;
2. The connectors are headed studs, so specimens with bolts were discarded;
3. The loading modes are monotonic loading and cyclic loading, which is closer to the actual engineering loading;
4. The materials of the concrete slab are not limited to ordinary concrete, but UHPC and HPC are also collected.

In order to avoid the influence of the concrete material properties by the different sizes of the specimens, the compressive strength of concrete is uniformly converted into the compressive strength $f_c{}'$ of the concrete cylinder according to GB 50010-2010 [75], and the conversion relationship is shown in Table 1. The details of the push-out test datasets are shown in Table 2.

Table 1. Compressive strength conversion for concrete coupons with different sizes and shapes.

Sample shape and size	Cube			300 mm × 150 mm Cylinder				
	Width/mm			Strength grade				
	100	150	200	C20-C40	C50	C60	C70	C80
Conversion coefficient	1.05	1.0	0.95	0.80	0.83	0.86	0.875	0.89

Table 2. Dataset of headed studs embedded in concrete push-out test specimens.

Ref.	Number	f_u /MPa	D /mm	h /mm	E_c /GPa	$f_c{}'$ /MPa	n	P_{stud} /kN
Hu et al. [38]	10	455, 495	19	60, 80, 110	35	45.2–45.9	2, 4, 6	17.3–35.1
Shim et al. [41]	18	625–900	25	155	33.5–41.0	35.3–64.5	8	139.4–240.0
Lin et al. [16]	8	430–465	22–30	200	37.7	60.5	4	233.9–352.4
Wang et al. [42]	13	326–515	16–22	50–280	34.5	46.2	4	82.5–206.5
Wang et al. [43]	6	436, 486	22, 30	70–120	33, 48	37.3, 119.0	4	128.4–215.5
Han et al. [44]	3	400	13	90	33.7	36.1	2	156.0–163.3
Luo et al. [45]	16	472	13, 22	47, 80	45	23.5–132.4	2–18	41.2–217.0
Chen [46]	4	400	16, 19	80	45.3	94.8	8	102.1–155.7
Kim et al. [47]	15	466, 484	16, 22	50–100	32, 45	35, 200	8	103–212
Kim et al. [48]	12	466, 484	16, 22	50–100	45	200	8	102.8–211.9
Luo [49]	4	462	16, 19	90	31.0	31.3	4	95.3–120.5
Zeng et al. [50]	4	400	10, 16	45	42.6	160.7	8	53.8–114.2
Lam et al. [51]	4	589	19	100	23.6	20–50	2	71.6–130.4
Zhou. [52]	20	450	16–25	150	41.4–49.2	82.3–146.4	8	92.1–189.8
Wei [53]	6	469	13, 16	100	32.5	33.6	8	75.2, 102.1
Chen [54]	9	477, 495	16, 19	80, 110	35.2–35.4	45.2–45.9	4–12	95.6–145.4
Wang et al. [55]	8	445	13, 16	40–80	34.0, 42.8	42.8, 145.3	16	69.9–136.7
Wang [56]	26	444	13–19	65–105	28.8–34.3	30.5–50.8	2	61.1–118.9
Cao et al. [57]	3	400	13	35	42.6	130.5	8	57.1–62.2
An et al. [58]	8	519	19	75	27–34	30.8–91.2	8	111.5–161.0
Yamamoto et al. [59]	8	491–569	16–22	10	30.3	29.6	4	92.2–145.7
Mainstone et al. [60]	10	600	19	102	29.0–32.3	26.6–34.0	4	94.4–119.1
Ollgaard et al. [7]	21	488, 489	16, 19	76	15.1–25.8	18.4–35.0	8	75.2–144.6
Menzies [61]	6	600	19	102	25.5, 34.4	16.6, 40.8	4	96.1–126.5
Oehlers [62]	6	611	19	96	26.1–27.1	24.9–30.9	2	122–142
Hiragi et al. [63]	4	485	19	70, 100	33.6–38.3	38.3–56.4	4	138.1–169.0
Roik et al. [64]	20	460, 472	19, 22	100	33.0–38.9	36.7–59.0	8	133.6–177.9
Hicks [65]	4	466	19	95	31.7–32.7	31.9–35.1	2, 4	90.4–118.1
Easterling [66]	3	447	19	102	34.7	42.1	4	104.9–119.2
Feldmann et al. [67]	22	537, 546	19–25	80, 100	39.1–43.6	102.5–111.0	1, 8	133.8–318.9
Viest [6]	12	436–507	13–32	102	30.1–33.5	27.5–37.8	4	61.8–222.4
Wang et al. [68]	9	465–675	22, 25	215–215	37.1	70.3	4	236.5–272.7
Hanswille et al. [69]	10	464	25	125	29.5	23.7, 41.3	8	179.6–238.0
Bullo et al. [70]	18	495	19, 25	75, 120	33.1–45.6	32.5–94.4	-	98.8–293.2
Döinghaus [71]	26	452–557	19–25	80–120	43	86.1–115.8	1, 8	139.8–254.4
Xue et al. [72]	5	475	22	200	34.5	69.7	6	181.2–208.8
Jähring et al. [73]	32	549–580	19–25	125	30.0–40.9	45.4–112.1	4	156.5–285.1
Hanswille et al. [74]	15	528	22	125	33–39	42.8–56.2	8	173.3–216.0

3.2. Anomaly Detection

When we use ML algorithms to model datasets, the presence of noise and outliers tends to have a larger impact on the accuracy of the final model. The size of the dataset used in this article is 428. If there are several or more than a dozen outliers in these data, the proportion of outliers in the overall data set cannot be ignored, and the existence of outliers will affect the training of the model. At the same time, taking into account the possible human operation errors, literature hard brushing errors, or measurement errors that may occur during the experiment, it is necessary to perform anomaly detection on the established dataset to ensure the high quality of the data points of the data set and improve the final model. In this study, isolation forest [76] is used to detect and remove anomalous data.

Isolation forest is an unsupervised algorithm that evaluates data through number of integrated decision trees and scores each data point in the interval of 0–1, which is shown in Figure 2. We use the following criteria to evaluate the datasets using the scores [76]:

5. If instances return a score very close to 1, then they are highly likely to be anomalies;
6. If instances have a score much smaller than 0.5, then they are quite safe to be regarded as normal instances;
7. If all the instances return a score $\approx$ of 0.5, then the entire sample does not really have any distinct anomaly.

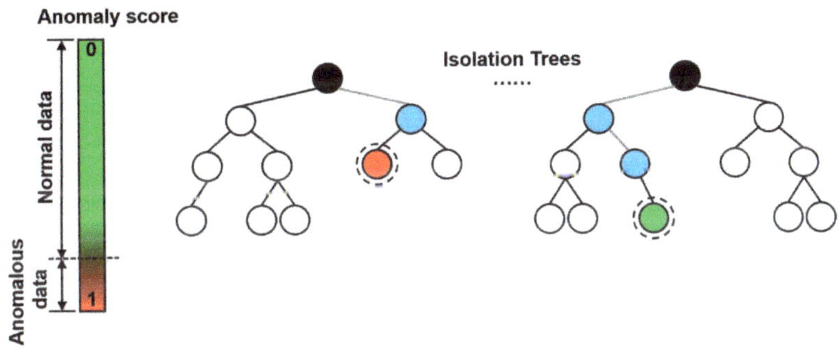

Figure 2. Illustration of anomaly detection with isolation forest.

The isolation forest algorithm was implemented in the Statistics and Machine Learning Toolbox of MATLAB. The evaluation scores of all specimens are shown in Figure 3a. The minimum value of the score is 0.320, and the maximum value is 0.686, which is generally distributed around 0.5. There are no obvious outliers in the overall datasets according to the appeal evaluation criteria. In order to determine the quality of the datasets, this study uses the K-Means clustering method to analyze the score (K = 2). The specific method is to use the clustering method to calculate the center points of the two types of score and obtain the two class label centers (normal data and anomalous data), as shown in Figure 3b. The abnormal data category accounted for 35.9%, and the center value was 0.520; the anomalous data category accounted for 64.1%, and the center value was 0.405. According to No.3 of the above evaluation criteria, the difference between the central values of the two categories is very small, and both are around 0.5, especially the central value of the anomalous data category is only 0.520, which is far away from 1. Therefore, the datasets used in this study can be regarded as having no obvious abnormal data, and all 428 sets of data will be used in the following; 85% of the dataset (364 data instances) were randomly selected as the training set, and 15% of the dataset (64 data instances) were used as the test set.

3.3. Pearson Correlation Coefficient Analysis

Data dimensions have a large impact on the efficiency and performance of ML model training. Therefore, parametric analysis of the data and consideration of dimensionality reduction is required before training ML models. Figure 4 shows the Pearson correlation coefficient analysis results for 5 parameters (f_u, D, f_c', h, and E_c). It is worth noting that the correlation coefficient between E_c and f_c' is as high as 0.798, while we find that E_c and f_c' are always put together in multinational design codes and represent the stiffness of concrete. Therefore, $K_c = E_c f_c'$ is used as the parameter to characterize the performance of concrete material to replace the original two parameters. The correlation coefficient between h and D is 0.560. Studies have shown that the length-to-diameter ratio of the stud has an impact on the shear strength [42], so this study considers the length-to-diameter ratio $\frac{h}{D}$ instead of the original two parameters h and D. In order to better reflect the influence of the parameters, the output representation becomes the stress form, namely $\sigma = \frac{P_{stud}}{A_s}$. Finally, the original seven features are reduced to three. For each specimen sample, the input feature parameters of ML models are f_u, K_c, and $\frac{h}{D}$, the output result is σ.

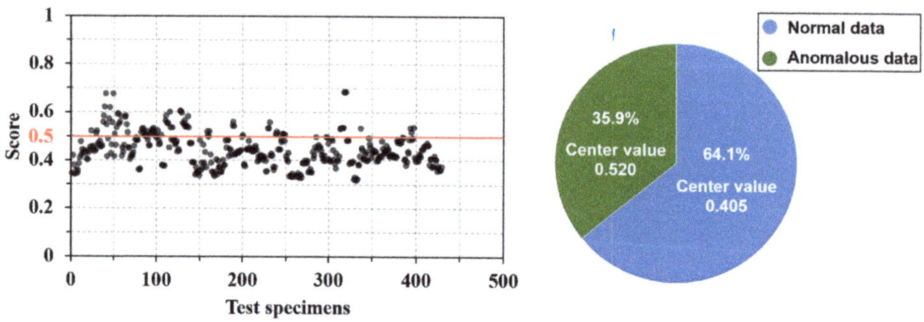

(a) (b)

Figure 3. Anomaly detection: (**a**) the score of datasets specimens and (**b**) K-means cluster analysis of score.

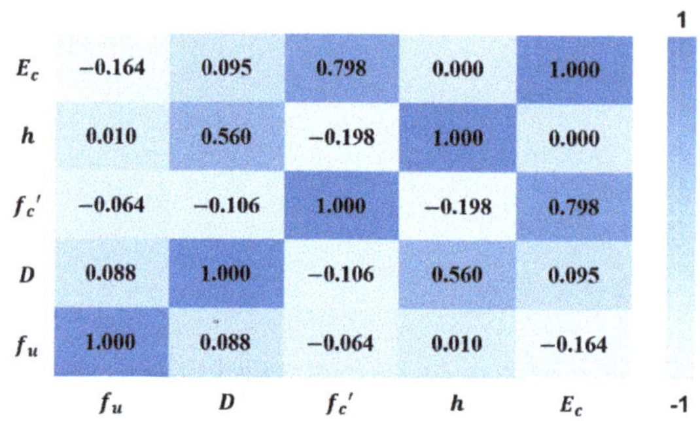

Figure 4. Heatmap of correlation coefficient.

3.4. Performance Metrics

Three performance metrics are used to analyze the performance of the prediction model, namely, root mean square error (RMSE), coefficient of determination (R^2), and mean absolute percent error (MAPE): RMSE is measured by is the deviation between the predicted value and the true value; R^2 is a statistical point of view that evaluates the goodness of fit of the model as a whole; MAPE is a relative error measure, which determines the scale as a percentage unit rather than a variable unit, and uses the absolute value to avoid positive and negative errors that cancel each other out. The mathematical expressions of the three indicators are as follows:

$$\text{RMSE} = \sqrt{\frac{\sum_{i=1}^{n}(\sigma_{stud\_Mi} - \sigma_{stud\_ti})^2}{n}} \tag{5}$$

$$\text{R}^2 = 1 - \frac{\sum_{i=1}^{n}(\sigma_{stud\_Mi} - \sigma_{stud\_ti})^2}{\sum_{i=1}^{n}\left(\sigma_{stud\_ti} - \bar{\sigma}_{stud\_t}\right)^2} \tag{6}$$

$$\text{MAPE} = \frac{\sum_{i=1}^{n} \left| \frac{\sigma_{stud\_Mi} - \sigma_{stud\_ti}}{\sigma_{stud\_ti}} \right|}{n} \times 100\% \quad (7)$$

where σ_{stud_Mi} and σ_{stud_ti} are the predicted and experimental values of the resistance of the studs, σ_{stud}, respectively; and n is the number of specimens considered.

4. ML Algorithms

In recent years, the ML algorithm has shown extremely strong prediction and fitting performance as a prediction method that can replace the traditional equation, and the application of ML in the field of civil engineering has shown considerable promise. This study selects five advanced and efficient ML algorithms: artificial neural network (ANN), support vector machines (SVM), decision tress (DT), random forest (RF), and gradient boosting decision tree (GBDT). These ML algorithms were chosen for this study because they are considered to be the most popular and widely used algorithms in structural engineering and have been shown to be reliable in predicting stud shear resistance ([38,77]). The five algorithms were elaborated on in reference [28]. The description of these algorithms is not duplicated in this study.

Hyperparameter optimization is a key step in developing ML models [78]. At present, the K-fold cross-validation method is mainly used to avoid overfitting in the training process. Additionally, the K-fold cross-validation method can be used together with grid search, random search, Bayesian optimization, and a manual search using the trial-and-error method to achieve hyperparameter tuning ([38,79–81]). K-fold cross-validation randomly splits the data into K groups of disjoint subsets of equal size, then traverses the K subsets in turn, each time using the current subset as the validation set and all the remaining samples as the training set for model training and evaluation; finally, the average value of K evaluation indicators is used as the final evaluation indicator, as shown in Figure 5. In this study, the 5-fold cross-validation was adopted to ensure that the resampled subset is large enough to represent the training dataset [38]. In the method of finding the best hyperparameter combination, the efficiency of grid search and random search is low. While Bayesian optimization can consider the previous parameter information, continuously update the prior, and has a small number of iterations and parameter adjustments. Therefore, 5-fold cross-validation combined with Bayesian optimization is selected for hyperparameter tuning in this study, as shown in Figure 6. ANN consists of an input layer (Input), a hidden layer, and an output layer (Output) [82], as shown in Figure 7a. It is found that the number of hidden layers of ANN is set to one layer, the number of hidden neurons is 60, and the activation function is the sigmoid function. Different from other ML algorithms, SVM regression is considered a nonparametric technique because it relies on kernel functions. SVM does not require the model output $f(x)$ to be exactly the same as the real output y when solving the regression problem. The loss is not calculated if the data are within the interval band, if and only if the absolute value of the gap between $f(x)$ and y is greater than ϵ. SVM optimizes the model by maximizing the width of the interval band and minimizing the total loss [83]; see Figure 7b. In this study, Radial Basis Function (RBF) is used as the kernel function of SVM, as considered that it can realize nonlinear mapping has fewer parameters which will affect the complexity of the model. In addition, this study adopts decision trees and two ensemble learning algorithms, namely RF and GBDT. DT is the base learner of RF and GBDT, which represents a mapping relationship between object attributes and object values. The DT model trained in this study has a depth of 14 layers, and the number of nodes is 14. On the basis of building a bagging ensemble with a decision tree as the base learner, RF further introduces random attribute selection in the training process of the decision tree and uses multiple decision tree models to obtain higher prediction accuracy than a single decision tree [84], see Figure 7c. The RF model consists of 10 decision trees and up to 20 layers. Unlike RF, GBDT optimizes by iterating over the residuals of the decision trees ([85,86]). In this study, we use 471 decision trees to build the GBDT model, the maximum depth is 5, and the learning rate is 0.3141. The gradient boosting decision tree algorithm was implemented in MATLAB, where more information can be found in the help menu of MATLAB [87].

Figure 5. K-fold cross-validation.

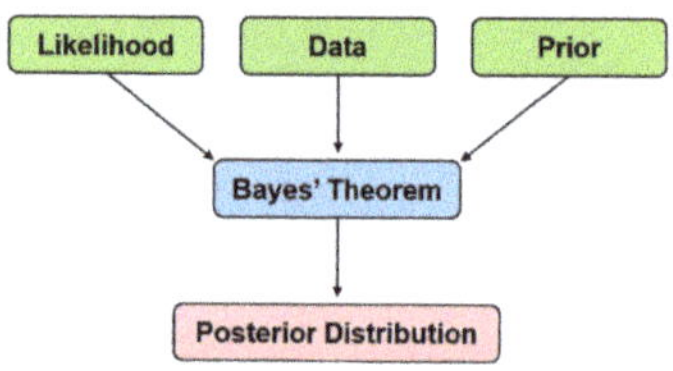

Figure 6. Bayesian optimization approach.

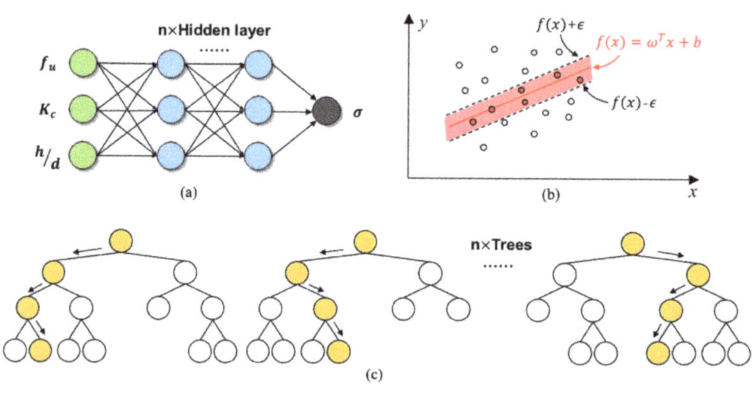

Figure 7. Structural diagram of ML models: (**a**) ANN, (**b**) SVM, and (**c**) random forest.

5. Results and Discussion

Figure 8 shows the prediction results of the collected 3 design codes and 5 ML models for 428 specimens. All points should lie on the diagonal dotted line when the predicted result is equal to the true value of the specimen sample. The vertical distance from the dotted line to a point is the prediction error of that point. The model in which the prediction points are scattered around the line indicates that the error is smaller and the prediction accuracy is higher. The black solid line represents the prediction trend of the model, and the smaller the angle between the solid line and the dotted line, the higher the prediction accuracy.

The RSME in Figure 9 is percentageized in order to visually compare the other three performance metrics on the same order of magnitude. The results show that AASHTO has the highest accuracy among the three design codes, but the error is still very large: RSME = 130.02, $R^2 = -0.45$, and MAPE = 0.21, followed by GB50017: RSME = 179.57, $R^2 = -1.76$, and MAPE = 0.30. The error of Eurocode-4:RSME = 203.61, $R^2 = -2.55$, and

MAPE = 0.35. Most of the predicted values of Eurocode-4 and GB50017 are generally less than the test values, which is manifested in that the point is below the dotted line. It is worth noting that the R^2 of the three design codes are all negative, which means that their prediction accuracy is not as good as taking the average directly [38]. The results show that the prediction accuracy of the existing codes is insufficient, and there is a certain error compared with the real test results of the specimens. In the ML models, the GBDT model has the highest accuracy:RSME = 22.66, R^2 = 0.96, and MAPE = 0.03, Figure 8g shows the prediction results are roughly distributed around the green line, and the error with the test results is effectively controlled, followed by the RF model:RSME = 31.31, R^2 = 0.92, and MAPE = 0.05, while their base learner DT has lower prediction accuracy: RSME = 41.81, R^2 = 0.85, and MAPE = 0.07. The prediction accuracy of SVM is the lowest among ML models: RSME = 44.03, R^2 = 0.83, and MAPE = 0.06, but the accuracy is still much higher than the design codes. Among ML models, ensemble models that include the RF model and GBDT model have higher prediction accuracy than other types of ML models. Compared with the design codes, ML models show better prediction performance, and the error is greatly reduced. Compared with AASHTO, which with the highest accuracy, the GBDT model has an 82.57% decrease in RSME, an increase in R^2 by 313.33%, and a decrease in MAPE by 85.71%. The comparison between ML models and the equation in AASHTO shows that the equation of the design codes is too conservative in giving accurate predictions, as most of the points lie below the diagonal dashed line. This may be because UHPC and HPC were not included in the data considered in the development of design codes. ML models can accurately predict the resistance of headed studs, and the dataset is of high quality and sufficient quantity to predict σ.

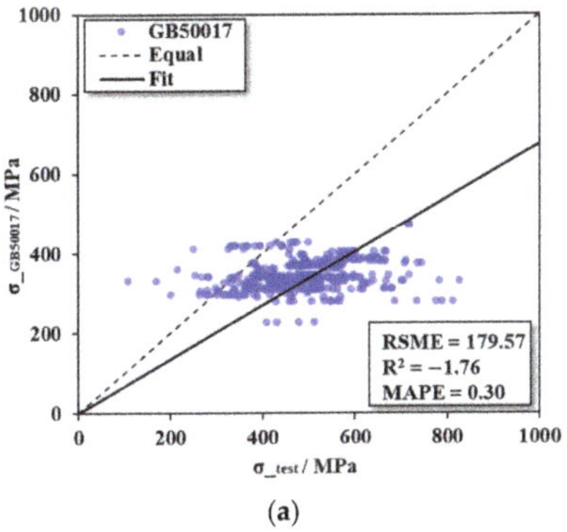

(a)

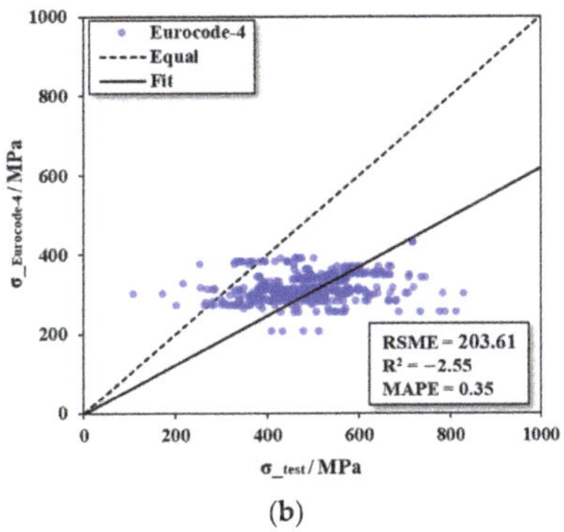

(b)

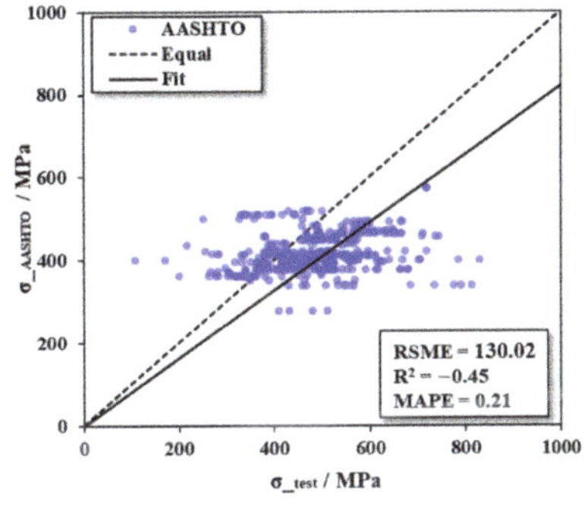

(c)

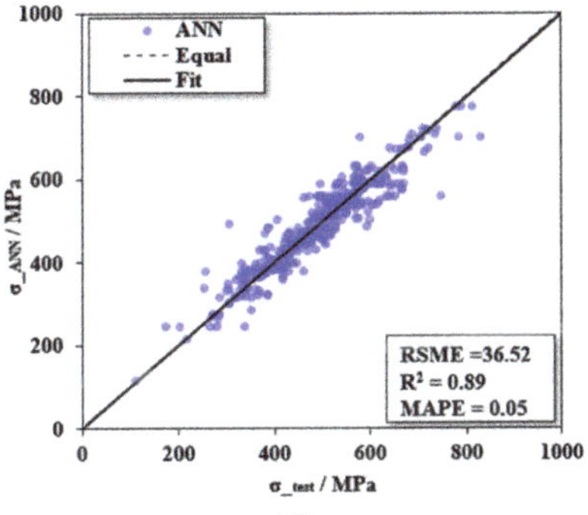

(d)

Figure 8. *Cont.*

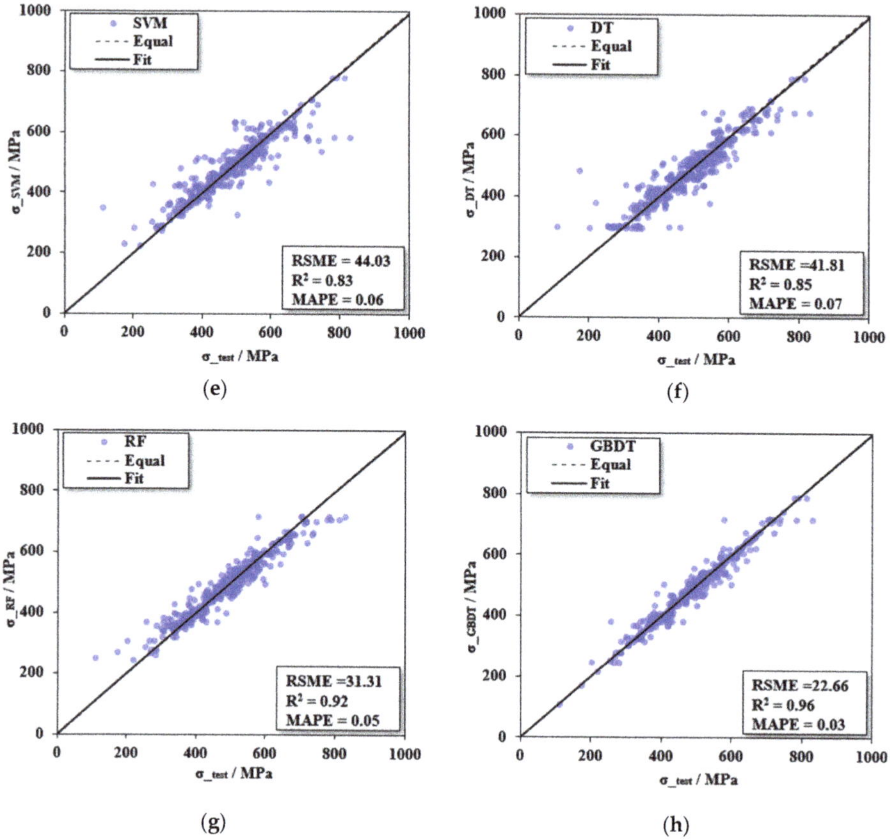

Figure 8. Comparison between experimental results and prediction: (**a**) GB50017, (**b**) Eurocode-4, (**c**) AASHTO, (**d**) ANN, (**e**) SVM, (**f**) DT, (**g**) RF, and (**h**) GBDT.

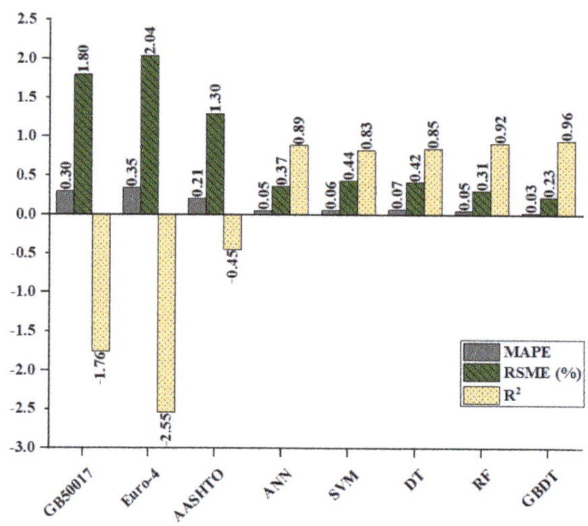

Figure 9. Comparison between experimental and predicted results.

6. Visualization and Interpretation of GBDT Model

The ML models can predict the resistance of headed studs with high accuracy, and their accuracy is much higher than the design codes of various countries. However, we cannot fully understand the decisions of ML models from human thinking patterns, which makes most ML models a kind"of "black"box". We can not know exactly when and why they will go wrong even though they show excellent predictive ability when dealing with large amounts of data, and it would be fatal if the error occurred in actual engineering. It is necessary to ensure that important decisions are made by humans rather than machines. The interpretability of the model is more important than the mere prediction of the results for decisions that ultimately need to be made by humans. Recently, shapley additive explanation (SHAP) [88], individual conditional expectation (ICE) plot, and partial dependence plot (PDP) [37] have shown great promise in interpreting the output of ML models and studying the importance of each factor and its interaction on the response variable, but SHAP may produce non-intuitive feature assignments. Therefore, in this study, individual conditional expectation (ICE) plots and partial dependence plots (PDP) are used to visualize and analyze the interpretability of the GBDT model, which has the highest accuracy.

PDP can show the marginal effect of one or two features on the prediction results of ML models and visualize the relationship between the prediction target and the feature [34], such as a linear relationship, monotonic relationship, or more complex relationship, but PDP is a visualization of the average relationship between the prediction results and features. It may obscure the influence of features that are only shown on some samples as a global analysis that does not consider individuals. ICE is a refined analysis based on the PDP, showing the dependence of the prediction results on the features of each sample, eliminating the influence of non-uniform effects, and visualizing the dependence of the prediction results on the features for each sample separately, each sample one line. In order to compare the curves of each sample and observe the cumulative effect of each feature parameter, the central ICE (C-ICE) is used to fix the prediction starting point of different samples at zero. The PDP and C-ICE are computed and plotted for the GBDT model with the highest prediction accuracy and visualize the relationship between the characteristic parameters f_u, K_c, $\frac{h}{D}$, and σ.

Figure 10 shows C-ICE plots of output σ for each of the three parameters f_u, K_c, and $\frac{h}{D}$. The dots in the figure represent each predicted value. Figure 10a show that the overall σ gradually increased with the increase in f_u, but there was a brief decrease around $f_u = 425$ MPa. This may be due to the small amount of f_u less than 425 MPa in the dataset, which leads to the larger prediction values of output σ. As shown in Figure 10b, The effect of K_c on σ is similar to that of f_u, and the trend is clearer: K_c is positively correlated with σ, and the effect is obvious when K_c is small. The growth of σ flattens and then starts to decline when K_c is over $5.36 \times 10^6 \text{MPa}^2$. From Figure 10c, The influence of σ by $\frac{h}{D}$ is more obvious when the $\frac{h}{D}$ is small, σ reaches a peak when $\frac{h}{D} = 10$ and no longer increases significantly after that, but due to a lack of follow-up data, the trend is downward. The existence of a cap on the influence of $\frac{h}{D}$ may be because the increase in the length-to-diameter ratio of the stud will increase the stress-bearing area of the concrete, the force generated by the stud will also increase accordingly, and the degree of reduction of the stud pull-shear coupling will also increase. As the $\frac{h}{D}$ of the stud increases, the stud can be better anchored in the concrete, thereby reducing the separation phenomenon of the steel–concrete composite interface and making the stud closer to the pure shear state, which can give full play to the material properties of the stud and improve the stud resistance ([42,47,48]). However, since the shear strength of the stud will not exceed the shear strength of the material of the stud itself, there is an upper limit on the influence of the stud height on the shear strength of the stud. This is consistent, as shown in Figure 10c.

Figure 11 shows 3-D and 2-D PDP plots for three parameters, and the results show the interactions between every two features. It can be seen that, on the whole, σ is positively correlated with the three parameters. Figure 11a indicates the dependence of σ on the tensile strength of the stud and the length-to-diameter ratio of the stud, which represents the effect of the key material and geometric variable of the stud, respectively. For a $\frac{h}{D}$ of less than 10, σ increases and reaches its peak as f_u increases, whereas for a $\frac{h}{D}$ of greater than 10, the tensile strength of stud f_u between approximately 411 and 511 MPa has a negative correlation with σ. It can be seen from Figure 11b that K_c and $\frac{h}{D}$ have apparent positive correlations with the resistance, but there is a clear upper limit on the effect of both.The stud resistance is relatively insensitive to the changes for K_c of greater than 5.36×10^6 MPa^2 and $\frac{h}{D}$ of greater than 10. As can be seen from Figure 11c, the stud resistance σ reaches its peak when f_u is 610 MPa and K_c is 4.55×10^6MPa^2, and f_u from around 400 to 425 MPa is negatively correlated with σ; however, as can be seen from Figure 11a, most of the data lie in the range 425 MPa $\leq f_u \leq$ 610 MPa. In the Eurocode-4 [9], $\frac{h}{D}$ only affects the results when $3 \leq \frac{h}{D} \leq 4$. However, the above results show that the stud aspect ratio $\frac{h}{D}$ could affect the resistance until 10, which may have implications in the development of future design models.

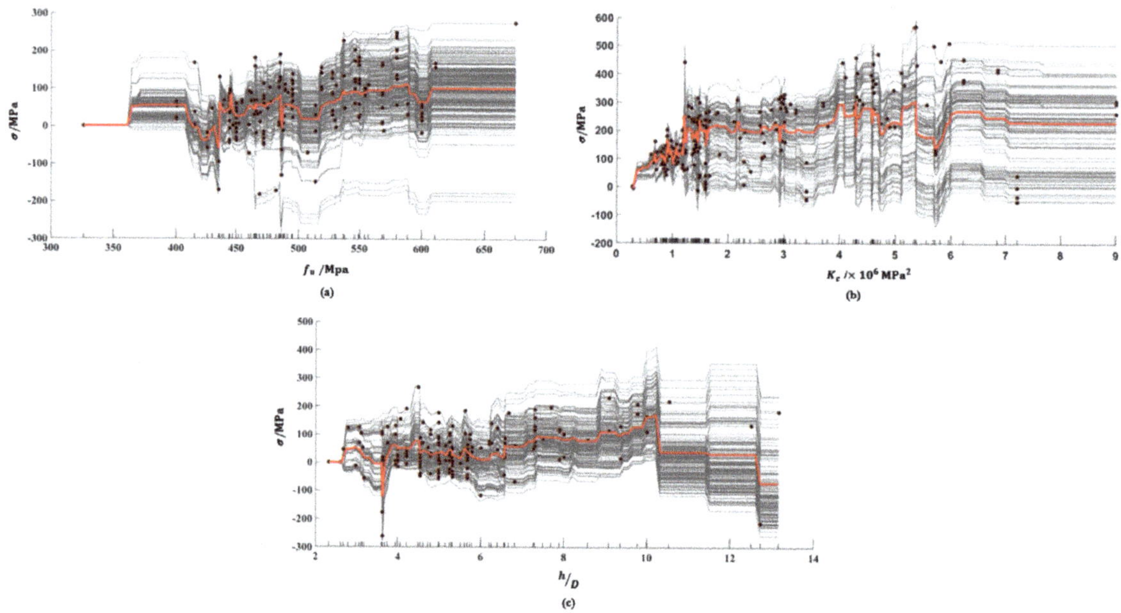

Figure 10. C-ICE for three parameters: (**a**) f_u, (**b**) K_c, and (**c**) $\frac{h}{D}$.

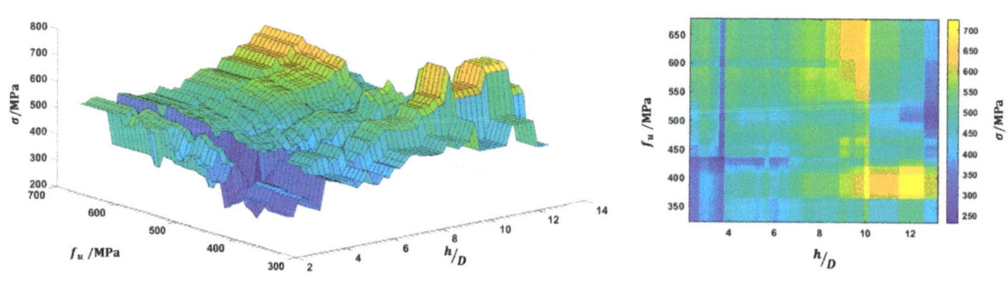

(**a**)

Figure 11. *Cont.*

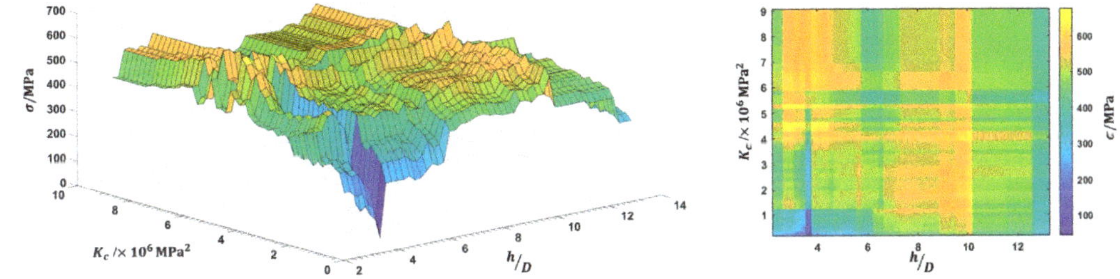

(b)

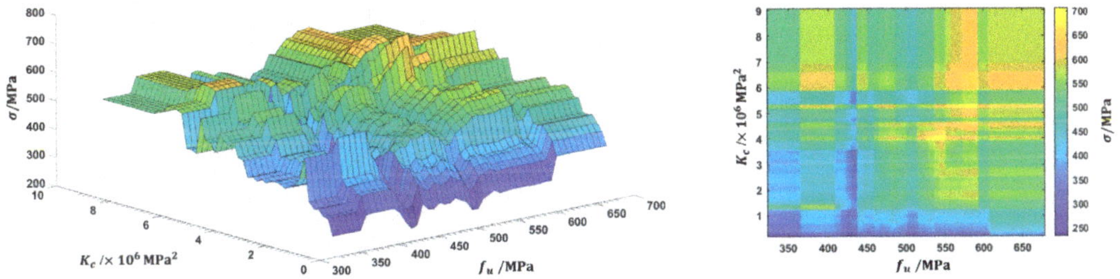

(c)

Figure 11. Three-dimensional and two-dimensional PDP plots for (**a**) f_u and $\frac{h}{D}$, (**b**) K_c and $\frac{h}{D}$, and (**c**) K_c and f_u.

7. Conclusions

In this study, a total of 428 push-out test specimens were collected from 37 published pieces of literature. Five machine learning (ML) models are derived to predict the shear resistance of headed studs in steel–concrete structures. The following conclusions are drawn:

- The R^2 values of the equations from design codes of various countries were all negative when predicting datasets, including HPC and UHPC, which means these equations cannot be used for headed studs in HPC and UHPC.
- The prediction accuracy of ML models was much higher than the international design codes. The gradient boosting decision tree (GBDT) model had the overall highest accuracy and was compared with AASHTO, which had the highest accuracy among the three design codes. The values of RSME and MAPE of the GBDT model were around 80% lower than that of the AASHTO equation.
- The visualization and interpretability analysis of the GBDT model showed that the length-to-diameter ratio of the stud had a substantial influence on the shear resistance of headed studs, which may be related to the effect of the length on the pull-out effect of the stud.
- The length-to-diameter ratio of the stud was suggested to be taken into account in the equations of future design codes, and there may be an upper limit on the positive effect of material properties on the shear resistance of headed studs, which requires future supplementation of high-strength material tests to determine.

Based on the research in this study, future research directions in this field are suggested to achieve better accuracy and to reveal more physical meaning behind parameters: it is expected that with the addition of new experimental data in the future, the prediction accuracy of ML models could be further improved. ML combined with data-driven analysis can provide a complementary research method to traditional experiments. This study investigated the prediction accuracy of limited to five ML models; there are many other ML algorithms that can be used to predict the stud shear resistance, including the simplest and most advanced ML algorithms, which are worth exploring in future research. Researchers can improve the design codes through the trained ML model. The dataset provided in this study, including UHPC and HPC, may be helpful in future studies to understand the influence of concrete types on the shear connection of headed studs.

Author Contributions: Conceptualization, F.Z. and X.Z.; methodology, F.Z. and X.Z.; validation, F.Z., C.W., and X.Z.; formal analysis, F.Z.; investigation, L.W. and X.Z.; resources, X.Z.; data curation, F.Z., C.W., and X.Z.; writing—original draft preparation, F.Z. and X.Z.; writing—review and editing, Q.W., D.C., Y.W., and X.Z.; visualization, F.Z.; supervision, L.W. and X.Z.; project administration, F.Z. and X.Z. All authors have read and agreed to the published version of the manuscript.

Funding: This study was funded by the Basic Research Program from the Natural Science Foundation of Jiangsu Province, China (No. BK20210620); and Natural Science Research of Jiangsu Higher Education Institutions of China (No. 21KJB560007).

Data Availability Statement: Some or all data, models, or codes generated or used during this study are available from the corresponding author by request.

Conflicts of Interest: The authors declare no conflict of interest.

References

1. Nie, J.; Cai, C.S. Steel-concrete composite beams considering shear slip effects. *J. Struct. Eng.* **2003**, *129*, 495–506. [CrossRef]
2. Nie, J.; Wang, J.; Gou, S.; Zhu, Y.; Fan, J. Technological development and engineering applications of novel steel-concrete composite structures. *Front. Struct. Civ. Eng.* **2019**, *13*, 1–14. [CrossRef]
3. Ashraf, M.; Hasan, M.J.; Al-Deen, S. Semi-rigid behaviour of stainless steel beam-to-column bolted connections. *Sustain. Struct.* **2021**, *1*, 000002. [CrossRef]
4. Qi, J.; Hu, Y.; Wang, J.; Li, W. Behavior and strength of headed stud shear connectors in ultra-high performance concrete of composite bridges. *Front. Struct. Civ. Eng.* **2019**, *13*, 1138–1149. [CrossRef]
5. Johnson, R.P. *Composite Structures of Steel and Concrete: Beams, Slabs, Columns and Frames for Buildings*; John Wiley Sons: New York, NY, USA, 2018; 288p.
6. Viest, I.M. Investigation of stud shear connectors for composite concrete and steel T.-beams. *ACI J.* **1956**, *27*, 875–981.
7. Ollgaard, J.G.; Slutter, R.G.; Fisher, J.W. Shear strength of stud connectors in lightweight and normal weight concrete. *AISC Eng. J.* **1971**, *71-10*, 55-34.
8. Civjan, S.A.; Singh, P. Behavior of shear studs subjected to fully reversed cyclic loading. *J. Struct. Eng.* **2003**, *129*, 1466–1474. [CrossRef]
9. *Eurocode-4*; Design of Composite Steel and Concrete Structures. European Committee for Standardization: Brussels, Belgium, 2004.
10. AASHTO. *AASHTO LRFD Bridge Design Codes*; American Association of State Highway and Transportation Officials: Washington, DC, USA, 2012; p. 20001.
11. *GB50017–2017*; Code for Design of Steel Structures. Ministry of Housing and Urban-Rural Development of China: Beijing, China, 2017. (In Chinese)
12. Model Code. *Joint Committee IASBSE/CEB/FIP/ECCS Composite Structures (Model Code)*; Construction Press: London, UK, 1981.
13. Lee, P.G.; Shim, C.S.; Chang, S.P. Static and fatigue behavior of large stud shear connectors for steel–concrete composite bridges. *J. Construct. Steel Res.* **2005**, *61*, 1270–1285. [CrossRef]
14. Hicks, S.J. Design shear resistance of headed studs embedded in solid slabs and encasements. *J. Construct. Steel Res.* **2017**, *139*, 339–352. [CrossRef]
15. Oehlers, D.J.; Foley, L. The fatigue strength of stud shear connections in composite beams. *Proc. Inst. Civ. Eng.* **1985**, *79*, 349–364. [CrossRef]
16. Lin, Z.F.; Liu, Y.Q. Experimental study on shear behavior of large stud connectors. *J. Tongji* **2015**, *43*, 1788–1793. (In Chinese)
17. Duan, M.; Zou, X.; Bao, Y.; Li, G.; Chen, Y.; Li, Z. Experimental investigation of headed studs in steel-ultra-high performance concrete (UHPC) composite sections. *Eng. Struct.* **2022**, *270*, 114875. [CrossRef]
18. Soleimani-Babakamali, M.H.; Esteghamati, M.Z. Estimating seismic demand models of a building inventory from nonlinear static analysis using deep learning methods. *Eng. Struct.* **2022**, *266*, 114576. [CrossRef]
19. Esteghamati, M.Z.; Flint, M.M. Developing data-driven surrogate models for holistic performance-based assessment of mid-rise RC frame buildings at early design. *Eng. Struct.* **2021**, *245*, 112971. [CrossRef]
20. Al-Bashiti, M.K.; Naser, M.Z. Machine learning for wildfire classification: Exploring blackbox, eXplainable, symbolic, and SMOTE methods. *Nat. Hazard Res.* **2022**, *2*, 154–165. [CrossRef]
21. Avci-Karatas, C. Application of machine learning in prediction of shear capacity of headed steel studs in steel-concrete composite structures. *Int. J. Steel Struct.* **2022**, *22*, 539–556. [CrossRef]

22. Cao, Y.; Wakil, K.; Alyousef, R.; Jermsittiparsert, K.; Ho, L.; Alabduljabbar, H.; Alaskar, A.; Alrshoudi, F.; Mohamed, A.M. Application of extreme learning machine in behavior of beam to column connections. *Structures* **2020**, *25*, 861–867. [CrossRef]
23. Gholampour, A.; Mansouri, I.; Kisi, O.; Ozbakkaloglu, T. Evaluation of mechanical properties of concretes containing coarse recycled concrete aggregates using multivariate adaptive regression splines (MARS), M5 model tree (M5Tree), and least squares support vector regression (LSSVR) models. *Neural Comput. Appl.* **2020**, *32*, 295–308. [CrossRef]
24. Huang, C.; Huang, S. Predicting capacity model and seismic fragility estimation for RC bridge based on artificial neural network. *Structures* **2020**, *27*, 1930–1939. [CrossRef]
25. Wang, X.; Liu, H.; Liu, Y. Auto-tuning deep forest for shear stiffness prediction of headed stud connectors. *Structures* **2022**, *43*, 1463–1477. [CrossRef]
26. Mahjoubi, S.; Meng, W.; Bao, Y. Logic-guided neural network for predicting steel-concrete interfacial behaviors. *Exp. Syst. Appl.* **2022**, *198*, 116820. [CrossRef]
27. Zhou, Y.; Zheng, S.; Huang, Z.; Sui, L.; Chen, Y. Explicit neural network model for predicting FRP-concrete interfacial bond strength based on a large dataset. *Compos. Struct.* **2020**, *240*, 111998. [CrossRef]
28. Zhang, F.; Wang, C.; Liu, J.; Zou, X.; Sneed, L.H.; Bao, Y.; Wang, L. Prediction of FRP-concrete interfacial bond strength based on machine learning. *Eng. Struct.* **2023**, *274*, 115156. [CrossRef]
29. Lee, S.; Lee, C. Prediction of shear strength of FRP-reinforced concrete flexural members without stirrups using artificial neural networks. *Eng. Struct.* **2014**, *61*, 99–112. [CrossRef]
30. Vinuesa, R.; Brunton, S.L. Enhancing computational fluid dynamics with machine learning. *Nat. Comput. Sci.* **2022**, *2*, 358–366. [CrossRef]
31. Tran, V.Q.; Dang, V.Q.; Ho, L.S. Evaluating compressive strength of concrete made with recycled concrete aggregates using machine learning approach. *Construct. Build. Mater.* **2022**, *323*, 126578. [CrossRef]
32. Yuan, X.; Tian, Y.; Ahmad, W.; Ahmad, A.; Usanova, K.I.; Mohamed, A.M.; Khallaf, R. Machine Learning Prediction Models to Evaluate the Strength of Recycled Aggregate Concrete. *Materials* **2022**, *15*, 2823. [CrossRef] [PubMed]
33. Asteris, P.G.; Skentou, A.D.; Bardhan, A.; Samui, P.; Pilakoutas, K. Predicting concrete compressive strength using hybrid ensembling of surrogate machine learning models. *Cem. Concr. Res.* **2021**, *145*, 106449. [CrossRef]
34. Cakiroglu, C.; Islam, K.; Bekdaş, G.; Isikdag, U.; Mangalathu, S. Explainable machine learning models for predicting the axial compression capacity of concrete filled steel tubular columns. *Construct. Build. Mater.* **2022**, *356*, 129227. [CrossRef]
35. Mangalathu, S.; Jeon, J.S. Machine learning–based failure mode recognition of circular reinforced concrete bridge columns: Comparative study. *J. Struct. Eng.* **2019**, *145*, 04019104. [CrossRef]
36. Feng, D.C.; Liu, Z.T.; Wang, X.D.; Jiang, Z.M.; Liang, S.X. Failure mode classification and bearing capacity prediction for reinforced concrete columns based on ensemble machine learning algorithm. *Adv. Eng. Inf.* **2020**, *45*, 101126.
37. Goldstein, A.; Kapelner, A.; Bleich, J.; Pitkin, E. Peeking inside the black box: Visualizing statistical learning with plots of individual conditional expectation. *J. Comput. Graph. Stat.* **2015**, *24*, 44–65. [CrossRef]
38. Setvati, M.R.; Hicks, S.J. Machine learning models for predicting resistance of headed studs embedded in concrete. *Eng. Struct.* **2022**, *254*, 113803. [CrossRef]
39. Hu, Y.; Qiu, M.; Chen, L.; Zhong, R.; Wang, J. Experimental and analytical study of the shear strength and stiffness of studs embedded in high strength concrete. *Eng. Struct.* **2021**, *236*, 111792. [CrossRef]
40. Nie, J.; Sheng, J.; Yuan, Y.; Lin, W.; Wang, W. Study on actual bearing capacity of shear connectors in steel-concrete composite beams. *J. Build. Struct.* **1996**, *17*, 21–29. (In Chinese)
41. Shim, C.S.; Lee, P.G.; Yoon, T.Y. Static behavior of large stud shear connectors. *Eng. Struct.* **2004**, *26*, 1853–1860. [CrossRef]
42. Wang, W.H. *Experimental and Analytical Study on Shear Properties of Headed Stud Connector*; Zhejiang University: Hangzhou, China, 2018. (In Chinese)
43. Wang, J.; Qi, J.; Tong, T.; Xu, Q.; Xiu, H. Static behavior of large stud shear connectors in steel-UHPC composite structures. *Eng. Struct.* **2019**, *178*, 534–542. [CrossRef]
44. Han, Q.; Wang, Y.; Xu, J.; Xing, Y. Static behavior of stud shear connectors in elastic concrete–steel composite beams. *J. Construct. Steel Res.* **2015**, *113*, 115–126.
45. Luo, Y.; Hoki, K.; Hayashi, K.; Nakashima, M. Behavior and strength of headed stud–SFRCC shear connection I: Experimental study. *J. Struct. Eng.* **2016**, *142*, 04015112. [CrossRef]
46. Chen, Z. *Research on Mechanical Properties and Bearing Capacity Analysis of Shear Connectors in Steel-UHPC Composite Structures*; Changan University: Xian, China, 2021. (In Chinese)
47. Kim, J.S.; Kwark, J.; Joh, C.; Yoo, S.W.; Lee, K.C. Headed stud shear connector for thin ultrahigh-performance concrete bridge deck. *J. Construct. Steel Res.* **2015**, *108*, 23–30.
48. Kim, J.S.; Park, S.H.; Joh, C.B.; Kwark, J.D.; Choi, E.S. Push-out test on shear connectors embedded in UHPC. *Appl. Mech. Mater.* **2013**, *351*, 50–54. [CrossRef]
49. Luo, Y.Z. *Research on Bolted Shear Connections of Steel-Concrete Composite Beams*; Central South University: Changsha, China, 2008. (In Chinese)
50. Zeng, D.; Liu, Y.; Cao, L. Shear performance of innovative shear connectors in steel-UHPC composite structure. *J. Zhejiang Univ.* **2021**, *55*, 1714–1724+1771. (In Chinese)
51. Lam, D.; El-Lobody, E. Behavior of headed stud shear connectors in composite beam. *J. Struct. Eng.* **2005**, *131*, 96–107. [CrossRef]

52. Zhou, X.D. *Experimental Study on Mechanical Properties of Large Diameter Shear Stud Connecters in Steel-UHPC Composite Structure*; Nanjing Forestry University: Nanjing, China, 2018. (In Chinese)
53. Wei, Z. *Push-Out Tests on Stud Shear Connector of Prefabricated Steel-Concrete Composite Beams*; Zhejiang University: Hangzhou, China, 2019. (In Chinese)
54. Chen, L. *Experimental Study of Static and Fatigue Properties of Interface Connection of Steel-Concrete Composite Beam Bridges*; Southeast University: Nanjing, China, 2014. (In Chinese)
55. Wang, W.F.; Chen, Z.J.; Zheng, X.H.; Xiong, Y. Experimental research on shear bearing capacity of Steel-RPC composite beam shear studs. *Guangdong Archit. Civ. Eng.* **2018**, *25*, 4. (In Chinese)
56. Wang, Y. *Experimental and Theoretical Research on Externally Prestressed Steel-Concrete Composite Beams*; Tongji University: Shanghai, China, 2004. (In Chinese)
57. Cao, J.; Shao, X.; Deng, L.; Gan, Y. Static and fatigue behavior of short-headed studs embedded in a thin ultrahigh-performance concrete layer. *J. Bridge Eng.* **2017**, *22*, 04017005. [CrossRef]
58. An, L.; Cederwall, K. Push-out tests on studs in high strength and normal strength concrete. *J. Construct. Steel Res.* **1996**, *36*, 15–29. [CrossRef]
59. Yamamoto, M.; Nakamura, S. *The Study on Shear Connectors*; The Public Works Research Institute, Construction Ministry Japan: Tokyo, Japan, 1962; Volume 5.
60. Mainstone, R.J.; Menzies, J.B. Shear connectors in steel-concrete composite beams for bridges. *Concrete* **1967**, *1*, 291–302.
61. Menzies, J.B. CP 117 and shear connectors in steel-concrete composite beams made with normal-density or lightweight concrete. *Struct. Eng.* **1971**, *49*, 137–154.
62. Oehlers, D.J. *Results on 101 Push-Specimens and Composite Beams*; Research Report CE 8; Department of Civil Engineering, University of Warwick: Coventry, UK, 1981.
63. Hiragi, H.; Miyoshi, E.; Kurita, A.; Ugai, M.; Akao, S. Static strength of Stud shear connectors in SRC Structures. *Trans. Jpn. Concr. Inst.* **1981**, *3*, 453–460.
64. Roik, K.; Bürkner, K.E. Beitrag zur Tragfähigkeit von Kopfbolzendübeln in Verbundträgern mit Stahlprofilblechen. *Bauingenieur* **1981**, *56*, 97–101. (In German)
65. Hicks, S.J. *Longitudinal Shear Resistance of Steel and Concrete Composite Beams*; University of Cambridge: Cambridge, UK, 1997.
66. Easterling, W.S.; Murray, T.M.; Rambo-Roddenberry, M. *Behaviour and Strength of Welded Stud Shear Connectors Data Report*; Civil and Environmental Engineering, Virginia Polytechnic Institute and State University: Blacksburg, VA, USA, 2002.
67. Feldmann, M.; Hechler, O.; Hegger, J.; Rauscher, S. Neue Untersuchungen zum Ermüdungsverhalten von Verbundträgern aus hochfesten Werkstoffen mit Kopfbolzendübeln und Puzzleleiste. *Stahlbau* **2007**, *76*, 826–844. (In German) [CrossRef]
68. Wang, Q.; Liu, Y.; Luo, J.; Lebet, J.P. Experimental study on stud shear connectors with large diameter and high strength. In Proceedings of the 2011 International Conference on Electric Technology and Civil Engineering, Lushan, China, 22–24 April 2011; IEEE: Piscataway, NJ, USA, 2011; pp. 340–343.
69. Hanswille, G.; Jost, K.; Schmitt, C.; Trillmich, R. Experimentelle Untersuchungen zur Tragfähigkeit von Kopfbolzendübeln mit großen Schaftdurchmessern. *Stahlbau* **1998**, *67*, 555–560. (In German)
70. Bullo, S.; Di Marco, R. Effects of high-performance concrete on stud shear connector behaviour. In Proceedings of the Nordic Steel Construction Conference, Malmö, Sweden, 19–21 June 1995; pp. 577–584.
71. Döinghaus, P. *Zum Zusammenwirken Hochfester Baustoffe in Verbundtragern*; Technische Hochschule: Lübeck, Germany, 2002. (In German)
72. Xue, D.; Liu, Y.; Yu, Z.; He, J. Static behavior of multi-stud shear connectors for steel-concrete composite bridge. *J. Construct. Steel Res.* **2012**, *74*, 1–7. [CrossRef]
73. Jähring, A. *Zum Tragverhalten von Kopfbolzendübeln in Hochfestem Beton*; Technische Universität München: Munich, Germany, 2008. (In German)
74. Hanswille, G.; Porsch, M.; Üstündag, C. Versuchsbericht über die Durchführung von 77 Push-Out-Versuchen. In *Forschungsprojekt: Modellierung von Schädigungsmechanismen zur Beurteilung der Lebensdauer von Verbundkonstruktionen aus Stahl und Beton*; Institut für Konstruktiven Ingenieurbau: Berlin, Germany, 2006; p. 7. (In German)
75. *GB 50010-2010*; Code for Design of Concrete Structures. Building Industry Press: Beijing, China, 2010. (in Chinese)
76. Liu, F.T.; Ting, K.M.; Zhou, Z.H. Isolation forest. In Proceedings of the 2008 Eighth IEEE International Conference on Data Mining, Northwest Washington, DC, USA, 15–19 December 2008; pp. 413–422.
77. Degtyarev, V.; Hicks, S. Reliability-based design shear resistance of headed studs in solid slabs predicted by machine learning models. *Archit. Struct. Construct.* **2022**. [CrossRef]
78. Kutty, A.A.; Wakjira, T.G.; Kucukvar, M.; Abdella, G.M.; Onat, N.C. Urban resilience and livability performance of European smart cities: A novel machine learning approach. *J. Clean. Prod.* **2022**, *378*, 134203. [CrossRef]
79. Abdella, G.M.; Kucukvar, M.; Kutty, A.A.; Abdelsalam, A.G.; Sen, B.; Bulak, M.E.; Onat, N.C. A novel approach for developing composite eco-efficiency indicators: The case for US food consumption. *J. Clean. Prod.* **2021**, *299*, 126931. [CrossRef]
80. Abdella, G.M.; Shaaban, K. Modeling the impact of weather conditions on pedestrian injury counts using LASSO-based poisson model. *Arab. J. Sci. Eng.* **2021**, *46*, 4719–4730. [CrossRef]
81. Wakjira, T.G.; Alam, M.S.; Ebead, U. Plastic hinge length of rectangular RC columns using ensemble machine learning model. *Eng. Struct.* **2021**, *244*, 112808.

82. Zhou, Z.H. *Machine Learning*; Tsinghua University Press: Beijing, China, 2016; pp. 138–139.
83. Vapnik, V. *The Nature of Statistical Learning Theory*; Springer Science and Business Media: Berlin, Germany, 1999; 188p.
84. Ho, T.K. The random subspace method for constructing decision forests. *IEEE Trans. Pattern Anal. Mach. Intell.* **1998**, *20*, 832–844.
85. Friedman, J.H. Greedy function approximation: A gradient boosting machine. *Ann. Stat.* **2001**, *29*, 1189–1232. [CrossRef]
86. Breiman, L. Bagging predictors. *Mach. Learn.* **1996**, *24*, 123–140. [CrossRef]
87. MATLAB Statistics and Machine Learning Toolbox. MathWorks. 2022. Available online: https://www.mathworks.com/products/statistics.html (accessed on 28 November 2022).
88. Lundberg, S.M.; Lee, S.I. A unified approach to interpreting model predictions. *Adv. Neural Inf. Proc. Syst.* **2017**, *30*, 4765–4774.

buildings

MDPI

Article

Cyclic Behavior of Multiple Hardening Precast Concrete Shear Walls

Hongbo Jiang [1,*], Jian Sun [2], Hongxing Qiu [2], Dafu Cao [1], Wenjie Ge [1], Qiang Fang [2], Hengwei Cui [1] and Kongyang Chen [3]

1 College of Civil Science and Engineering, Yangzhou University, Yangzhou 225127, China
2 Key Laboratory of Concrete and Prestressed Concrete Structures of Ministry of Education, Southeast University, Nanjing 211189, China
3 Suzhou Electric Power Design Institute Co., Ltd., Suzhou 215011, China
* Correspondence: jhb@yzu.edu.cn

Abstract: Precast Concrete (PC) shear walls are becoming popular in building structures. With "wet" connection techniques, PC shear walls often behave like conventional cast-in-place walls, where hardening occurs after yielding. In this study, two PC shear walls assembled by the "dry" connection technique, and one cast-in-place shear wall, were tested by means of quasi-static cyclic loading. The main purpose of the experiment was to systematically investigate the cyclic response of PC shear walls with varying types of vertical connection in the form of a friction-bearing device. The results showed that vertical bearing in devices, which mainly stems from the longitudinal elongation of PC wall panels, could enlarge the axial force of end column so that it provided an additional resistance moment. The PC shear wall with weak connection achieved ductile failure and second ascending branches on load-displacement relationship, i.e., secondary hardening, and the wall with strong vertical connection performed great moment capacity as well as tertiary hardening. Compared to cast-in-place walls, the peak load and cumulative hysteretic energy of PC shear walls increased by about 60% and 100%, respectively. A conceptual analysis of the multiple hardening phenomenon is presented based on experimental results.

Keywords: precast concrete; shear wall; longitudinal elongation; vertical connection; multiple hardening; conceptual analysis

1. Introduction

In the past 20 years, developing countries such as China and Chile have made great progress. People flock to cities to live, and new constructions of middle- and high-rise building are constantly emerging. Reinforced Concrete (RC) shear walls, which provide strong, stiff, and cost effective lateral-load resisting members to resist earthquake load and wind load, have become the first choice in practical engineering [1]. In recent years, due to the rise in labor costs, construction quality requirements, and carbon emission problems, Precast Concrete (PC) construction has become a popular alternative to cast-in-place construction. The PC shear wall has attracted a lot of research in recent years, see Figure 1. With the development of building industrialization, PC construction can also be adopted in seismic regions [2]. Once proper connection is incorporated, one large shear wall can be disassembled into several well-sized PC wall members and then reassembled on site, so as to facilitate economical manufacture, transportation, and construction.

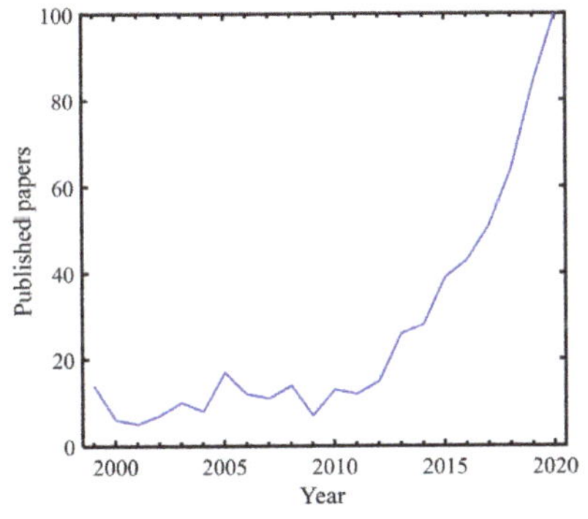

Figure 1. Numbers of published papers searched by "PC shear wall" in Web of Science database [3].

One issue about PC walls appears when connecting precast panels. Two kinds of connection are crucial to walls, namely horizontal connection and vertical connection, as shown in Figure 2. Most assembled construction techniques are based on "wet" methods [4–6], i.e., placing fresh concrete or cementitious grout after connecting steel bars with mechanical couplers/lap splices/welding [7,8]. As a consequence of labor-intensive construction methods, the "wet" method has lagged behind the efficiency demands of building industrialization. To avoid these drawbacks, some researchers combined the connection technique with the "dry" method, such as steel shear key [9], damper [10,11], buckling restrained plates [12], and damping devices at the toes of the wall [13]. However, site-welding operations are required for the assembly of wall panels with these convenient equipment, which bring other potentially dangerous and harmful factors, pointing to the bolt-based assembly method as a more attractive alternative. Another issue is design philosophy. Along with cast-in-place work in joints, a simplified process is to design the PC members as a cast-in-place one. Hence, the characteristics of cast-in-place members (such as accumulated damage, low ductility, and low resilience) are also retained [14,15]. The lateral load–top displacement relationship is often used to assess the seismic behavior of a shear wall, see Figure 3. Generally, most slender shear walls experience the elasticity, hardening, and softening phase. In particular, after flexural yielding, different walls may fail in different ways, including fracture/buckle of longitudinal reinforcing bars, crushing of the boundary element zone, and concrete crushing in the web [16]. With proper horizontal and vertical connection, PC technology provides a chance to delay the degeneration process and improve the ductility.

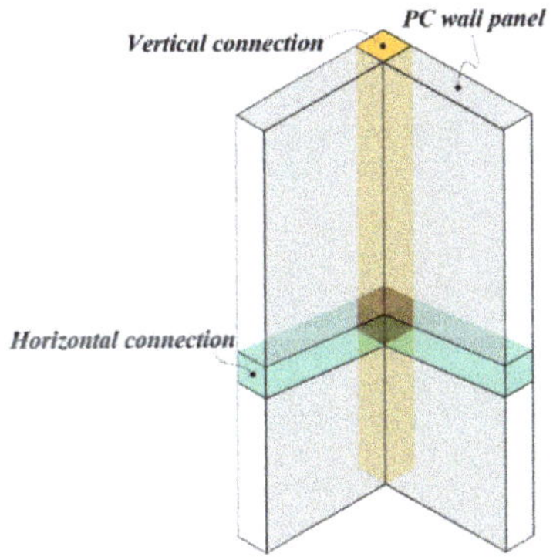

Figure 2. Horizontal and vertical connection in PC wall system.

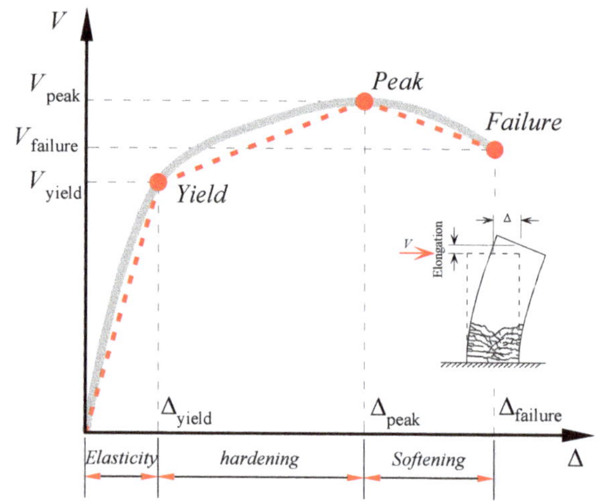

Figure 3. Lateral load–top displacement relationship (grey solid line) and classic backbone (red circle markers and dashed line) of cast-in-place shear wall.

As a common damage phenomenon depicted in Figure 3, the axial elongation is generally considered to be caused by accumulated damage in the plastic hinge zone [17–20]. Axial elongation, elongation of tensile side, and elongation of compressive side are collectively called "longitudinal elongation" here. In ductile RC shear walls, the longitudinal elongation grows quickly after the plastic hinge is formed at the bottom of the wall [21,22], which provides insight for a new connecting philosophy and innovative working mechanism for PC walls. Limited existing research could be retrieved on the elongation of RC shear walls [18,23–25], not to mention the effective utilization of this kind of deformation. Recently, a bolt-based PC shear wall was developed and experimentally validated to take into account the wall elongation [22]. Figure 4 depicts the schematics of the proposed PC shear wall system, in which steel square columns representing the end column and friction-bearing devices with limited slippage are placed. As expected, the longitudinal elongation could cause the slipping of friction-bearing devices during cyclic loading. The advantage of the new system is great moment capacity and ductility, however, in the early study [22], the walls failed with concrete spalling around the devices at the end of test.

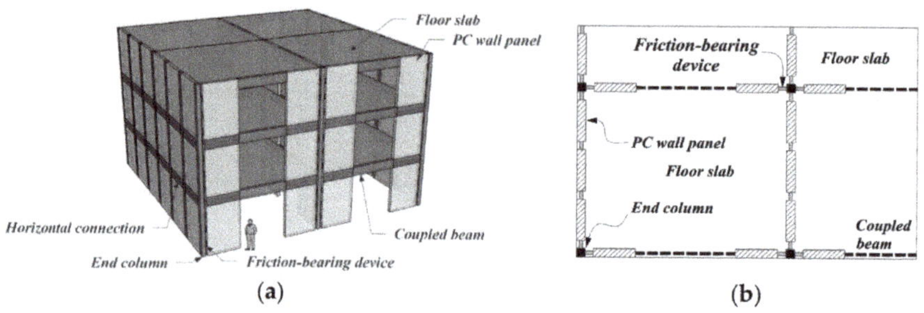

Figure 4. Proposed PC shear wall system [22]. (**a**) 3D view; (**b**) Plan view.

To better understand the behaviour of the new-type PC shear wall, this paper describes an experimental investigation and discussion of PC shear walls. Different from the previous experimental study [22], a rigid column base and various vertical connections were adopted, see Figure 5. Rigid column base is much more common in engineering practice and can be considered to enhance the stability and moment resistance of the column. Experimental results of walls with weak or strong vertical connection tested to failure are reported in terms of multiple hardening phenomena. As well as the secondary hardening observed in the previous study, a new tertiary hardening was presented and contrasted herein. A discussion is provided to explain these uncommon phenomena.

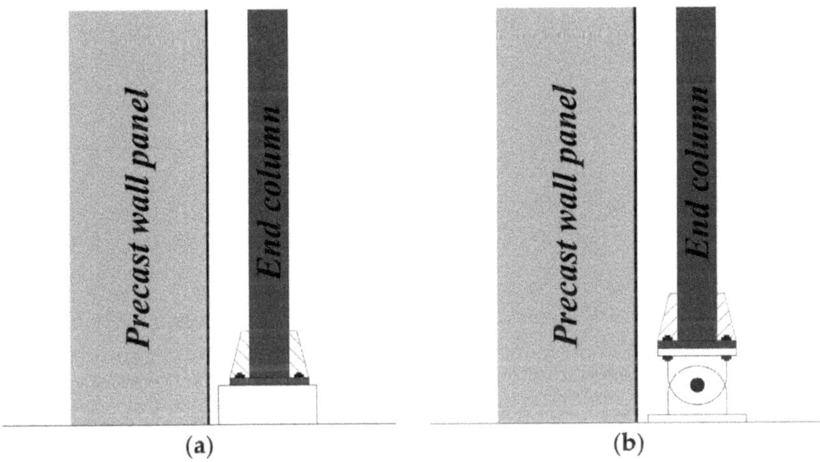

Figure 5. Comparison of column base between current and previous test. (**a**) Column base of testing wall in this paper; (**b**) Column base of previous wall [22].

2. Experimental Program

The testing walls are reported in this section, together with the loading scheme.

2.1. Testing Walls

Three testing walls, named W0 (cast-in-place construction), W1-R, and W2-R (precast construction) were designed and built for quasi-static reversed cyclic loading tests. As depicted in Figures 4 and 6, the testing PC shear walls are basic members extracted from the overall PC system, with a loading beam and a short wall segment cast at the top of wall panel and foundation block, respectively. As control variables, the cast-in-place wall and wall panels were designed in same section dimensions and reinforcement details. The wall panels were measured 2250 × 1000 × 120 mm (i.e., aspect ratio = 2.25) and reinforced with three diameters of steel bars.

Similar to the previous proof-of-concept test [22], the wall panel is designed as a 1/2-scale RC shear wall with boundary elements on the left and right side in accordance with Chinese standard GB 50010-2010 [26]. To facilitate the assembly of the horizontal connection, steel shoes were placed at the bottom end of the wall panel before casting, with vertical steel bars being welded on the inner surface of the steel shoe. To facilitate assembly of the vertical connection, two kinds of vertical connection were utilized for the PC shear walls. W1-R was cast with built-in bolts at both sides of the wall panel, resulting in a weak vertical connection; a thicker friction-bearing device and vertical steel shoes were utilized in W2-R, resulting in a strong vertical connection. The details of the horizontal and vertical connections are depicted and summarized in Figures 7 and 8, and Table 1. Two PC shear walls were designed with comparable limited slippage (12 mm) and varying magnitude of friction in the devices. As emphasized before, different from the previous experimental study [22], which contained a pin base under the end column, this time the PC shear walls contained a rigid column base, see Figure 5.

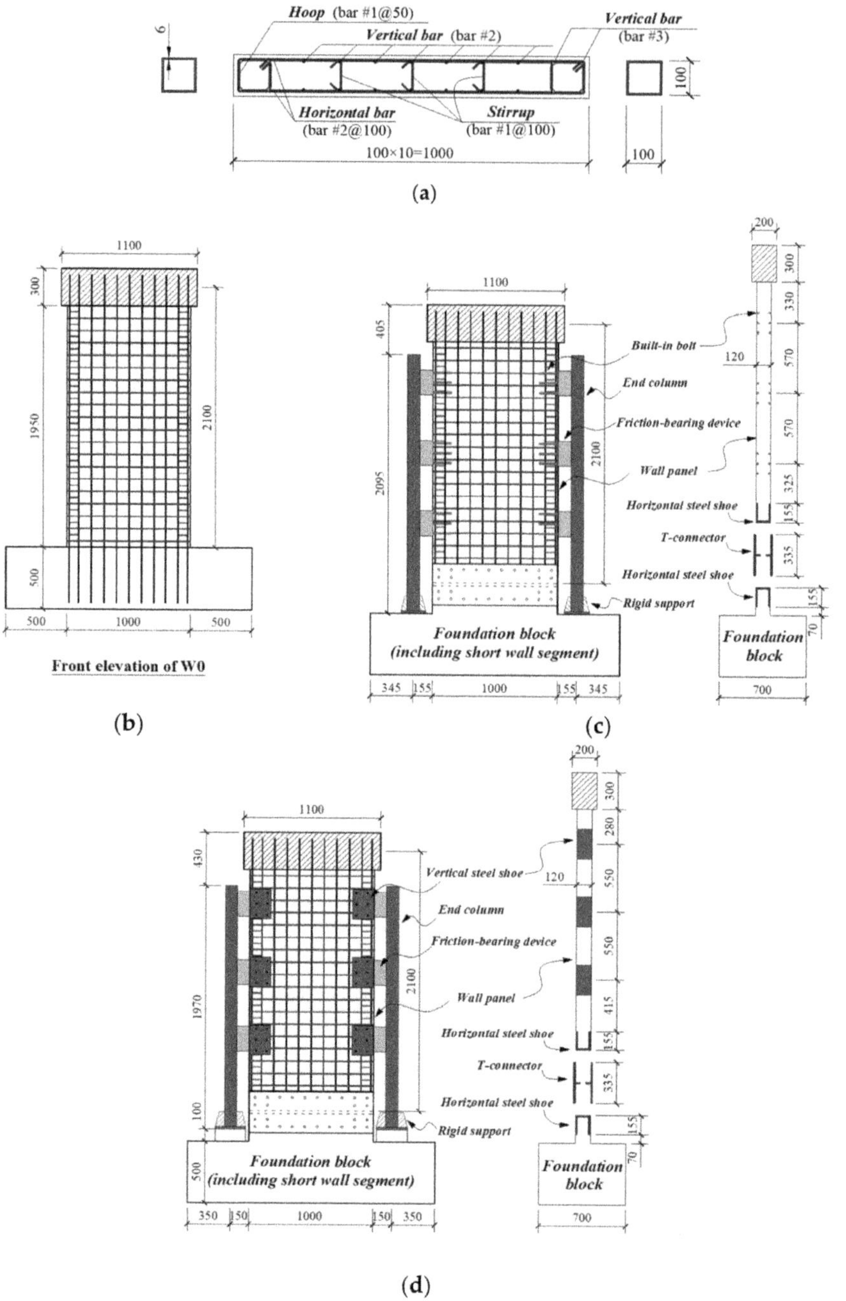

Figure 6. Details of walls (unit: mm). (**a**) Cross section of PC shear wall; (**b**) W0; (**c**) W1-R; (**d**) W2-R.

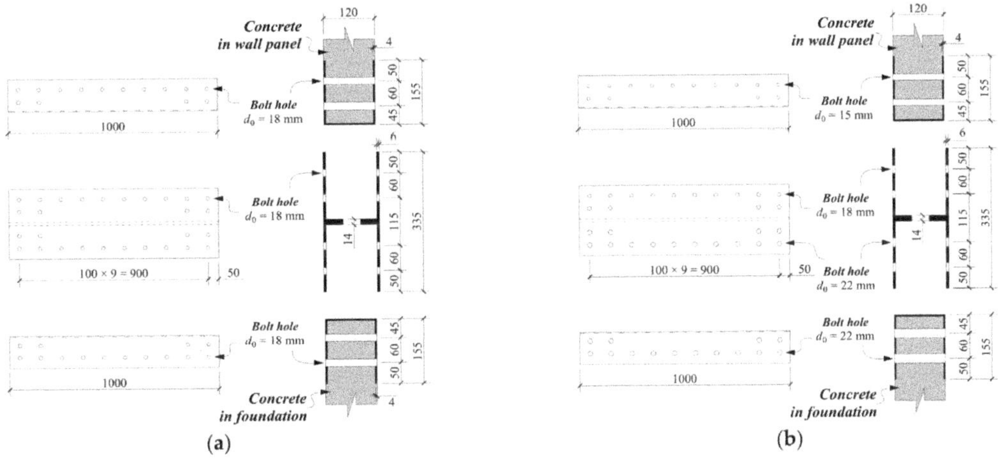

Figure 7. Details of horizontal connection in PC walls (unit: mm). (**a**) W1-R; (**b**) W2-R.

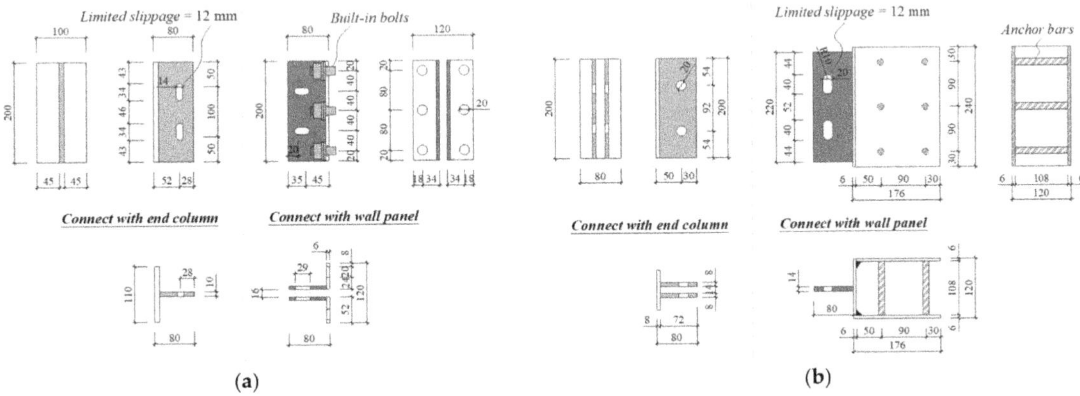

Figure 8. Details of vertical connection in PC walls (unit: mm). (**a**) W1-R; (**b**) W2-R.

Table 1. Details of vertical connection.

Testing PC Walls	Column Base	Friction-Bearing Device		
		High Strength Bolt	Friction Applied by One Bolt	Friction Applied by One Device
W1-R	Rigid	M12	11.4 kN	22.8 kN
W2-R	Rigid	M18	20.0 kN	40.0 kN

During the member casting, standard concrete cubes (150 × 150 × 150 mm) and prismoids (150 × 150 × 300 mm) were cast according to Chinese standard GB/T 50081-2002 [27]. Steel coupons for types of steel bar and steel plate were also prepared and tested in accordance with GB/T 228.1-2010 [28]. W0 and W1-R were cast in one batch with measured concrete compressive strength f_c = 27.0 MPa, and for W2-R another batch measured at f_c = 27.5 MPa. Testing results of steel are summarized in Table 2.

2.2. Loading Scheme

Three walls were tested by a horizontal reversed quasi-static cyclic loading and constant axial force P. The axial force and horizontal load were both applied on the loading beam, the former by a hydraulic jack and the latter by an MTS actuator. A sliding support was seated between the reaction frame and hydraulic jack so that both vertical and horizontal load could be applied simultaneously. The axial load applied on W0 and W1-R was 235 kN, and on W2-R it was 220 kN. The corresponding axial load ratio, $P/f_c A_g$, for the above two axial loads were 0.073 and 0.067, respectively, where A_g is gross area of the concrete section. The light axial load was chosen for two reasons, being that heavy axial load may restrain the elongation, and safety considerations for an innovative wall test. The foundation block and a pair of out-of-place triangle trusses were fixed to the strong floor to provide a stable testing boundary. Typical testing photo is illustrated in Figure 9a.

Table 2. Testing results of steel properties.

Material	Diameter/Thickness (mm)	Yield Strength (MPa)	Ultimate Strength (MPa)
Bar #1	6.7 (6.5)	370.9 (314.4)	523.0 (527.5)
Bar #2	8.0 (8.0)	338.8 (308.1)	509.4 (488.1)
Bar #3	9.5 (8.0)	490.0 (308.1)	544.3 (488.1)
Plate	6.0 (6.0)	372.0 (337.1)	431.8 (380.5)

Note: The data out of brackets are properties of W0 and W1-R. The data in brackets are properties of W2-R.

(a)

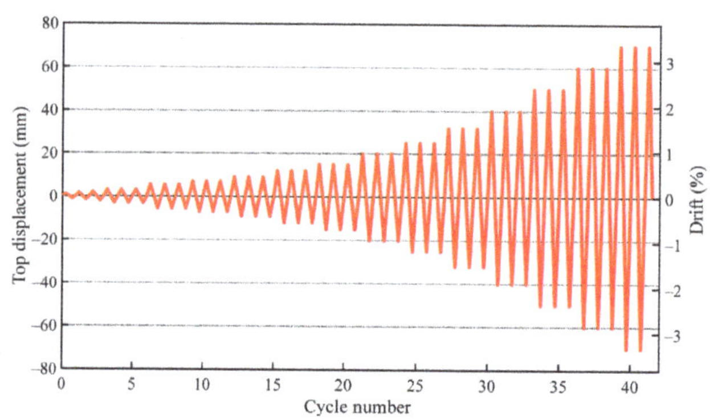

(b)

Figure 9. Testing setup and loading protocol. (**a**) Photo of W2-R; (**b**) Loading protocol.

The displacement-controlled protocol of horizontal load is described in Figure 9b, with one cycle per step before 2.0 mm top displacement, and after that, three cycles per step until failure [22]. The failure is defined as the recovery force degrading to less than 85% of the peak lateral load, as generally adopted in seismic tests [29,30]. Additionally, the drift ratio was stated as Δ/H_e, where Δ is the top displacement (i.e., horizontal displacement at the lateral loading point), and H_e is the effective height 2100 mm shown in Figure 6. Apart from the top displacement and lateral load measured by the sensors in the actuator, displacement gauges were placed at the foundation block and friction-bearing devices to monitor the potential slippage. Numbers of strain gauges were mounted on the steel bars and end columns to measure the real-time steel strain. Detailed locations of gauges are reported in Figure 10.

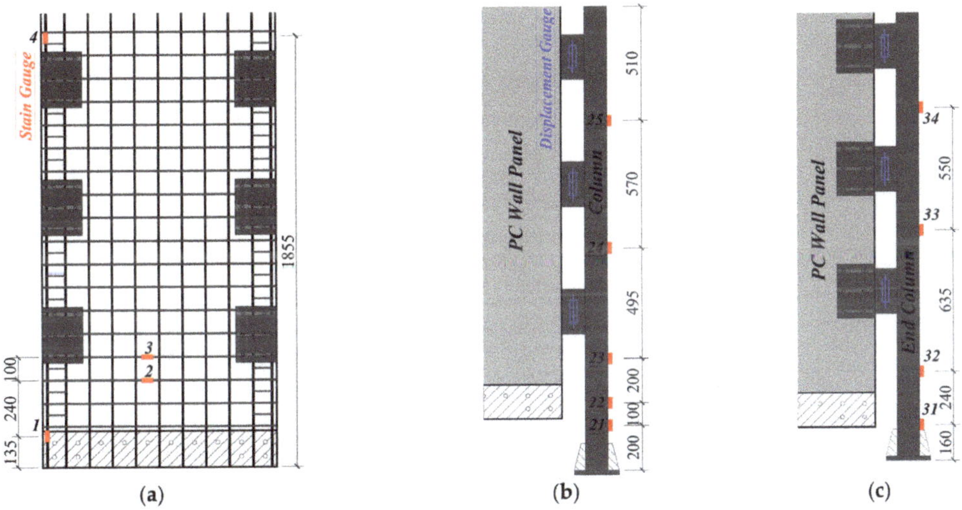

Figure 10. Instrumentation of wall. (**a**) Gauges mounted on wall; (**b**) Gauges mounted on W1-R end column; (**c**) Gauges mounted on W2-R end column.

3. Experimental Results

The response pertaining to friction-bearing devices, damage, and failure modes, load–displacement relationship, measured strains, stiffness reduction, and energy dissipation are reported in this section.

3.1. Working Mechanism and Response of Friction-Bearing Devices

See Figures 11 and 12, "vertical bearing" (i.e., the bolt bearing in the friction-bearing devices) may occur during the cyclic loading, which stems from the magnitude of limited slippage. Both PC shear walls were designed with same limited slippage, 12 mm. The vertical slippage in devices was recorded by displacement gauges, whose responses are shown in Figure 13.

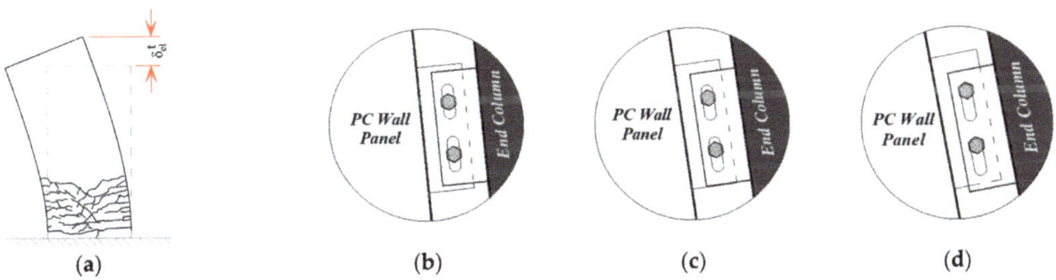

Figure 11. Vertical bearing of Tensile side (VT). (**a**) Longitudinal elongation of tensile side; (**b**) Before slip; (**c**) Slipping; (**d**) Bolt bearing.

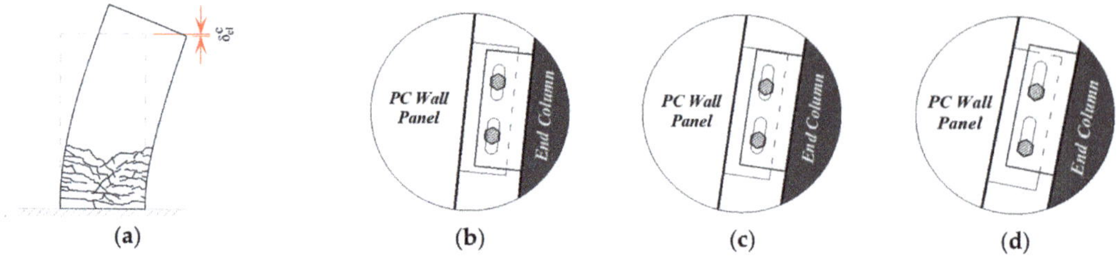

Figure 12. Vertical bearing of Compressive side (VC). (**a**) Longitudinal elongation of compressive side; (**b**) Before slip; (**c**) Slipping; (**d**) Bolt bearing.

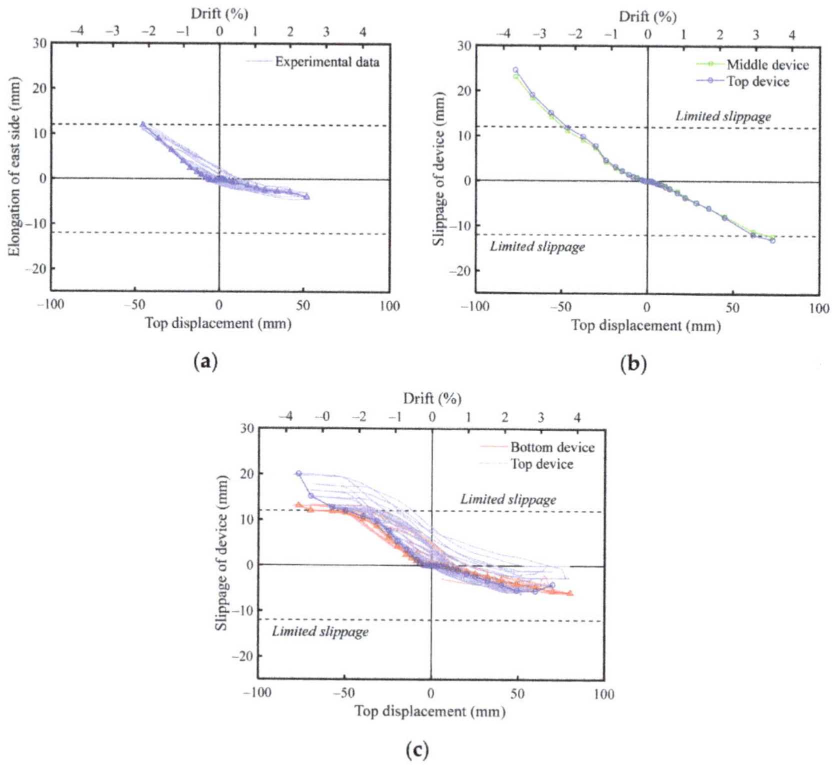

Figure 13. Results of vertical slippage and elongation. (**a**) Elongation in W0; (**b**) Vertical slippage in W1-R; (**c**) Vertical slippage in W2-R.

Table 3 summarizes the lateral top displacements when vertical bearing occurred, and two PC shear walls behaved similarly. Wall W1-R and W2-R attained "Vertical bearing of Tensile side" ("VT" for short) at a drift ratio of 1.33% (top displacement 28 mm) and 1.48% (top displacement 31 mm), respectively. The rough value of lateral top displacement when VC occurred reached an approximate drift ratio of 3.05% (top displacement 64 mm) for W1-R and 2.86% (top displacement 60 mm) for W2-R. The occurrence time of vertical bearing was comparable to the previous study [22], which indicated that the pin/rigid column base had relatively small influence on the working mechanism of vertical bearing. Because the slipping response for all three walls were comparable, the friction in devices was considered to have no significant influence on the elongational phenomenon. It can be inferred the effect of friction on the left and right sides balances itself out.

Table 3. Top displacement when "vertical bearing" occurred.

Test Walls	"Vertical Bearing of Tensile Side"		"Vertical Bearing of Compressive Side"	
	Positive (mm)	Negative (mm)	Positive (mm)	Negative (mm)
W1-R	28	32	64	64
W2 R	31	30	60	60

Another interesting and important behavior is that, once VT occurred, strong vertical connection (W2-R) almost stopped the device from slipping, while weak vertical connection (W1-R) did not. The continuous increase in slippage in W1-R is due to the low robustness of weak friction-bearing devices. Furthermore, the slippage of the top device in W2-R increased again when the top displacement was over 60 mm, resulting from the severe damage of the upper wall part.

3.2. Damage and Failure Mode

Early in the loading, all three walls behaved in a comparable manner, with flexural cracks initially opening above the horizontal connection as the 0.10% drift ratio for wall W0 and 0.24% drift ratio for other two PC shear walls, with later cracks gradually extending diagonally downwards. Before the occurrence of vertical bearing (20 mm, 0.95% drift ratio), we found less diagonal and X-shaped cracks in the two PC shear walls, similar to previous PC walls using pins as the column base [22], demonstrating that bending deformation was the major factor in PC wall panels compared with cast-in-place walls. Furthermore, the less cracks also mean that shear force demand in the PC walls was observably decreased by vertical connection in the initial loading period.

After these initial cracks appeared, no more cracks were found in W0. However, the PC shear walls continued to crack along with vertical bearing. VT occurred during the load step to 1.52% drift ratio (top displacement 32 mm), together with the diagonally oriented cracks around friction-bearing devices.

Finally, three walls failed in different ways, as depicted in Figure 14. W0 failed in flexure-shear mixed mode, with X-shape cracks from the bottom end to the top, and concrete crushing at the wall toes at the end. Additionally, VC occurred together with distinguishing failure in two PC shear walls. For PC shear wall W1-R, which assembled with weak vertical connection, concrete spalling and crushing were observed around the friction-bearing devices as well as the toes of the wall panels. For PC shear wall W2-R, being equipped with strong vertical connection prevented the crushing failure around the vertical connections, but the wall failed with diagonal tension cracking and concrete crushing in the upper wall part. Figure 15 shows the cracking observation of all three walls after testing.

The failure mode of wall W2-R is unusual but understandable. See Figure 16, the six devices divide the wall panel into four wall parts: upper, middle 1, middle 2, and bottom. The VT and VC of strong vertical connection restrained the elongation of the wall panels strictly, leading to high additional axial load in the middle and bottom wall parts. This constraint further increased the flexural and shear capacities of the middle and bottom wall parts, followed by the slow deterioration of the bottom plastic hinge region, and the bottom of wall panel ceased cracking, see Figure 14e,f. The upper wall part became a nominal "squat wall" and experienced shear failure.

3.3. Lateral Load–Top Displacement Relationship

The measured lateral load versus top displacement relationships of three walls are drawn in Figure 17a–c, and the comparison of backbone curves is reported in Figure 17d. The lateral top displacement when VT and VC occurred are marked with the black and red dotted lines, respectively. The hysteretic loops of W1-R and W2-R are fatter than those of W0 with incremental lateral top displacement. Furthermore, the vertical connection and accessory friction failed to rise the residual displacements associated with the cyclic loading.

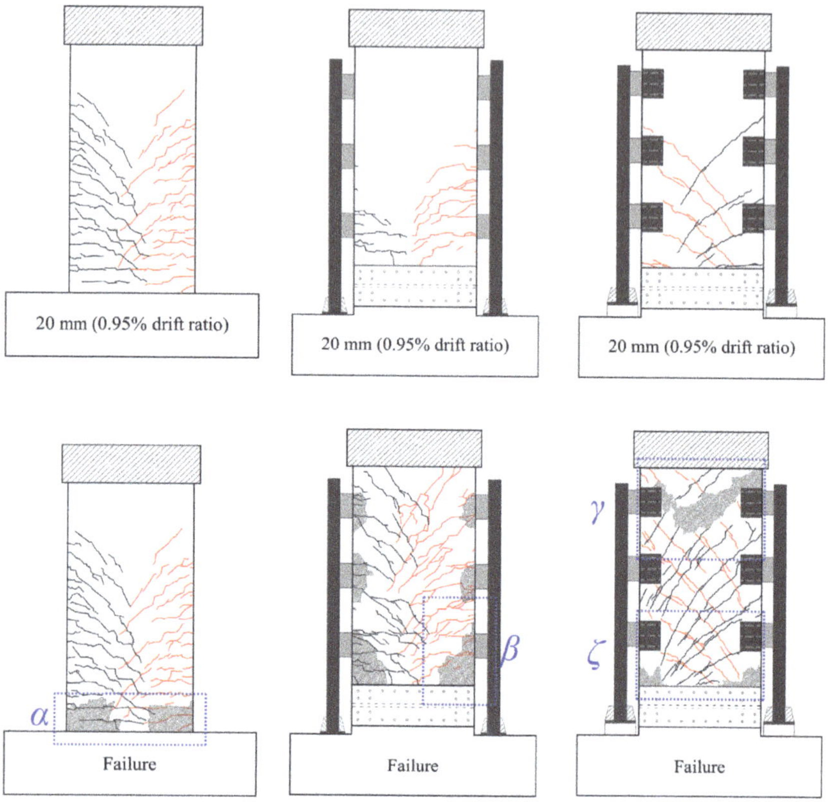

(a) (b) (c)

Figure 14. Crack distribution of testing walls (black and red lines represent the cracks as subjecting to different loading direction). (**a**) W0; (**b**) W1-R; (**c**) W2-R.

(a) (b) (c)

Figure 15. *Cont.*

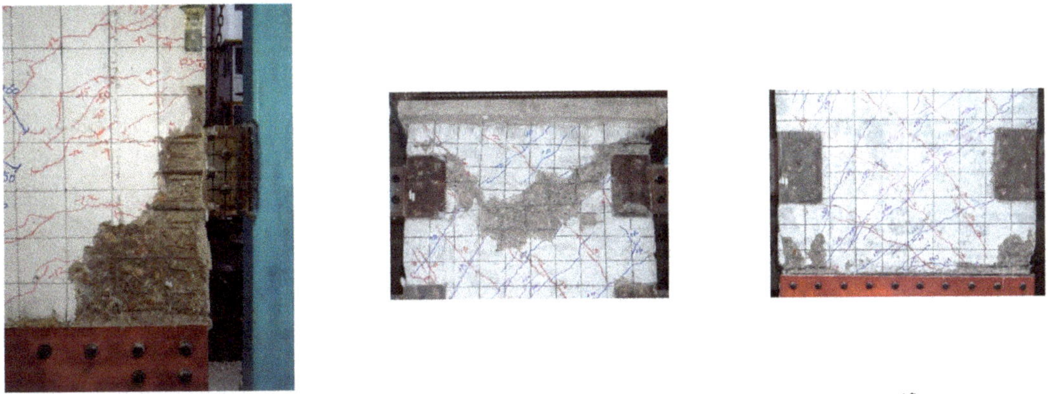

(**d**) (**e**) (**f**)

Figure 15. Cracking observation after test (blue and red lines represent the cracks as subjecting to different loading direction). (**a**) Global damage in W0; (**b**) Global damage in W1-R; (**c**) Global damage in W2-R; (**d**) Local damage in W1-R (detail from zone β, Figure 14d); (**e**) Local damage at the top of W2-R (detail from zone γ, Figure 14f); (**f**) Local damage at the bottom of W2-R (detail from zone ζ, Figure 14f).

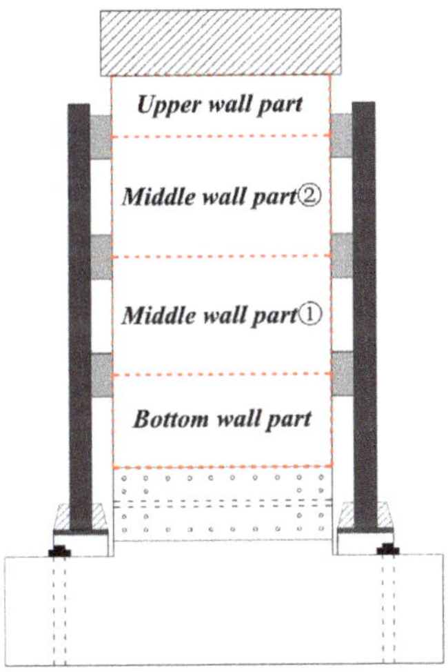

Figure 16. Wall parts divided by friction-bearing devices in PC shear wall.

Four key points representing the behavioral phases were identified in Figure 17. The same as the definition of failure in the preceding part of the paper, the ultimate point was defined as 85% peak lateral load of response [29,30]. The cast-in-place wall W0 achieved peak lateral load at a drift ratio of 1.03% with 203.2 kN. Both PC shear walls achieved great moment capacity and drift capacity (i.e., drift ratio of ultimate point). W1-R and W2-R achieved peak loads at a drift ratio of 2.92% (315.1 kN) and 3.81% (365.1 kN), respectively. The peak loads of PC shear walls increased by about 60% when compared to cast-in-place W0. Compared with wall W1-R, the peak lateral loads of wall W2-R with strong vertical connection increased by 16% (Positive direction), and top displacement of ultimate point decreased by 9% (Positive direction). Wall W1-R was more ductile than wall W2-R, which was in accordance with the response characteristics about peak lateral loads and top displacement. In fact, the weak vertical connection could be considered to be a kind of ductile connection to some extent, although it failed it resulted in wall W1-R being more ductile without sudden load loss. For safety reasons, wall W2-R, in which the upper wall part failed in shear, was not tested to 85% peak load in positive direction.

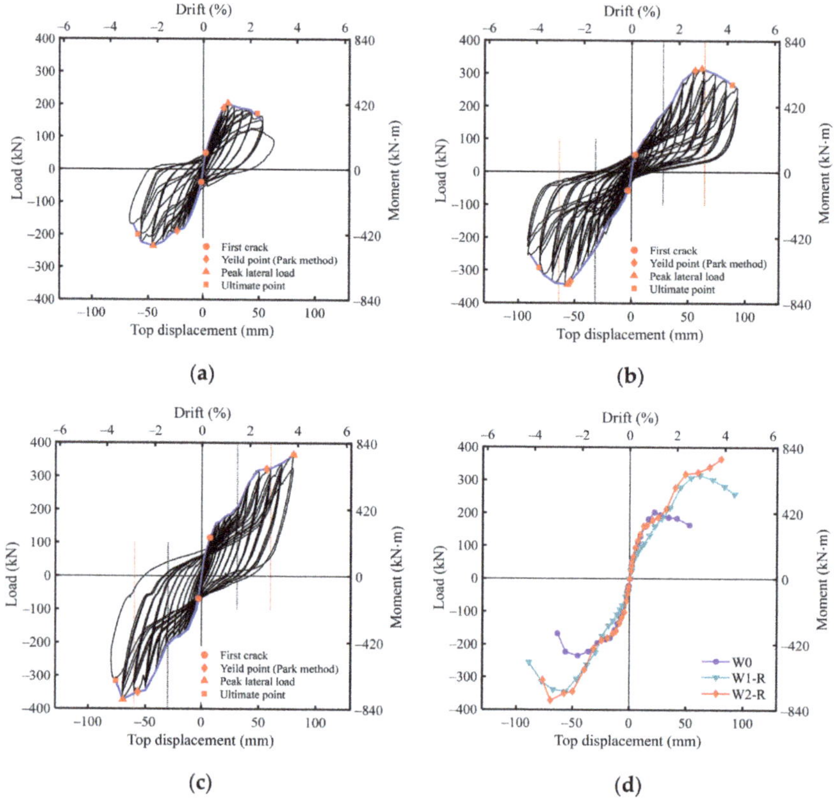

Figure 17. Measured lateral load–top displacement responses (black dotted lines represent VT, red dotted lines represent VC). (**a**) W0; (**b**) W1-R; (**c**) W2-R; (**d**) Comparison of backbone curves.

VT could be able to provide additional resistant moment, the wall W1-R and W2-R here reproved the fact and further demonstrated the influence of weak/strong connection. For wall W1-R, VT enlarged the internal force of the vertical connection, concrete around the built-in bolts cracked gradually, then, VC which led to upward compression in devices meantime widened the existing cracks. Eventually, W1-R reached its peak lateral load. As can be observed in Figure 17, the second ascending branches were governed by VT and peaked when VC occurred. For wall W2-R, both VT and VC increased, which caused the overall response to be extra ascending branches, however, due to hardening of the middle and bottom wall parts influenced by VC, the upper wall part failed finally. In this special way, it is interesting to note that W1-R achieved secondary hardening, while W2-R achieved tertiary hardening.

3.4. Measured Strain

The strain of vertical bars above the horizontal connection is crucial data to estimate the working phases of shear wall. The measured strains in vertical bars are depicted in Figure 18. With well welding between vertical bars and steel shoes, the bars of all walls yielded at a comparable top displacement, approximately 20 mm (0.95% drift ratio). Apparently the hysteretic loops of W1-R and W2-R are slimmer than those of W0. It can be inferred that, before the occurrence of vertical bearing, the vertical connection dissipated most of the energy, while the PC wall panel itself only dissipated a small part.

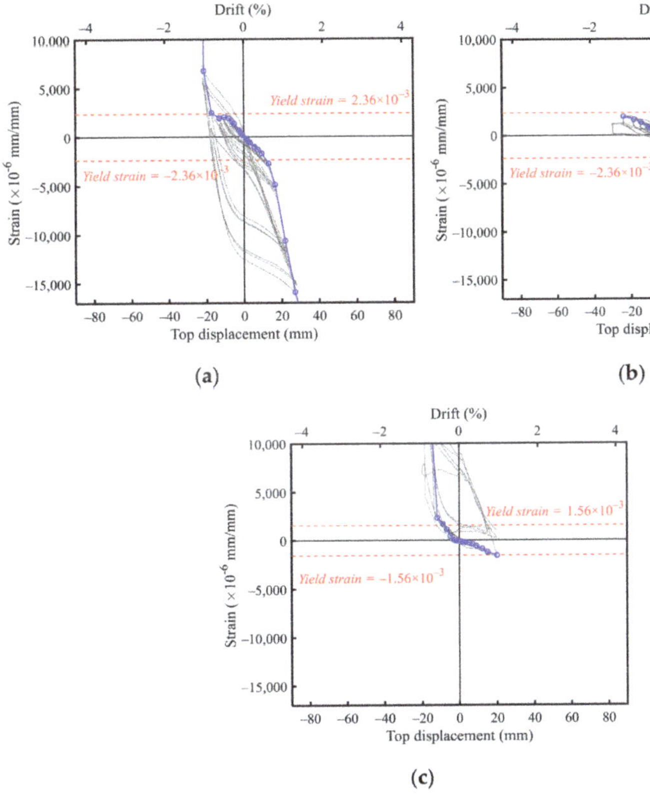

Figure 18. Strain response of vertical bars (No. 1 gauge in Figure 10). (**a**) W0; (**b**) W1-R; (**c**) W2-R.

The measured strains on the horizontal bars are depicted in Figure 19. Only the horizontal bar in W0 yielded slightly until the last load steps, while the bars in W1-R and W2-R were elastic during the test. This phenomenon indicates again that the shear force demand in PC walls was decreased by vertical connection, as mentioned in Section 3.2.

The strain of the vertical bar in W2-R upper wall part is also recorded, as depicted in Figure 20. The vertical bar yielded during the load step to 1.90% drift ratio (top displacement 40 mm), which was later than VT and earlier than VC. Though no horizontal strain gages were pasted on the upper wall parts, concerning the damage phenomenon, we could reconfirm that the failure mode of wall W2-R was a mixed mode of shear and flexure in the upper wall part.

The measured results of strain gages along the height of the end column are reported in Figure 21. In general, due to the flexure and vertical connection, the strain decreased with increasing height. Because the global deformation of the PC shear wall was dominated by wall panel, the strain of the end column increased rapidly once the plastic hinge region of wall panel occurred. As the test continued, the bottom of the end column yielded as the drift ratio reached 1.67% (top displacement 35 mm). As expected, strain increased faster once VT occurred, which means that VT would enlarge the axial force of the end column. The vertical bearing and associated axial force of the end column provided additional resistance moment for the shear walls.

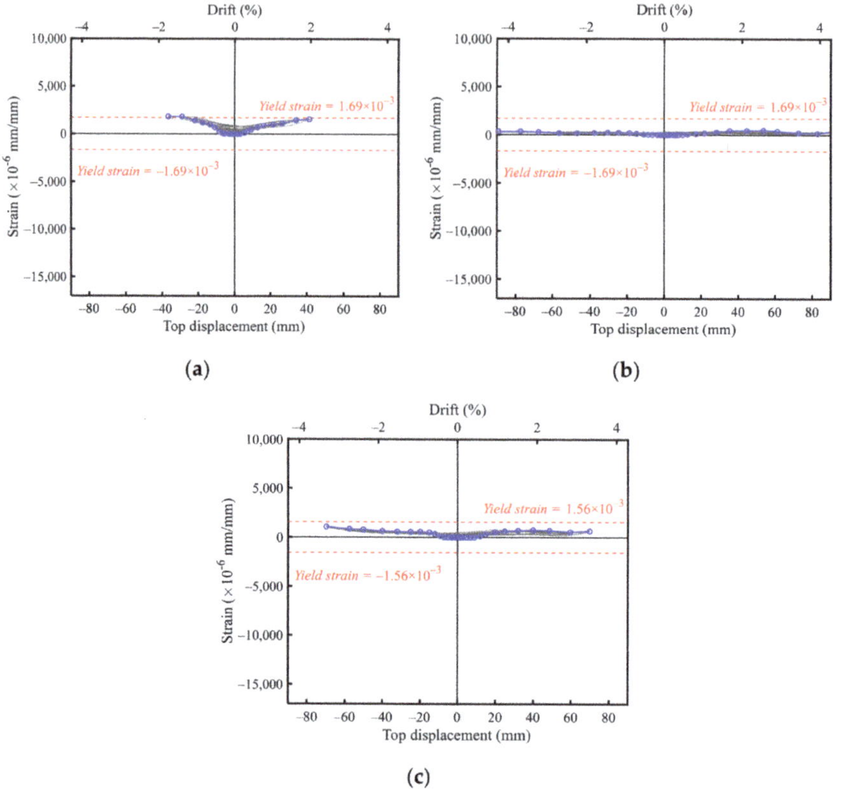

Figure 19. Strain response of horizontal bars. (**a**) W0 (No. 2 gauge in Figure 10); (**b**) W1-R (No. 2 gauge in Figure 10); (**c**) W2-R (No. 3 gauge in Figure 10).

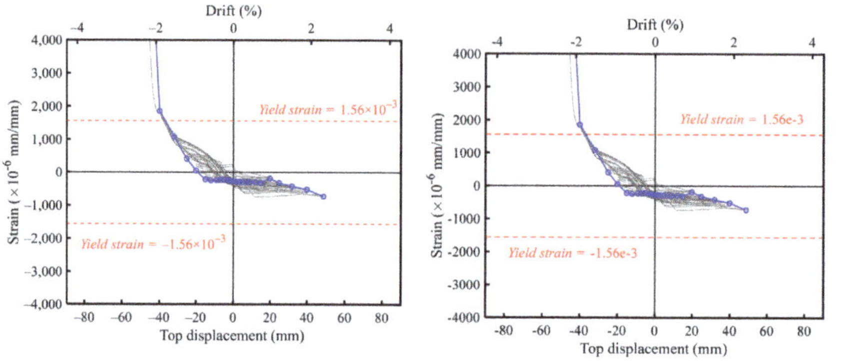

Figure 20. Strain response of vertical bars for upper wall part of W2-R (No. 4 gauge in Figure 10).

3.5. Stiffness Reduction and Energy Dissipation

Secant stiffness and energy dissipation are frequently used parameters to evaluate the seismic behaviour of PC members. The secant stiffness is defined as the slope of line which links the maximum loads of first cycle at positive and negative loading directions. The accumulated enclosed area of each hysteresis hoop is used to assess energy dissipation ability in various loading stages. Both two criteria, varying with the drift ratio of the walls, are illustrated in Figure 22.

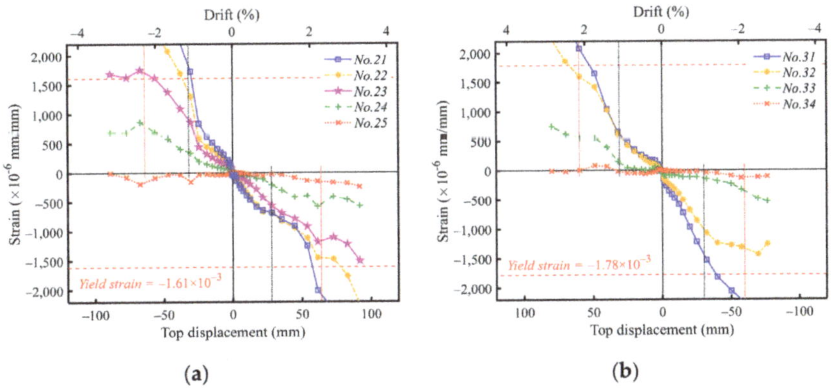

Figure 21. Comparison of strain results on end column. (**a**) W1-R; (**b**) W2-R.

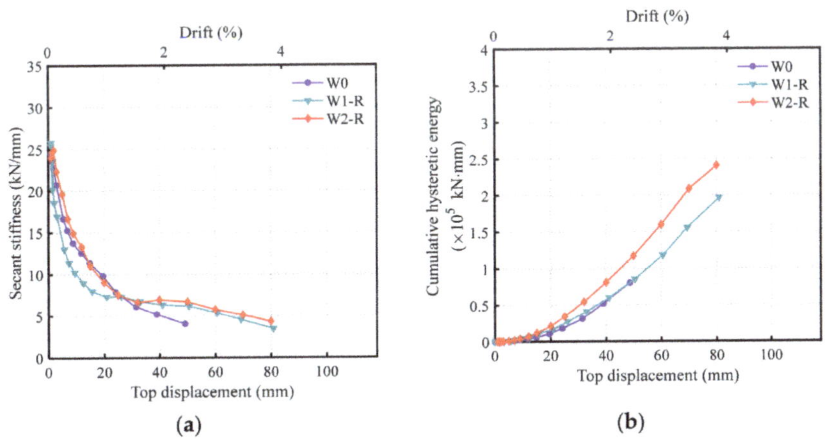

Figure 22. Comparison of stiffness degradation and cumulative energy dissipation. (**a**) Stiffness degradation; (**b**) Cumulative hysteretic energy.

Before VT occurred, the stiffness degradation of W0 and W2-R behaved comparably, but the curve of W1-R was lower. We inferred that more shrinkage cracks existed around the built-in bolt before loading, and that the weak vertical connection deteriorated the initial stiffness of W1-R. See Figure 22a, after the occurrence of VT, the tangent stiffness of two PC shear walls W1-R and W2-R were comparably flat, while the cast-in-place wall W0 kept falling. In detail, note that the curve of W2-R is slightly higher than W1-R after VT occurred, the reason is that strong vertical connection could provide stricter constraint; later VT brought additional stiffness and moment capacity and successfully delayed the degradation of the PC shear wall. Furthermore, VC had almost no obvious effect on the degradation.

Benefitting from the elevated drift capacities, both PC shear walls dissipated greater energy than W0 at the end of the test. The cumulative hysteretic energy of PC shear walls was increased by approximately two times compared with wall W0 at the end of the test. By comparing the cumulative hysteretic energy of walls, it was established that W2-R dissipated the most energy by benefitting a larger magnitude of friction in strong friction-bearing device.

4. Discussion of Multiple Hardening

As described in previous sections, different failure modes were observed in PC walls. The secondary hardening phenomenon occurred in the PC wall with weak vertical connection (wall W1-R), concrete crushing around vertical connection and wall panel toes were found at the end of the test. The wall with strong vertical connection (wall W2-R) achieved the highest moment capacity and tertiary hardening. In this section, conceptual analysis of the multiple hardening phenomena is presented.

See Figure 23a, according to the Modified Compression Field Theory (MCFT) [31] and Rotating Angle Soften Truss Model (RA-STM) [32,33], the elongation enlarges the smeared tensile strain perpendicular to the crack direction with smeared diagonal cracks in the plastic hinge region. The effective compressive strength of concrete decreases as elongation increases, resulting in a reduction in the shear strength in plastic hinge. Most well-designed slender cast-in-place shear walls failed in a flexure or flexure-shear mode after flexural yielding, as depicted in Figure 23b. Different from the cast-in-place wall, PC walls in this study were equipped with vertical connection. Because of the elongation and associated VT, the elongation of the middle and bottom wall parts were restrained, see Figure 23c,d. Hence, flexural/shear strength of bottom/middle wall parts were enhanced, while the upper wall part was not.

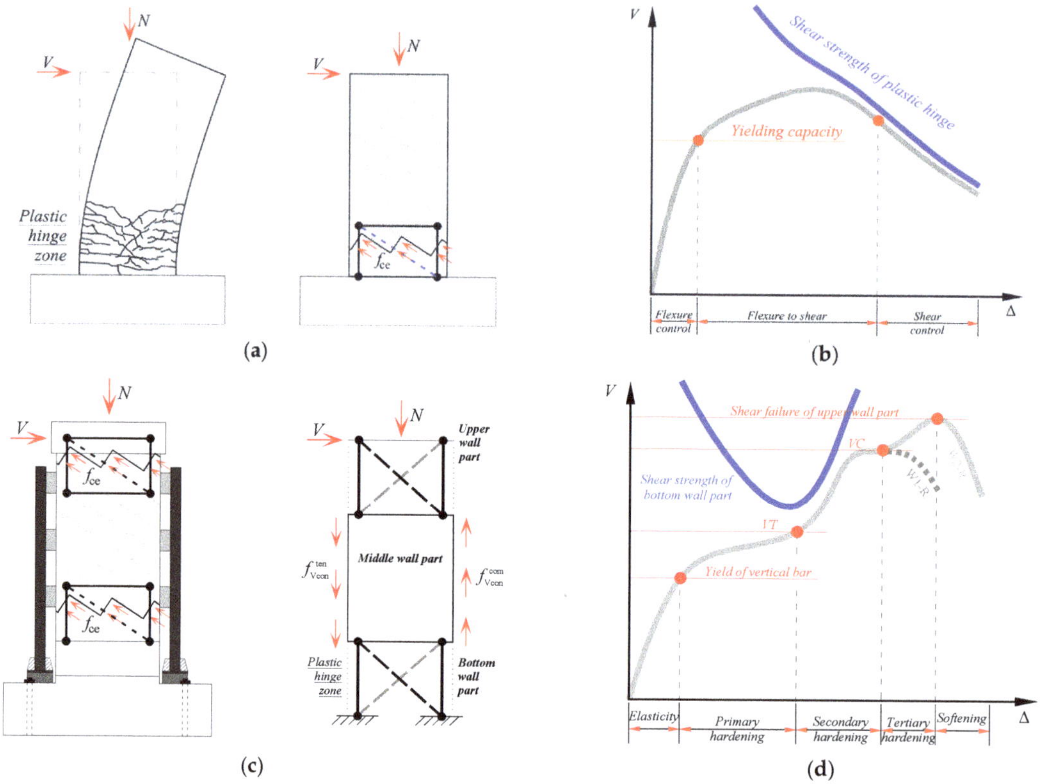

Figure 23. Conceptual analysis shear walls. (**a**) Load path and truss mechanism of slender cast-in-place wall; (**b**) Schematic of load–displacement relationship of slender wall [32,33]; (**c**) Load path and truss mechanism of PC wall; (**d**) Schematic of load–displacement relationship of PC wall.

Following above logic, for PC wall W1-R, the time of VT was consistent with the time when the concrete around the vertical connection began to crack, and the backbone curve showed "secondary hardening"; however, the vertical connection was too weak to provide subsequent additional resistance moment, the time of the failure of vertical connection was consistent with VC, and later lateral load declined. For another PC wall W2-R, whose vertical connection was strengthened, the VT also caused the backbone curve to have "secondary hardening". However, with strong vertical connection, the moment capacity of W2-R was re-elevated by VC, and "tertiary hardening" occurred gradually. Therefore, the failure model of W1-R was defined as a connection failure, while the one of W2-R was defined as shear failure at the upper wall part.

Predictably, as the limited slippage lengthens, shear strength of the plastic hinge region may drop quickly. Thus, for the PC wall with long limited slippage vertical connection, shear failure in the plastic hinge region may occur earlier than vertical bearing. In the existing parallel studies [22,34,35], PC shear walls with long limited slippage (22 mm) failed with web crushing of the bottom wall panel, which reproved the validity of the conceptual analysis in this chapter; the photo is shown in Figure 24.

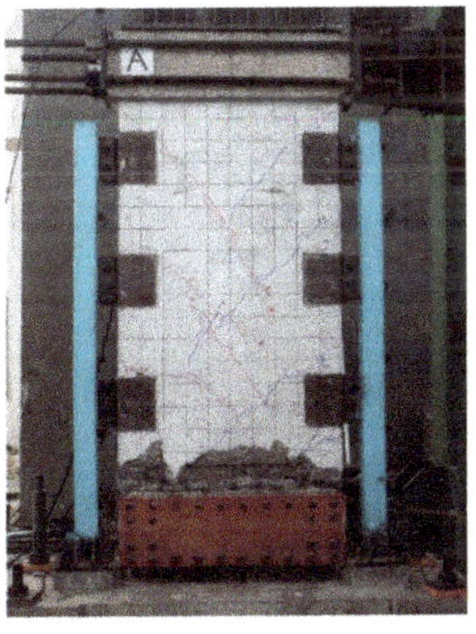

Figure 24. Web crushing failure of PC wall with long limited slippage vertical connection (blue and red lines represent the cracks as subjecting to different loading direction) [22,34,35].

5. Conclusions

An experimental programme consisting of quasi-static cyclic loading on shear walls was conducted. Two PC shear walls were designed with both horizontal and vertical connections. In the vertical connection, the friction-bearing devices and the end columns were used to assemble PC wall panels. The cyclic responses of PC shear walls with varying types of friction-bearing devices, namely weak connection and strong connection, were systematically investigated. The following conclusions can be summarized according to the current study.

(1) The proposed PC shear walls first achieved "Vertical bearing of Tensile side" ("VT" for short), and then achieved "Vertical bearing of Compressive side" ("VC" for short). Once VT occurred, strong vertical connection almost stopped the vertical connection from slipping, while weak vertical connection did not.
(2) The PC shear walls cracked slower than cast-in-place walls. Eventually, different failure modes were observed on three walls: W0 (cast-in-place wall) failed in flexure; W1-R (PC shear wall with weak vertical connection) failed with concrete crushing occurring around the friction-bearing devices and toes of the PC wall panels; and W2-R (PC shear wall with strong vertical connection) prevented the failure mode like W1-R, and later failed in a mixed mode of shear and flexure at the upper wall part.
(3) The VT delayed the stiffness degradation of PC shear walls, thus leading to additional drift capacity and moment capacity. For wall W1-R, the time of VT was consistent with the time when the backbone curve achieved "secondary hardening"; the time of VC was consistent with the failure of vertical connection. For wall W2-R, benefitting

from the strengthened vertical connection, the wall also achieved "tertiary hardening". In detail, the moment capacity of PC shear walls was increased by more than 60% when compared with the testing cast-in-place W0. The moment capacity of wall W2-R was higher than wall W1-R.

(4) The magnitude of friction in the devices had a great influence on the energy dissipation, but not on the stiffness reduction and elongation. The cumulative hysteretic energy of the PC shear wall was increased by approximately two times compared with the cast-in-place W0.

Author Contributions: Conceptualization, H.J., J.S. and H.Q.; methodology, H.J. and J.S.; software, H.J. and H.C.; validation, H.J., Q.F. and D.C.; formal analysis, H.J. and W.G.; investigation, H.J. and Q.F.; data curation, H.J.; writing—original draft preparation, H.J.; writing—review and editing, H.J., H.C., W.G. and K.C.; visualization, H.C. and K.C.; supervision, H.Q. and D.C.; project administration, J.S. and H.J.; funding acquisition, J.S. and H.J. All authors have read and agreed to the published version of the manuscript.

Funding: This research was funded by the Lvyangjinfeng Talent Program of Yangzhou (Grant No. YZ-LYJFJH2021YXBS110); the Deputy Manager of Science and Technology Program of Jiangsu Province (Grant No. FZ20211212); the High-Level Talents Support Program of Yangzhou University (No. 137012283); Jiangsu Provincial Double-Innovation Doctor Program; the Natural Science Foundation of Jiangsu Province (Grant No. BK20170668).

Institutional Review Board Statement: Not applicable.

Informed Consent Statement: Not applicable.

Data Availability Statement: Correspondence and requests for materials should be addressed to H.J.

Conflicts of Interest: The authors declare no conflict of interest.

References

1. Carrillo, J.; Vargas, D.; Sánchez, M. Stiffness degradation model of thin and lightly reinforced concrete walls for housing. *Eng. Struct.* **2018**, *168*, 179–190. [CrossRef]
2. Kurama, Y.C.; Sritharan, S.; Fleischman, R.B.; Restrepo, J.I.; Henry, R.S.; Cleland, N.M.; Ghosh, S.K.; Bonelli, P. Seismic-Resistant Precast Concrete Structures: State of the Art. *J. Struct. Eng.* **2018**, *144*, 03118001. [CrossRef]
3. Clarivate Analytics. Web of Science. Available online: https://www.webofscience.com (accessed on 1 May 2022).
4. Dang, L.J.; Liang, S.T.; Zhu, X.J.; Zhang, M.; Song, Y.M. Seismic performance of precast concrete wall with vertical energy-dissipating connection. *Struct. Des. Tall Spec.* **2021**, *30*, e1820. [CrossRef]
5. Wu, D.; Liang, S.; Shen, M.; Guo, Z.; Zhu, X.; Sun, C. Experimental estimation of seismic properties of new precast shear wall spatial structure model. *Eng. Struct.* **2019**, *183*, 319–339. [CrossRef]
6. Yan, J.-B.; Yan, Y.-Y.; Wang, T. Cyclic tests on novel steel-concrete-steel sandwich shear walls with boundary CFST columns. *J. Constr. Steel Res.* **2020**, *164*, 105760. [CrossRef]
7. *JGJ 1-2014*; Technical Specification for Precast Concrete Structures. China Architecture & Building Press: Beijing, China, 2014.
8. *GB/T 51231-2016*; Technical Standard for Assembled Buildings with Concrete Structure. China Architecture & Building Press: Beijing, China, 2017.
9. Shen, S.-D.; Cui, Y.; Pan, P.; Gong, R.-H.; Miao, Q.-S.; Li, W.-F. Experimental Study of RC Prefabricated Shear Walls with Shear Keys Affected by a Slotted Floor Slab. *J. Aerosp. Eng.* **2019**, *32*, 04019013. [CrossRef]
10. Wang, Y.L.; Zhang, Y.M.; Cui, H.J.; Cai, Z.J.; Xu, L.X. Experimental study on fabricated shear wall structure with energy-dissipating vertical joint. *Struct. Concr.* **2020**, *21*, 2654–2668. [CrossRef]
11. Zhang, H.; Li, H.N.; Chen, X.Y.; Li, C. Experimental investigation on the hysteretic behavior of precast concrete walls with energy-dissipated dry connections. *Struct. Concr.* **2020**, *21*, 2836–2853. [CrossRef]
12. Li, X.H.; Wu, G.; Kurama, Y.C.; Cui, H.R. Experimental comparisons of repairable precast concrete shear walls with a monolithic cast-in-place wall. *Eng. Struct.* **2020**, *216*, 110671. [CrossRef]
13. Xu, L.H.; Xiao, S.J.; Li, Z.X. Behaviors and Modeling of New Self-Centering RC Wall with Improved Disc Spring Devices. *J. Eng. Mech.* **2020**, *146*, 04020102. [CrossRef]
14. Han, Q.H.; Wang, D.Y.; Zhang, Y.S.; Tao, W.J.; Zhu, Y. Experimental investigation and simplified stiffness degradation model of concrete shear wall with steel connectors. *Eng. Struct.* **2020**, *220*, 110943. [CrossRef]
15. Mangalathu, S.; Jang, H.; Hwang, S.-H.; Jeon, J.-S. Data-driven machine-learning-based seismic failure mode identification of reinforced concrete shear walls. *Eng. Struct.* **2020**, *208*, 110331. [CrossRef]

16. Eom, T.-S.; Park, H.-G.; Kim, J.-Y.; Lee, H.-S. Web Crushing and Deformation Capacity of Low-Rise Walls Subjected to Cyclic Loading. *ACI Struct. J.* **2013**, *110*, 575–584.
17. Paulay, T.; Priestley, M.J.N. Stability of ductile structural walls. *ACI Struct. J.* **1993**, *90*, 385–392.
18. Cheng, X.W.; Ji, X.D.; Henry, R.S.; Xu, M.C. Coupled axial tension-flexure behavior of slender reinforced concrete walls. *Eng. Struct.* **2019**, *188*, 261–276. [CrossRef]
19. Shegay, A.; Dashti, F.; Hogan, L.; Lu, Y.; Niroomandi, A.; Seifi, P.; Zhang, T.; Dhakal, R.; Elwood, K.; Henry, R.; et al. Research programme on seismic performance of reinforced concrete walls: Key recommendations. *Bull. N. Z. Soc. Earthq.* **2020**, *53*, 54–69. [CrossRef]
20. Masoudi, M.; Khajevand, S. Revisiting flexural overstrength in RC beam-and-slab floor systems for seismic design and evaluation. *Bull. Earthq. Eng.* **2020**, *18*, 5309–5341. [CrossRef]
21. *NZS 3101:2006*; Concrete Structures Standard. Standards New Zealand: Wellington, New Zealand, 2017.
22. Jiang, H.; Qiu, H.; Sun, J.; Del Rey Castillo, E.; Ingham, J.M. Influence of friction-bearing devices on seismic behavior of PC shear walls with end columns. *Eng. Struct.* **2020**, *210*, 110293. [CrossRef]
23. Mehrabani, R.V.; Sigrist, V. Elongation of Reinforced Concrete Plastic Hinges Subjected to Reversed Cyclic Loading. *J. Struct. Eng.* **2015**, *141*, 04014188. [CrossRef]
24. Henry, R.S.; Dizhur, D.; Elwood, K.J.; Hare, J.; Brunsdon, D. Damage to concrete buildings with precast floors during the 2016 Kaikoura earthquake. *Bull. N. Z. Soc. Earthq. Eng.* **2017**, *50*, 174–186. [CrossRef]
25. Dashti, F.; Dhakal, R.P.; Pampanin, S. Out-of-Plane Response of In-Plane-Loaded Ductile Structural Walls: State-of-the-Art and Classification of the Observed Mechanisms. *J. Earthq. Eng.* **2020**, *26*, 1325–1346. [CrossRef]
26. *GB 50010-2010*; Code for Design of Concrete Structures. China Architecture & Building Press: Beijing, China, 2016.
27. *GB/T 50081-2002*; Standard for Test Method of Mechanical Properties on Ordinary Concrete. China Architecture & Building Press: Beijing, China, 2003.
28. *GB/T 228.1-2010*; Metallic Materials-Tensile Testing-Part 1: Method of Test at Room Temperature. Standards Press of China: Beijing, China, 2011.
29. *GB/T 50152-2012*; Standard for Test Method of Concrete Structures. China Architecture & Building Press: Beijing, China, 2012.
30. *JGJ/T 101-2015*; Specification for Seismic Test of Buildings. China Architecture & Building Press: Beijing, China, 2015.
31. Vecchio, F.J.; Collins, M.P. The modified compression-field theory for reinforced concrete elements subjected to shear. *ACI J.* **1986**, *83*, 219–231.
32. Hsu, T.T.C. Softened truss model theory for shear and torsion. *ACI Struct. J.* **1988**, *85*, 624–635.
33. Lee, J.Y.; Watanabe, F. Shear deterioration of reinforced concrete beams subjected to reversed cyclic loading. *ACI Struct. J.* **2003**, *100*, 480–489.
34. Jiang, H. Behaviour of Precast Shear Wall with End Column Subjected to Seismic Loading. Ph.D. Thesis, Southeast University, Nanjing, China, 2020.
35. Fang, Q.; Qiu, H.X.; Sun, J.; Dal Lago, B.; Jiang, H.B. Performance study of precast reinforced concrete shear walls with steel columns containing friction-bearing devices. *Arch. Civ. Mech. Eng.* **2021**, *21*, 110. [CrossRef]

MDPI AG
Grosspeteranlage 5
4052 Basel
Switzerland
Tel.: +41 61 683 77 34

Buildings Editorial Office
E-mail: buildings@mdpi.com
www.mdpi.com/journal/buildings

www.ingramcontent.com/pod-product-compliance
Lightning Source LLC
Chambersburg PA
CBHW061843210926
51850CB00006B/959
* 9 7 8 3 7 2 5 8 5 8 9 9 6 *